BIOLOGICAL ASSESSMENT OF NATURAL AND ANTHROPOGENIC ECOSYSTEMS

Trends in Diagnosis of Environmental Stress

BIOLOGICAL ASSESSMENT OF NATURAL AND ANTHROPOGENIC ECOSYSTEMS

Trends in Diagnosis of Environmental Stress

Edited by
Eugene M. Lisitsyn, DSc
Larissa I. Weisfeld, PhD
Anatoly I. Opalko, PhD

Reviewers and Advisory Board Members:
Gennady E. Zaikov, DSc, and
Alexander N. Goloshchapov, PhD

A∧P | **APPLE ACADEMIC PRESS**

First edition published 2022

Apple Academic Press Inc.
1265 Goldenrod Circle, NE,
Palm Bay, FL 32905 USA
4164 Lakeshore Road, Burlington,
ON, L7L 1A4 Canada

CRC Press
6000 Broken Sound Parkway NW,
Suite 300, Boca Raton, FL 33487-2742 USA
2 Park Square, Milton Park,
Abingdon, Oxon, OX14 4RN UK

© 2022 Apple Academic Press, Inc.

Apple Academic Press exclusively co-publishes with CRC Press, an imprint of Taylor & Francis Group, LLC

Library and Archives Canada Cataloguing in Publication

Title: Biological assessment of natural and anthropogenic ecosystems : trends in diagnosis of environmental stress / edited by Eugene M. Lisitsyn, DSc, Larissa I. Weisfeld, PhD, Anatoly I. Opalko, PhD.

Names: Lisitsyn, Eugene M., editor. | Weisfeld, Larissa I., editor. | Opalko, Anatoly I., editor.

Description: First edition. | Includes bibliographical references and index.

Identifiers: Canadiana (print) 20210239883 | Canadiana (ebook) 20210239905 | ISBN 9781771889773 (hardcover) | ISBN 9781774639344 (softcover) | ISBN 9781003145424 (ebook)

Subjects: LCSH: Indicators (Biology)

Classification: LCC QH541.15.I5 B53 2022 | DDC 363.73/63—dc23

Library of Congress Cataloging-in-Publication Data

....

CIP data on file with US Library of Congress

....

ISBN: 978-1-77188-977-3 (hbk)
ISBN: 978-1-77463-934-4 (pbk)
ISBN: 978-1-00314-542-4 (ebk)

About the Editors

Eugene M. Lisitsyn, DSc

Eugene M. Lisitsyn, DSc of Biological Sciences, is an Assistant Professor at the N.V. Rudnitsky North-East Agricultural Research Institute, Russian Academy of Sciences in Kirov, Russia, and a professor of ecology and zoology at Vyatka State Agricultural Academy, also in Kirov, Russia. He is a member of the N.I. Vavilov Society of Geneticists and Breeders. Dr. Lisitsyn is the author of over 230 publications in scientific journals and conference proceedings, as well as the cobreeder of oats variety Krechet. His research interests concern the basic problems of plant adaptation to environmental stressors such as soil acidity, aluminum and heavy metals, and other ecological problems. He has published chapters in several books in Apple Academic Press, including *Barley: Production, Cultivation and Uses*; *Biological Systems, Biodiversity and Stability of Plant Communities*; *Chemical and Structure Modification of Polymers*; *Materials Chemistry: A Multidisciplinary Approach to Innovative Methods*; *Temperate Crop Science and Breeding: Ecological and Genetic Studies*; *Antioxidants in Systems of Varying Complexity: Chemical, Biochemical and Biological Aspects*.

Larissa I. Weisfeld, PhD

Chief Specialist, Emanuel Institute of Biochemical Physics,
Russian Academy of Sciences, Moscow, Russia

Larissa I. Weisfeld, PhD, is a Chief Specialist at the Emanuel Institute of Biochemical Physics of the Russian Academy of Sciences, Moscow, Russia; and a member of the All-Russia Vavilov Society of Geneticists and Breeders. She is also the author of more than 300 articles published in scientific journals and conference proceedings and holds seven patents for inventions. She is the co-author of four new winter wheat cultivars registered in the State Register of the Russian Federation. Her main field of interest concerns basic problems of chemical mutagenesis, mutational selection, and the mechanism of action of *p*-aminobenzoic acid. She has worked as a scientific editor in the publishing house Nauka ("Sciences" in Russian) (Moscow, Russia) and with the journals Genetics and Ontogenesis. She is the author of several book chapters and has co-edited several books with Apple Academic Press, including *Ecological Consequences of Increasing Crop Productivity: Plant Breeding and Biotic Diversity (2015); Biological Systems, Biodiversity, and Stability of Plant Communities (2015); Temperate Crop Science and Breeding Ecological and Genetic Study (2016); Heavy Metals and Other Pollutants in the Environment: Biological Aspects (2017); Chemistry and Technology of Plant Substances: Chemical and Biochemical Aspects (2017); Temperate Horticulture for Sustainable Development and Environment: Ecological Aspects (2019)*, and *Antioxidants in Systems of Varying Complexity (2020)*.

Anatoly I. Opalko, PhD

Anatoly I. Opalko, PhD in Agriculture, Professor, is Leading Researcher of the Genetics, Plant Breeding and Reproductive Biology Division at the National Dendrological Park "Sofiyivka" of the National Academy of Sciences of Ukraine. He is also Head of the Cherkassy Regional Branch of the Vavilov Society of Geneticists and Breeders of Ukraine. He is a prolific author, researcher, and lecturer. He has received several awards for his work, including the badge of honor, "Excellence in Agricultural Education," and the badge of honor of the National Academy of Sciences of Ukraine "for professional achievement." He has also received the Nikolai Cholodny Prize in Botany and Plant Physiology. He is member of many professional organizations and is on the editorial boards of the Ukrainian and International Biological and Agricultural Science Journals. He is the author and co-editor of the books with Apple Academic Press: *Ecological Consequences of Increasing Crop Productivity: Plant Breeding and Biotic Diversity; Biological Systems, Biodiversity, and Stability of Plant Communities; and Temperate Crop Science and Breeding*. He was also a reviewer and advisory board member for the book *Heavy Metals and Other Pollutants in the Environment: Biological Aspects;* and *Temperate Horticulture for Sustainable Development and Environment: Ecological Aspects*.

Contents

Contributors

Rafail A. Afanas'ev
DSc of Agricultural Sciences, Professor, Project Leader, M. D. Pryanishnikov All-Russian Scientific Research Institute of Agrochemistry, 31A Pryanishnikov St., Moscow, 127550, Russia. Tel.: +74999764757, +74999761531, +79191040585. E-mail: rafail-afanasev@mail.ru

Alexander Y. Bome
PhD of Agricultural Sciences, Professor, Sr. Manager, Exeter Produce and Storage Ltd., 149A Thames Rd. W, Exeter, ON, Canada N0M 1S3. Tel.: +1 5196368612. E-mail: alex_bo1@aol.com

Nina A. Bome
DSc of Agriculture Sciences, Professor, Tyumen State University, Head of Department of Botany, Biotechnology and Landscape Architecture, Institute of Biology, Tyumen State University, 6 Volodarsky St., Tyumen 625003, Russia. Tel.: +79129236177. E-mail: bomena@mail.ru

Juliya V. Burmenko
PhD of Biological Sciences, Senior Scientist, Department of Fruit and Berry Crops, All-Russian Selection and Technology Institute of Horticulture and Nursery, 4 Zagor'evskaya, St., Moscow 115598, Russia. Tel.: +79803922524. E-mail: burmenko_j@mail.ru

Nikolai G. Ivanov
Postgraduate Student, Department of Botany, Biotechnology and Landscape Architecture, Institute of Biology, Tyumen State University, Department of Botany, Biotechnology and Landscape Architecture, 6 Volodarsky St., Tyumen 625003, Russia. Tel.: +79123932814. E-mail: tgu_bf@mail.ru

Vladislav N. Kalaev
Professor, DSc of Biological Sciences, Professor, Voronezh State University, Department of Genetics, Cytology and Bioengineering, 1 University Sq., Voronezh 394018, Russia. Tel.: +79103450072. E-mail: dr_huixs@mail.ru

Natalia N. Kolokolova
Associate Professor, PhD of Agricultural Sciences, Department of Botany, Biotechnology and Landscape Architecture, Institute of Biology, University of Tyumen, 6 Volodarsky St., Tyumen 625003, Russia. Tel.: +79129234499. E-mail: campanella2004@mail.ru

Eugene M. Lisitsyn
DSc of Biological Sciences, Assistant Professor, Head of Department of Plant Edaphic Resistance, N.V. Rudnitsky Federal Agricultural Research Center of North-East, 116a Lenin St., Kirov 610007, Russia; Professor of Department of Ecology and Zoology, Vyatka State Agricultural Academy, 133 Oktyabrsky Av., Kirov 610017, Russia. Tel.: +79123649822. E-mail: edaphic@mail.ru

Genrietta E. Merzlaya
Project Leader, DSc of Agricultural Sciences, Professor, MD Pryanishnikov All-Russian Scientific Research Institute of Agrochemistry, 31A Pryanishnikov St., Moscow 127550, Russia. Tel.: +74999761191, +74999761531, +79623694197. E-mail: lab.organic@mail.ru

Lee A. Newman
Associate Professor, Professor, DSc of Biological Sciences, Environmental and Science Biology,
State University of New York College of Environmental Science, Syracuse, NY 13210, USA.
Tel.: +1 315 470 4937. E-mail: lanewman@esf.edu

Elena S. Novoselova
Postgraduate, Department of Ecology and Zoology, Vyatka State Agricultural Academy,
133 Oktyabrsky Av. Kirov, 610017, Russia. Tel.: +78332574381. E-mail: gonina-elena@mail.ru

Anatoly I. Opalko
Full Professor, PhD of Agricultural Sciences, Leading Researcher of the Genetics,
Plant Breeding and Reproductive Biology Division in National Dendrological Park
"Sofiyivka" of National Academy of Science of Ukraine, 12-a Kyivska St. Uman,
Cherkassy 20300, Ukraine. Tel.: +380506116881. E-mail: opalko_a@ukr.net

Olga A. Opalko
PhD of Agricultural Sciences, Associate Professor, Senior Scientist of the Genetics,
Plant Breeding and Reproductive Biology Division in National Dendrological Park
"Sofiyivka" of National Academy of Science of Ukraine, 12-a Kyivska St.,
Uman, Cherkassy 20300, Ukraine. Tel.: +380664569116. E-mail: opalko_o@ukr.net

Irina V. Pak
DSc of Biological Sciences, Professor, Head of the Department of Ecology and Genetics,
Institute of Biology, Tyumen State University, 3 Pirogov St., Tyumen 625003, Russia.
Tel.: +789129292286. E-mail: pakiv57@mail.ru

Irina I. Rudneva
DSc of Biological Sciences, Professor, A.O. Kovalevsky Institute of the Biology of the Southern
Seas RAS, Laboratory of Ecotoxicology, Head of the Laboratory, 2 Nakhimov Av., Sevastopol,
Cream 299011, Russia. Tel.: +7978749 17 04. E-mail: svg-41@mail.ru

Rizvan Dilman Ogly Rustamov
Postgraduate Student, Department of Ecology and Genetics, Institute of Biology,
Tyumen State University, 3 Pirogov St. Tyumen, 625043 Russia. Tel.: +79523459031.
E-mail: kafedraekogen@mail.ru.

Marina V. Semenova
PhD of Biological Sciences, Department of Botany, Biotechnology and Landscape Architecture,
Institute of Biology, Tyumen State University, Tyumen 625003, Russia; Associated Professor,
Department of Botany, Biotechnology and Landscape architecture, 6 Volodarski St.,
Tyumen 625003, Russia. Tel.: +79129925333. E-mail: lanewman@esf.edu

Valentin G. Shaida
Researcher, AO Kovalevsky Institute of the Biology of the Southern Seas RAS, Laboratory of
Ecotoxicology, 2 Nakhimov Av., Sevastopol, Crimea 299011, Russia. Tel.: +79787491708.
E-mail: svg-41@mail.r

Lyudmila N. Shikhova
DSc of Agricultural Sciences, Assistant Professor, Professor of Department of Ecology and Zoology,
Vyatka State Agricultural Academy, 133 Oktyabrsky Av., Kirov 610017, Russia.
Tel.: +79127213759. E-mail: shikhova-l@mail.ru

Michael O. Smirnov
Senior Scientist, PhD of Biological Sciences, MD Pryanishnikov All-Russian Scientific Research
Institute of Agrochemistry, 31A Pryanishnikov St., Moscow 127550, Russia.
Tel.: +74999764757, +74953337071, +79057966323. E-mail: User53530@yandex.ru

Vladimir N. Sorokopudov
Dsc of Agriculture, Professor, Department of Landscape Gardening and Greenkeeping,
Professor, Russian State Agrarian University-Moscow Timiryazev Agricultural Academy,
49 Timiryazevskaya st., Moscow 127550 Russia
Tel.: +79999235654, E-mail: sorokopud2301@mail.ru

Nikolay V. Tetyannikov
PhD of Agricultural Sciences, All-Russian Horticultural Institute for Breeding, Agrotechnology and
Nursery, Laboratory of Field Crops, Researcher, 4 Zagoryevskaya St., Moscow 115598, Russia.
Tel.: +79224790808. E-mail: tetyannikovnv@ya.ru

Oleg V. Trofimov
PhD of Biological Sciences, Associate Professor of the Department of Ecology and Genetics,
Institute of Biology, Tyumen State University, 3 Pirogov St., Tyumen 625003, Russia.
Tel.: +789068753543. E-mail: oleg_v_trofimov@mail.ru

Maral U. Utebayev
Postgraduate Student, A.I. Barayev Research and Production Centre of Grain Farming,
Department of Biochemistry and Technology of Quality, Head of the Department,
12 Barayev St., Shortandy-1 021601, The Republic of Kazakhstan.
Tel.: +77163123029, +77163123369, +77011697536. E-mail: phytochem@yandex.ru

Tatyana V. Vostrikova
Federal State Budgetary Scientific Institution "A.L. Mazlumov All-Russian Research Institute of
Sugar Beet and Sugar", Federal Agency of Scientific Organizaions, 86, Ramonsky District,
Voronezh Region, 396030, Russia. Tel.: +79191825573.
E-mail: tanyavostric@rambler.ru

Larissa I. Weisfeld
PhD of Biological Sciences, Chief Specialist of the Laboratory of Solar Photoconverters,
Emanuel Institute of Biochemical Physics, RAS (IBCP RAS), 4 Kosygin St., Moscow 119334,
Russia. Tel.: +79162278685. E-mail: liv11@yandex.ru

Olga A. Zemlyanukhina
Voronezh State University, Department of Genetics, Cytology and Bioengineering,
1 University Sq., Voronezh 394018, Russia

Ksenya A. Zubkova
Postgraduate, Department of Ecology and Zoology, Vyatka State Agricultural Academy,
133 Oktyabrsky Av., Kirov 610017, Russia. Tel.: +78332574381. E-mail: ks.zubkowa@gmail.com

Abbreviations

A	anaphase
AD	Latin *Anno Domini* (used to indicate that a date comes the specified number of years after the traditional date of Christ's birth)
AGEs	glycation end products
a.m.	Latin *ante meridiem* means "before noon", used after times of day between midnight and noon not expressed using the 24-hour
Ap	apoptotic cells
ANOVA	analysis of variance
AOA	antioxidant activity
AOPP	advanced oxidation protein products
APG	Angiosperm Phylogeny Group
APG IV	is the fourth version of a modern system of plant taxonomy for flowering plants (angiosperms)
a.s.	active substance (in medicine)
B	vitamin of the B group
BC	before Christ (a date before the Christian era)
BL	blebbed nuclei
BN	binucleated cells
°C	the degree or degrees Celsius
Cd	cadmium is chemical element with atomic number 48
CFU	colony-forming unit, the number of colony-forming cells in 1 ml of medium
ChL	chemiluminescence
cm	centimeter or cantimeter is equal to one hundredth of a meter
CMCase	carboxymethylcellulase
CO_2	carbon dioxide
Cu	copper, a chemical element with atomic number 29
cv.	cultivar
CV	coefficient of variation
cvs.	cultivars

DES   diethyl sulfate
DMS   dimethyl sulfate
DNA   deoxyribonucleic acid
EI   ethyleneimine
EMF   electromagnetic field
17β-estradiol (E_2)   estrogen, the sex hormone of female
et al.   an abbreviation for the Latin *"et alia"* or *"et alii"*
                             means and the rest; and others; and so forth
etc.   et cetera means "and others"
Fe(III)/Fe(II)   indicator system of reduction-oxidation mechanism
                             and kinetics studies
g/m^2 *gram* per square meter
Hz   value of the rate of periodical processors ($1Hz+1s^{-1}$)
IBCP RAS   Institute Biochemical Physics of Russian Academy of
                             Sciences
i.e.   Latin *id est*, means "that is to say"
kg/ha   kilograms per hectare
kg/m^2 kilograms per square meter
FARC   N.V. Rudnitsky Federal Agricultural Research Center
                             of the Northeast
Fr   fragmented cells
G, g   gram
g/m^2 gram/square meter (1 gram/(square meter = 0.001 kg/m^2)
H_2O_2 chemical compound hydrogen peroxide
Hz   hertz is the derived unit of frequency in the Interna-
                             tional System of Units (SI)
GSH   glutathione
IBCP RAS   Institute Biochemical Physics of Russian Academy of
                             Sciences
IF   inflorescence
LF-EMF   low-frequency electromagnetic fields
ka   The Abbreviation "ka" means "kilo-annum" Before
                             Present (era)
km/h   kilometre/ hour
LHC   light-harvesting complexes
$LSD_{.05}$ least significant difference between the treatment
                             variants at the 5% level for significance
                             (significantly with 95% probability)

LVs	leaves
m	meter (is the base unit of length in the International System of Units)
M	symbol: mol, mole (International System of Units)
M_1	there are plants of the first generation after treatment with the mutagen of the parent population
M_2	there is second generation of plants after selections from M_1 generation
M_3	there is a third generation of plants after selections in the M_2 population
m^2	square meter
MAC	maximum allowance concentration
MDA	malondialdehyde
MI	mitotic index = ratio of number of dividing cells to total number of calculated cells
MF	magnetic field
mg/kg	milligram/kilogram
mm	abbreviation for millimeter or millimeters
MN	micronuclei
MP	level pathologies of mitoses
mT	millitesla
N	haploid number of chromosomes
N	north pole of the magnetic field
N	nitrogen
N0	zero dose of nitrogen
$2n$	diploid chromosome number in a somatic cell
NAS	National Academy of Sciences
NAAS	National Academy of Agricultural Sciences
NaCl	sodium chloride
NaN_3	sodium azide
NB	nuclear buds
NBf	nuclear buds on filament cells
NDP	National Dendrological Park
NDVI	normalized differentiated vegetation index
NEM	nitrosoethylurea
NG	nitroso guanidine
NIR	near infrared is photometry indexes
Nm	nanometer (a one-billionth of a meter)

NMM	nitrosomethilurea
NO	nitric oxide
17-OHP	17-α hydroxy progesterone
P	*p*-value in statistic
P	prophase
PABA	*para*-aminobenzoic acid
pcs.	pieces or instances
pcs/sq. m	pieces of 1meter square
Pb	lead, a chemical element with atomic number 82
pH	is a scale used to specify how acidic or basic a water-based solution
p.m.	Latin *post meridiem* means "after noon", used after times of day between noon and midnight not expressed the 24-hour clock
PN	persistent nucleoli
Pro. Sp.	a hybridogenic taxon, published as a species
R	correlation coefficient for measuring the strength and direction of a linear relationship between two variables on a scatter plot
RAAS	Russian Academy of Agricultural Sciences
RAS	Russian Academy of Sciences
RC	reaction centers
RED	index of photometry red
RESP	Registered Education Savings Plans
RI	Research Institute
ROS	reactive oxygen species
S	south pole of the magnetic field
SI	International System of Units
SMF	static magnetic field
SO_2	sulfur dioxide
SOD	superoxide dismutase
spp.	abbreviation for species (plural)
sq km	square kilometer
STM	stems
T	tesla (Tl) the value of magnetic induction, mT = 0.01 T
T	telophase
t/ha	tonnes per hectare
TILLING	targeting induced local damage in genomes

U	unit of enzymatic activity
UAV	unmanned aerial vehicle (commonly known as a drone)
UV-B	ultraviolet radiation
UK	United Kingdom
UNUH	Uman National University of Horticulture
USA	United States of America
USDA	United States Department of Agriculture
USSR	Union of Soviet Socialist Republics
UV-B	ultraviolet irradiation
var.	variety
x	monoploid chromosome number
Zn	zinc, the first element in group 12 of the periodic table with atomic number 30
μT	microtesla

Introduction

One of the costs of scientific and technological progress is the degradation of natural ecosystems by reducing biodiversity and disrupting the normal functioning of natural communities. Anthropogenic impacts on the natural environment are multifactor: they are related to the use of nonrenewable mineral resources, environmental pollution, change in climate, landscapes, and so on.

The main factor supporting the existence of modern problems of a global nature is the priority of short-term interests over long-term goals, the priority of economic and political issues over the protection of nature.

A second problem closely related to the formation of a guideline for endless consumption growth is the increase in industrial emissions both in terms of total volume and in terms of the variety of harmful substances contained in their composition and their combinations. In the structure of production products, new substances for the biosphere, which it is unable to process, account for an increasing share. All this gives rise to a range of global problems, among which anthropogenic impacts on climate and the geological environment are prominent.

A truly dangerous global problem is loss in biodiversity, as species extermination (direct and indirect, through habitat destruction) by humans occurs faster than in past eras of mass extinction. A number of regional problems, such as the problems of certain large regions and the destruction of unique ecosystems, often reach global status.

Some of the most pressing human resource challenges—population growth, food shortages, environmental pollution (with human waste and because of disruption of production technologies), including global warming, plant and animal species extinction, and all associated sociological and political problems—are largely ecological.

Encyclopedia Britannica defines ecology as a science that studies the relationships between organisms and the environment. The concept of environment includes both other organisms and the physical environment. It includes relationships between individuals within a population and between individuals of different populations. Ernst Haeckel introduced the term "ecology" into science in his book "Generelle Morphologie der

Organismen" (1866). He defines ecology as "the animal's relationship to both its organic and inorganic environment." In relation to plant organisms, for the first time, the term "ecology" was used by Danish botanist Eugenius Warming in his book *Oecology of Plants: An Introduction to the Study of Plant Communities* (1895). Until the early 1900s, plant and animal ecology developed separately until American biologists emphasized the relationship of both plant and animal communities as a biotic whole.

Traditionally, ecology is divided into several main areas of research.

Evolutionary ecology studies the environmental factors that stimulate species adaptation. Studies of species evolution may aim to answer the question of how populations changed genetically over several generations but not necessarily attempt to learn what the underlying mechanisms might be. Evolutionary ecology is looking for these mechanisms.

Physiological ecology finds out how organisms survive in their environment. There is often an emphasis on extreme conditions such as very cold or very hot environments or aquatic environments with unusually high salt concentrations. Physiological ecology considers the special mechanisms that individuals of a species use to function and the restrictions on species imposed by the environment.

Behavioral ecology studies environmental factors that stimulate behavioral adaptation, such as how individuals find their food and avoid enemies, why some birds migrate and others are settled, and why some animals live in groups and others are mostly single.

Population ecology explores individual species, reasons for their settlement, and fluctuation in numbers, and **community ecology** studies interactions between several or many species. **Biogeography** studies the geographical distribution of organisms. The fundamental issue of these environmental directions concerns the "species set"—that is, what environmental factors determine how many species are present in a given area. Another set of questions includes examining how many trophic levels there are in a particular location and what factors limit that number.

Ecosystem ecology addresses large-scale environmental problems, often defined not by species but by measures such as biomass, energy flow, and cyclic nutrient transport. Human activity modifies global ecosystems by increasing carbon dioxide content in the atmosphere, enhances greenhouse effects, and causes excessive flow of fertilizers into rivers and then into the oceans, which significantly impairs the living conditions of the species living there.

Scientific research in all the above areas has been the most pressing task of biological science in recent decades. From the point of view of sustainable economic development, we need to present in detail the possible consequences of certain manifestations of technogenesis and learn to prevent them.

Since ecologists work with living systems, which are characterized by a high level of variation of different parameters, methods of exact sciences (physics, chemistry, and mathematics) are not as easily applied in ecology and are not as accurate as the results obtained in other sciences. Environmental measurements can never undergo the same ease of analysis as measurements in physics, chemistry, or some quantifiable fields of biology.

Despite these problems, various aspects of the environment can be identified by physical and chemical means ranging from simple chemical identification and physical measurements to the use of complex mechanical devices. The development of biostatistics, the development of an appropriate experimental framework and improved sampling methods now allow for a quantitative statistical approach to the study of environmental problems.

The environmental assessment system on the basis of chemical and analytical control data has its disadvantages, since in this case, it is necessary to determine the content of numerous components of pollutants in different natural environments, then compare their concentrations with the maximum permissible ones, and on this basis to conclude the "danger" or "safety" for the biota of the complex impact of all these factors. At the same time, chemical analysis allows to determine concentrations of only a relatively small number of potentially dangerous and already known mutagenic and toxic substances. At the same time, the question arises about the regulation of the content of pollutants: in different countries for the same toxicants, the values of the maximum permissible concentrations differ significantly; often these values vary depending on the physical and chemical properties of the environment or the direction of the use of resources.

Ecologists have long established that in real conditions biological objects are influenced by a complex of physical, chemical, and biological factors, the joint effects of which, depending on the nature, intensity, and order of the agents, cause fundamentally different types of cell/body response—additive, synergistic, antagonism. Therefore, the resulting response of the biological system to the combined effect cannot be predicted based solely on information on the effects of the separate action of the agents. Under these conditions, the use of the MPC indicator will either underestimate or reassess the possible effects of man-made stress on the ecosystem.

In contrast to chemical and analytical control, biological monitoring allows to correctly estimate and predict deviations in the state of biological systems from the response norm caused by the influence of manmade factors. Although biological monitoring does not allow the recorded effect to be linked to a certain active factor, it gives an integral assessment of the effects of the complex of pollutants on various members of the wildlife.

The joint application of chemical analysis and biological testing techniques allowed Russian environmental scientists to establish that the toxic effect of pollutants can be due to different factors, and its magnitude depends primarily on the chemical properties and biological significance of the active agents, rather than on the concentrations of the individual components. Thus, the need to find sensitive test objects and test systems that adequately reflect the level of man-made exposure comes to the fore.

These challenges require the development of new and improved approaches to environmental assessment and forecasting based on bioin-dication data. Bioindicators are organisms or communities of organisms whose reactions are observed to representative the situation, giving keys to the state of all physical or chemical variables, so that changes in the presence/absence, number, morphology, physiology, or behavior of this species indicate that these physical or chemical variables are outside their preferred limits.

Bioindicators are useful in three situations: (1) where pointed environmental factor cannot be measured, for example, in situations where environmental factors in the past are reconstructed, such as climate change, are studied in paleo-biomonitoring; (2) when this factor is difficult to measure, for example, pesticides and their residues or complex toxic wastewater containing several interacting chemicals; and (3) where the environmental factor is easy to measure but difficult to interpret, for example, whether the changes observed are of environmental significance.

Accordingly, the following categories of bioindicators are identified:

An **environmental indicator** is a species or group of species that predictably responds to an environmental disturbance or change. The system of environmental indicators is a set of indicators designed to diagnose the state of the environment for environmental policy-making.

Ecological indicator: These are species that are sensitive to pollution, habitat fragmentation, or other efforts. The response to this indicator is a response for the environmental community.

Biodiversity indicator: The species richness of the indicator taxon is used as an indicator of the species richness of the community. However, the definition has been extended to "measurable biodiversity parameters," including, for example, species richness, endemism, genetic parameters, population-specific parameters, and landscape parameters.

In practice, bioindication is used to answer the following three questions:

1. Is there a pollutant in the medium
2. Whether the pollutant is capable of accumulating in biological objects
3. Whether the pollutant has biological effect

Bioindication is possible at almost all levels of living matter organization:

Level of macromolecules—changes in concentration, structure, and functioning of biological molecules are taken into account

Cell and organ levels—changes in their structure and functioning are assessed

Level of organism—reactions of individual organisms of indicator species are taken into account

Population level—dynamics of indicator species numbers in disturbed and intact habitats are studied

Level of communities and ecosystems—takes into account species composition of communities, ratios of number of different species, as well as changes in functional indicators of whole communities (e.g., ratio of production processes to destruction).

The first use of organisms as indicators for environmental conditions dates back to the days of Aristotle, which placed freshwater fish in salty water to observe their reactions. Farmers have used plants as bioindicators for thousands of years. There is now an urgent need for robust environmental assessment procedures, as environmental policies (e.g., EU Habitat Directive) focus on the cost-effectiveness and applicability of bioindication systems on a large scale (at least pan-European).

Bioindication information is a significant complement to the environment information from chemical analysis. The combination of both approaches is increasingly part of environmental quality assessment. At the same time, the general ecological state of the environment as a whole can be assessed by the results of accounting and analysis of biological indicators, and direct assessment of physical and chemical characteristics will help to understand which anthropogenic factors most negatively affect the environment and what are the mechanisms of this impact.

In order to use a biological object as a bioindicator, the object must meet certain requirements: high sensitivity with low individual variability; genetic uniformity; the possibility of existing in a wide range of environmental conditions; and easy identification in nature. The results obtained using a particular bioindicator species or test system should be highly reproducible. The test object should have complexity in terms of the possibility of recording on it of biological effects differed in the mechanisms.

Huge experimental work and the involvement of scientific potential from other fields of natural science is needed to justify the correctness of the application of an object for bioindication purposes and to select exactly the criterion that will reflect changes in habitat properties related to man-made factors.

The discussions on the results of the multiyear environmental studies carried out by the International Bioindicator Commission symposia specifically designed to coordinate research on the biological indication and monitoring of environmental pollutants highlighted not only the practical but also the theoretical importance of such research. It is that such studies improve our knowledge of the mechanisms and patterns of response of whole ecosystems and individual species to the simultaneous effects of factors of different nature; contribute to the development of scientifically sound criteria and methods for assessing the risk of pollutants; enable the development of common methodological approaches used to protect the environment at the planetary level and advance the global issue of human–nature interaction, such as the impact of environmental mutagens on human inheritance.

Now ecologists face rather complex and important problems—identification of indicator species, assessment of applicability of certain criteria (biochemical, morphological, and anatomical), and methodological approaches, expansion of applications of bioindicators not only in nature protection but also in scientific research purposes, development of theoretical bases and methodology of analysis of response of biological systems to multifactor effects taking into account differences of pathogenic agents.

In connection with the above, the leading Russian specialists and their colleagues from some other states carried out theoretical and applied research. Most performed studies are devoted to diagnostics of environment status, as well as forecasting of possible risks for environment of diverse activity of entrepreneurs of different countries.

—**A. A. Soloviev**
DSc of Biological Sciences,
Professor of Russian Academy of Sciences,
All-Russian Research Institute of Agricultural Biotechnology,
Deputy Director,
Head of the Laboratory Marker and
Genomic Plant Breeding

Preface

Happiness and well-being not only of our grandchildren and great-grandchildren, but also great-grandchildren of our great-grandchildren will depend on bioecological outlooks of modern generation.

The phrase put forward in the epigraph was formulated as a result of a rethinking of the consequences of man's technogenic activities. Anthropogenic pressing on environment begins not today. As early as prehistoric times, the impact of *Homo sapiens* on habitat neighbors, including other *Homo* species, as well as on the entire animal and plant world, has been very significant; but in recent decades, the anthropological load on the environment has reached global-threatening proportions. The benefits of today's consumption society, which until recently were predominantly accessible to the inhabitants of the most advanced economies, are now taken for granted in many countries with disabilities, but for reasons unclear to them, the standard of living they lack. Such regional inequality causes envy, which provokes the desire to industrialize production and chemize agriculture quickly and at all costs, ignoring the risks of destructive loads on the environment, as well as to serve as a ground for the cultivation of aggression and terrorism, the desire to "take everything away, and divide."

At the same time, the risks of rapid industrialization and chemicalization related to the consequences of industrial pollution, including radioactive waste, as well as violations of the status of protected areas, disappearance of certain species et cetera, are perceived by the public of developed countries mainly negatively. As a result, both entrepreneurs and governments of those countries had to take into account the views of their states' civil society and adopt more or less effective preventive actions relating to the management of natural resources. At the same time, most people in developing countries approve deforestation to free territories for "intensive" agriculture or industrial and housing construction, linking such activities to their perennial aspirations for well-being. Such contradictions are exacerbated by the information pressure around environmental problems. In developed countries, professional protesters have emerged that bring natural conservation and antiglobalization initiatives to

the absurd with their manifestation in irrational forms, eliciting the retalia-
tory irrational opposition of many entrepreneurs and individual officials.
At the same time, the local, regional, and global impacts of the observed
imbalance of biotic interactions in the ecosystem can be minimized by
the use of an environmentally based sociotechnical management strategy,
which requires objective information on the state of the environment, not
identified by pseudo-activists on both sides.

It has long been known that environmental factors have a direct or
indirect effect on the main parameters of ecosystems—density of plant and
animal populations, dynamics of species structure, behavioral characteristics, and so on. Hence, the reverse pattern arises: the state of the ecosystem
elements can be judged on the state of the physical and chemical habitat.
The use of chemical methods of assessing of the environment quality is the
most informative, as it allows to determine the concentrations of various
substances in the environment, compare them with reference concentrations, and conclude the reasons for the changes. However, the researcher
does not always know which substance caused the observed changes;
whether it deals with a single factor or possibly simultaneous exposure to
a complex of several pollutants. In addition, chemical analysis does not
provide information on potential effects on biological objects. The presence of the pollutant in the environment does not per se mean that it will
have a significant impact on living systems.

And here the researcher comes to help the biological objects them-
selves, which by their reaction to living conditions help to determine
biologically significant anthropogenic loads. Diagnosis of early disorders
in the most sensitive components of biotic communities is defined by
the term "bioindication" (in Europe) or "bioassessment" (in the United
States). Biological methods of medium quality control do not require
preliminary identification of influencing factors; they are easy to use,
often express.

Without the use of such bioindicators, it is impossible to determine
correctly the state of biological resources, to develop a strategy for the
rational use of a region, to calculate the levels of maximum permissible
loads on the ecosystem, to identify areas of environmental disasters, and
so on. The application of animal and plant tests is necessary to assess
the effects of radiation, heavy metals, transformable chemicals and other
pollutants, as well as factors such as magnetic fields, noises, the effects of
solar activity, and other phenomena.

In recent years, there has been a steady tradition of assessing the environmental impacts of anthropogenic and extreme natural factors on water bodies by identifying changes in the development and production rates of plant cenoses, zooplankton, and zoobenthos. The criteria for assessing air pollution are quite well developed. For other habitats, however, work is still far from complete.

Despite major advances in bioindication and biomonitoring, there are a number of major environmental problems in general. The most common and significant of these are:

- The problem of complexity of biological systems response to pollutants and within its scope—selection of biological indicators—related to the need to take into account the peculiarities of arrival, accumulation, and excretion of different substances for different biological systems, as well as the need to take into account the peculiarities of individual development, metabolism, and reproduction of the biological object.
- Complexity of assessing effects of different concentrations of different agents of different nature (physical, chemical, and biological) combined in space and time.
- The problem of analysis of specificity, nonspecificity, and the relationship between specific and nonspecific, occurring in response of living systems to damaging effects.

Now ecologists face rather complex and important problems—identification of indicator species, assessment of applicability of certain criteria (biochemical, morphological, and anatomical) and methodological approaches, expansion of applications of bioindicators not only in nature protection but also in scientific research purposes, development of theoretical bases and methodology of analysis of response of biological systems to multifactor effects taking into account differences of pathogenic agents.

In connection with the above, the leading Russian specialists and their colleagues from some other states carried out theoretical and applied research. Most performed studies are devoted to diagnostics of environment status, as well as forecasting of possible risks for environment of diverse activity of entrepreneurs of different countries.

Compiled from the results of these studies, this book, book *Biological Assessment of Natural and Anthropogenic Ecosystems: Trends in Diagnosis of Environmental Stress,* consists of three related common ecological–biological orientation parts.

PART I of the book is titled "Bioindication of Natural and Anthropogenic Ecosystems"

The first two works of this part of the book are devoted to the use
of peatland ecosystems to assess not only existing, but also, importantly
in theoretical and practical terms, the prior environmental conditions
of phytocenosis. Peatlands are unique natural ecosystems that store the
history of phytocenosis and climate development in their location. Peat
is an easy-to-process material, so it is relatively simple in chronological
terms to determine environmental changes. The study of botanical and
spore-pollen composition of peat, as well as the stages of formation of
swamps, is necessary for better understanding of peat-forming processes
and preservation of peatlands. The study of botanical structure demonstrates the alternation of different climatic eras. At each stage, phytocenoses are formed, which are most adapted to these climatic factors,
which is visually reflected in the peat (organic) chronicle of the peatlands.
Application of botanical and paleopalinological analyses of peat allows
characterizing vegetative groups at different stages of peat formation.
Taking into account the biotic preferences of different plant species, it
is possible to restore to some extent the climatic and hydrological conditions of phytocenosis formation with their participation. According to the
Web of Science database, 2422 papers on peat geochemistry and 1355
papers on peat palynology have been published from 1900 to present. In
the same time period, only 247 papers were published using geochemical
and palynological analysis for peat records.

In the work of Ksenya A. Zubkova, Lyudmila N. Shikhova, and Eugene
M. Lisitsyn "Sporous-pollen Bio-indication of Conditions of Peatlands
Formation on European North-East of Russia" pollen, sporous, seeds, and
other parts of plants were used as a bioindicator of living conditions of
natural peatland ecosystems. The authors were able to identify patterns
of association of plant residues with climate change in the distant past.
Analysis of literary data for the last 20–25 years shows that while the
importance of peat ecosystems is beyond doubt on a global scale, detailed
environmental studies are in most cases carried out on upper peatlands
(ombotrophic systems = bogs), but lowland peatlands (minerotrophic
worlds = fens) and transition-type peatlands are much less frequently
studied. The main reason for this is the complexity of the factors affecting
the peat formation of the fens compared to the bogs. The following work
(Elena S. Novoselova, Lyudmila N. Shikhova, and Eugene M. Lisitsyn

"Possibility of Using Cutover Peatbogs in Bioindication of Environmental State") shows that various chemical elements, including heavy metals, accumulate, and persist in peat formation. The authors identified patterns of distribution of heavy metal ions in the soil component of peat biogeocenosis of fens and transition-type bogs due to their drainage and further economic use, as well as their effect on the relative composition of peat phytocenoses. Peatland drainage and their further economic development contribute to the transformation of these natural complexes. Economic use results in active mineralization of the top of the remaining peat column. As a result of these processes, the stability of the biogeochemical cycles of the individual elements is impaired. At the same time, the obtained results allow to consider these peatlands as indicators of the state of the natural environment. Good safety of peat deposits allows to study processes of accumulation and migration of chemical elements.

One of the factors of anthropogenic influence on natural objects can be magnetic field (MF). They are successfully used to pretreat seeds to increase seeding and crop growth strength. There is not much information available about the mechanism arising in magnetically processed organisms, and there is no unique theory. With people making extensive use of the potential effects of high-intensity static magnetic fields when developing new technologies, the biological effects of MF studies are very important. In Irina I. Rudneva and Valentine G. Shaida's chapter "Use of *Artemia salina* Biomarkers for the Evaluation of the Static Magnetic Fields Effects," the authors measured the hatching rate of cysts of artemia and the level of chemiluminescence (CHL) of their extracts. The results show that SMF can affect the metabolism of living systems through various mechanisms. For example, a static magnetic 50-mT field has a clear biological effect on the hatching percentage of *Artemia salina* cysts and the level of chemiluminescence of the nauplia extracts. Direct and indirect mechanisms of effect of magnetic fields on early development of *A. salina* were observed. Static MF irradiation alters the pro-oxidant/ antioxidant status of developing brine shrimp embryos. A separate magnet pole also plays a role in the reaction of the *A. salina* hatching process and the CHL level of nauplia. In general, the authors have shown that *Artemia* is a good tool in ecotoxicological research.

PART II of the book "Genetic Diversity: Experimental Induction and Evaluation of Ecosystems in the Modern Natural Environments" concerns the problems of expansion of biological diversity by artificial mutagenesis, changes occurring in the hereditary apparatus of cells, and the possibility of assessing the effect of such changes in the introduction of wild plants or the creation of new varieties of agricultural plants.

Abiotic and biotic environmental stresses can affect the morphophysiological and biochemical status of a plant, changing its metabolism, growth, and development. Changing climatic conditions present new requirements for the parameters of the designed agroecosystems. Agrocenosis with varietal mixture could use of environmental resources more complete and have greater resistance to stressful environmental factors. The solution to a problem to increase plants' stresses resistance is the creation and use of new breeding material. The driving force behind plant diversity is the process of mutation. Induction of artificial mutations significantly increases the diversity of crops, increases ecological fitness of plants having traits useful for agronomy. Chemical mutagenesis has advantages over radiation in increasing biodiversity for further plant breeding. The chapter of Nina A. Bome et al. "Theoretical and Applied Aspects of Mutations Induction for Improving Agricultural Plants" presents up-to-date data on methods for obtaining and efficient use of induced mutations to create new varieties or source material for breeding. This chapter is a result of cooperative work of scientists of the Tyumen State University, Institute of Biochemical Physics. A.M. Emanuel of the Russian Academy of Sciences (Russia), Scientific and Production Center for Grain Management named after A.I. Barayev (Kazakhstan) and School of Biological Sciences: College of Science and Technology, Flinders University (Australia).

Irina V. Pak et al. ("Enzymatic Activity of Bacteria *Bacillus subtilis*— Assessment Using Phosphemidum") performed the evaluation of fermentative activity of exogenic enzymes of *Bacillus subtilis*, which are the base of numerous probiotic preparations. The following fermentative activities of *B. subtilis* bacteria were determined after the treatment by phosphemidum: amylolytic, protease, lipase, xilanase, carboxymetylcellulase, and β-gluconase. The high fermentative effectiveness was shown under phosphemidum concentrations of 0.0001% and 0.00001%. The authors have shown that with increasing age of the cultures, the activity of the studied enzymes decreases.

The group of Tatyana V. Vostrikova consists of scientists from the Voronezh State University and All-Russian Horticultural Institute for Breeding, Agrotechnology and Nursery (Moscow) presents a set of chapters that deal with cytogenetic changes in plant cells of different species such as *Rhododendron mucronulatum* Turcz., *Rhododendron dauricum* L., *Rhododendron ledebourii* Pojark., *Rhododendron sichotense* Pojark ("Anthropogenic Pollution Influence on the Antioxidant Activity in Leaves and on the Cytogenic Structures in the Seedlings of the Representatives of the *Rhododendron* genus" and "Cytogenetic Characteristics of Seed Seedlings of *Rhododendron ledebourii,* Introduced in the Botanical Garden of Voronezh State University"), and *Betula pendula* ("Cytogenetic Indices, Germination Ability, and Content of Total Protein in Seed Progeny of *Betula pendula* from Various Areas of Voronezh City with Different Levels of Anthropogenic pressure"). They identified positive correlation of antioxidant activity level with the number of vacuolated cells in the apical root meristem, as well as with the level of mitosis pathologies, allow using these indicators for bioindication. The seed progeny of the mutable group shows a decrease in mitotic activity, but an increase of the mitosis pathologies level, which indicates a high cytogenetic instability. The same parameters reveal the stability of maternal plants producing mutable and low-mutable seed progeny. The cytogenetic method may be used for identification of the seed progeny as the mode of separation of the parental plants producing seed progeny with a high level of stability of the genetic material. These investigations may be appropriate for the assessment of the seed quality according to cytogenetic characteristics in parties with unknown origin. The seed progeny of *Betula pendula* has increased indicators of germination ability and total protein in areas with low pollution levels as compared to control group (seeds collected in ecologically clean territory) and to the same parameters for areas with high levels of anthropogenic pressure.

In PART III: Diagnosis of Environmental Status and Plants Adaptive Potential, there are some chapters on use of biochemical and physiological parameters in ecological investigations.

Rafail A. Afanas'ev, Genrietta E. Merzlaya, and Michail O. Smirnov from "D. N. Pryanishnikov All-Russian Scientific Research Institute of Agrochemistry" (Moscow) in their chapter "Photoindication of the Nitrogen Nutrition of Plants" discussed the advantages of photometric diagnostics of plant nitrogen nutrition in comparison with chemical methods. They took experimental substantiation of photometric diagnostics while using ground and remote methods of diagnostic works in sowings of cereal and other agricultural crops. On the basis of obtained data authors made recommendations on the use of unmanned aerial vehicles (drones) for the diagnostics of agricultural crop nitrogen nutrition. The practical significance of these recommendations is that the development of drones for the agricultural interests can significantly reduce the cost of agricultural production by replacing the previous labor-intensive operations for the diagnostics and optimization of plant mineral nutrition with modern high-performance methods.

Vegetation monitoring of the road interchange slopes Tyumen—Krivodonovo since the commissioning (1996) to the present time was made by Nina A. Bome and colleagues (Institute of Biology, University of Tyumen (Russian Federation) and State University of New York College of Environmental Science and Forestry (Syracuse, NY, USA) in the chapter "Revegetation and Launching Self-restoration Process of the Disturbed Landscape along the Transport corridor in Western Siberia." To develop of a plant formation, they included grass mixture with four species plants (*Bromopsis inermis* L., *Festuca pratensis* L., *Brassica napus*, and *Amaranthus cruentus* L.) and wild plants. The increasing of herbaceous plant species and the emergence of trees and shrubs were found out. There were found major differences in a ratio of species, biomass, and phenological stages of plants in different layers of both the slopes. More favorable conditions for plants and the formation of seeds were on the southern slope of the site.

In the chapter "Pigment Content in Plant Leaves as Bio-indicator of Adaptability to Growing Conditions", Eugene M. Lisitsyn considered a problem of application of data on the content of pigments in plant leaves for (1) diagnostics of resistance to stressful abiotic factors; (2) differentiation of closely related taxonomic groups of plants on the example of two subspecies of oats: covered oats (*Avena sativa* subsp. *sativa* L.) and naked oats (*Avena sativa* subsp. *nudisativa* L.). It is shown that under influence of a stress of aluminum ions, conditions of harvesting of light energy change in high degree, whereas processes of its transformation into organic substances in the reaction centers are rather protected from a stressor. As a result, author concluded that the content of photosynthetic pigments can be used as a diagnostic indicator of resistance to environmental factors (air temperature, conditions of moistening, and existence of toxic ions of metals in the soil) at cultivars and breeding lines of cereal crops and as a diagnostic indicator of belonging to different genetic taxa.

And, at last, in work of Olga A. Opalko and Anatoly I. Opalko "Some Indirect Methods for Predicting the Rooting Ability of Apple Tree

(*Malus* spp.) Stem Cuttings" with predicting the rooting ability of apple tree (*Malus* spp.), stem cuttings performed by authors from National Dendrological Park "Sofiyivka" of NAS of Ukraine (Uman, Ukraine) have been found that radish seeds and bean cuttings can be used as biotesters with certain premonitions, and sprouted seeds of garden cress do not meet the requirements of the regenerative capacity tester of the stem cuttings of apple tree.

Thus, the authors hope that proposed results of the research given in this book, *Biological Assessment of Natural and Anthropogenic Ecosystems: Trends in Diagnosis of Environmental Stress*, will serve as a theoretical-practical base for the bioindication of ecosystems changes, designing new methodologies of assessment of environmental state. The authors, editors, and other participants of the project on the manual creation hope that the materials offered to the reader will help him orient in the most difficult and multiplane problem of environmental protection.

—Eugene M. Lisitsyn, DSc
Assistant Professor, N.V. Rudnitsky North-East Agricultural Research Institute, Russian Academy of Sciences, Kirov, Russia;
Professor of Ecology and Zoology,
Vyatka State Agricultural Academy, Kirov, Russia;
Professor of Cathedra of Ecology and Zoology,
Vyatka Agricultural Academy, Kirov, Russia

PART I

Bioindication of Natural and Anthropogenic Ecosystems

CHAPTER 1

Possibility of Using Cutover Peat Bogs in Bioindication of Environmental State

ELENA S. NOVOSELOVA[1], LYUDMILA N. SHIKHOVA[2], and
EUGENE M. LISITSYN[2*]

[1]*Vyatka State Agricultural Academy, 133 Oktyabrsky Av.,
Kirov, 610017, Russia*

[2]*Federal Agricultural Research Center of the North-East named N.V.
Rudnitsky, 166a Lenin Street, Kirov, 610007, Russia*

Corresponding author. E-mail: edaphic@mail.ru

ABSTRACT

Two peatlands in central part of Kirov region (Zenginsky and Karinsky) are investigated. During three field seasons, soil profiles were made and peat samples by layers were collected. The content of total and mobile forms of zinc (Zn), copper (Cu), lead (Pb), and cadmium (CD) were determined by method of an inversion volt-amperemetry. It is revealed that content of the studied elements in general does not exceed average values for various soil types of area. On light-drained undeveloped part of a Zenginsky peatland, the content of total Zn in different profile layers varied from 0.16 to 2.60 mg/kg, mobile—from trace values to 1.60 mg/kg. The content of Cu changed from trace values to 0.90 mg/kg for total and up to 0.12 mg/kg for mobile forms. The content of total Pb fluctuated from 0.37 to 5.60 mg/kg, mobile forms—from trace to 2.90 mg/kg (Fig. 1.1). The low content of Cd is noted (up to 0.59 mg/kg of total and 0.038 mg/kg mobile). Biogenous and accumulative type of accumulation in top layers of peat deposit is accurately traced for all four elements. Development of a peatland led to change in content and redistribution of chemical elements in newly formed profiles. The content of total Cu varied from trace values up to 34.00 mg/kg; Zn—up to 126.00 mg/kg, in the lower mineral part of a profile—up to 40.00 mg/kg. The content of total Pb in profiles varied from trace values up to 17.00 mg/kg; the content of mobile forms did not exceed 2.30 mg/kg. Content of total Cd did not exceed 1.00…1.20 mg/kg on average. Nevertheless, on some sites concentration of an element in separate layers reached 4.41…8.60 mg/kg. Content of mobile Cd is noted as trace. Residual layers of peat of the Karinsky peatland were also characterized by the insignificant content of heavy metals. Content of total Zn varied within 4.20…62.00 mg/kg, mobile form—from trace up to 1.46 mg/kg. Content of total Cu made 3.70…22.0 mg/kg, mobile forms—up to 0.69 mg/kg. Content of total Pb changed from trace values up to 4.10 mg/kg; mobile forms had trace concentration. Content of total and mobile Cd did not exceed 1.90 mg/kg. The botanical composition of peat substantially influenced the content of chemical elements: it is revealed accumulation of the studied metals in layers with prevalence of wood remains—*B. pube-scens* Ehrh., *P. sylvestris* L., and *P. abies* (L.) H. Karst. as well as *Typha sp.* and *Phragmites sp.* Economic use leads to an active mineralization of the top part of the remained peat mass. Profiles of the developed peat bogs are characterized by considerable fluctuations in content of heavy metals both in different profiles, and in the different horizons of the same profile. These facts indicate occurring of reorganization of soil processes and modes in the developed and drained soils.

Plants growing on dried peat soils accumulate small amounts of zinc, copper, lead, and cadmium in vegetative mass, which do not exceed the standards of maximum permissible concentrations and are close to average values of zinc, copper, lead, and cadmium content in plants growing on automorphic soils of Kirov region. Among the plant species under study, *Rosa majalis* accumulates zinc, lead, and cadmium the least. *Salix caprea* leaves revealed the highest content of all four elements studied.

1.1 INTRODUCTION

The term "**peatland (mire)**" has many definitions that differ significantly from each other. This lack of agreement among scientists is due to the complexity of this natural object and the great variety of types of peatlands. For example,[1] definite **peatland** as organic wetland ecosystems character-ized by an accumulation of peat. Peatlands contain more than 40 cm of peat accumulation on which organic soils (excluding folisols) develop;[2] indicates that "**peatlands** are areas with or without vegetation with a naturally accumulated peat layer at the surface". Further, peatlands where peat is currently being formed they call "**mire**"; bog and fen is two main types of mire. A **bog** is mire that due to its location relative to the surrounding landscape obtains its water mainly from precipitation and not influenced by groundwater; *Sphagnum*-dominated vegetation (ombrotrophic), while a **fen** is located in depressions or on slopes and is fed from both groundwater and rainfalls (minerotrophic); vegetation cover composed dominantly of graminoid species and brown mosses. Thus, a bog is always acidic and nutrient-poor, a fen may be slightly acidic, neutral, or alkaline, and either nutrient-poor or nutrient-rich.

D.H. Vitt[3] and C. Craft[4] also recognized peatlands (or mires) as ecosystems that contain deep accumulations of decomposed organic material (derived from the remains of dead and decaying plant material), or peat. L.L. Bourgeau-Chavez et al.[5] pointed out that peatlands having saturated soils, anaerobic conditions, and large deposits of partially decomposed organic plant material (peat).

Russian scientists N.I. Piyavchenko[6] gives the following definition: "The peatland is a geographical landscape naturally occurring and developing under the influence of interaction of environment and vegetation factors, which is determined by constant or periodic humidity and manifests itself in the hydrophilicity of ground vegetation cover, bog-type of soil formation process, and the accumulation of peat."

All peatland ecosystems are characterized by constant long-term stagnant or flowing moistening, which determines the specific nature of their development, the presence of different life forms of plants, a special type of soil formation, expressed in peat deposition. In Russian tradition, peatlands by nature of mineral nutrition are divided into three types—fen, transitional-type mire, and bog. Waters of fens are rich in mineral salts, with the ash content of the top substrate layer above 6–7%. Transition-type mires have average content of mineral substances with ash content from 4 to 6–7%. Feeding of bogs is usually carried out at the expense of mineral-poor atmospheric precipitation; the ash content of the substrate is less than 4%.

Almost all countries of the world have peatlands. Canada and Russia have the most extensive areas of peatlands although extensive areas are found in northern Europe, especially Fennoscandia, and in the tropics of

Indonesia. Different type peatlands of Russia make up to 369.1 million ha or 21% of the country.[7] Peat resources are the richest natural potential. But peat bogs are also unique natural formations that perform many of the most important biosphere functions: they preserve huge reserves of fresh water, deposit carbon, largely determine the water and hydrological regimes of territories, serve as natural filters liberating the Earth's atmosphere from toxic elements.

Peatland biogeocenoses are an important constituent of the natural complexes of the boreal zone. Peatlands are involved in regulating the territory's climate and water exchange. The formed organic substance is stored for long periods, reducing carbon dioxide release back into the atmosphere.[8] During the formation of the peatland, various chemical elements, including heavy metals, such as Pb, Cu, Zn, and Cd, are accumulated and preserved in the peat.[9,10]

Due to the specific properties of peat (presence of a large number of acidic functional groups in the composition of organic matter),[11] peatland systems are huge natural accumulators, which are able to accumulate and preserve many millennium different chemical elements and substances.[12]

Peat drainage and extraction lead to destruction and sharp change of the conditions of peatlands' existence,[13] to transformation of the bog soils toward the zone type of soil formation. The fate of heavy metals after the development of peatlands is not clear. On the one hand, active mineralization of organic matter as a result of drying should contribute to increasing mobility and migration capacity of most elements.[14] On the other hand, changes in hydrological, redox, acid-alkaline, and other regimes result in geochemical barriers on which sedimentation of heavy metals is possible. Therefore, the content and behavior of chemical elements in peatland soils will depend on the specific geochemical situation.

Analysis of literary data over the past 20–25 years has shown that while the importance of peatland ecosystems is beyond doubt on a global scale, detailed environmental studies are in most cases carried out on upper peatlands (ombrotrophic systems = bogs), but low-lying peatlands (minerotrophic mires = fens) and transition-type peatlands are studied much less frequently. The main reason for this is the complexity of factors affecting the peat formation of fens compared to bogs.[15]

Kirov region of Russian Federation belongs to the belt of intensive peat accumulation and is considered one of the richest in peat resources on the territories of the Volga-Vyatka economic region. The peatlands are most common on flat, smoothed, and lowered water-covered areas of the northern, north-eastern, and central parts of the region, on low-lying sections of river floodplains and terraces. In the region, there are peatlands of all classification types—fens (57%), bogs (23%), transition-type mires (10%), and mixed mires (10%). A lowland type of peat formation prevails. Even bogs tend to have a more or less significant lower layer of fen-type peat. Large peatlands mostly consist of areas of all types. In addition, quite a large number of small fen-type deposits can be found in the areas of distribution of bog-type peatlands. About 270 fields of peat deposits are currently disturbed to varying degrees due to the industrial development of the peat fund of the region.[16] Due to the appearance of large areas of cutover peatlands, plantations of pine, spruce, larch, oak, poplar, and currants have been planted on some peatland[17] (including Zenginsky) since 1961. In the 1970–1980, a large-scale attempt was made to use peatlands for the development of fodder production, but due to the shortage of mineral fertilizers and other economic reasons of the 1990s, the activity of agricultural use of these workings has sharply decreased.[18]

The purpose of this work is to identify the patterns of distribution of heavy metal ions in the soil component of peatland biogeocenoses of bogs and transition-type mires in connection with their drainage and further economic use of cutover peat bogs as well as their influence on specific composition of peatland phytocenoses.

1.2 MATERIALS AND METHODOLOGY

The Kirov region occupies the eastern part of the Russian Plain. On the available data, in the 70th years of the 20th century in the territory of the area, 1734 peat fields were revealed and explored, by the end of the 90th years in the area 1858 peat fields are revealed and described. The area of peat deposits of the region is almost 510 thousand ha.[19] The largest peat bog masses are drained and mastered. The largest distribution of the peatlands was on wide flat, smoothed, and lowered water-covered areas of the northern, north-eastern, and central parts of the region, on low-lying areas of river floodplains and terraces.[20] A feature of the peatlands of the Kirov region is that the underlying bedrocks in the southern part of the region are carbonate rocks of Paleozoic age, and in the northern part—carbonate-free deposits of Moscow and Dnieper ice.

Two peat bog massifs in the central part of the region (Zenginsky and Karinsky) were chosen for the study. The **Zenginsky peatland** is located in the Orichi district of the Kirov region on the first fluvial terrace above floodplain of the Vyatka River. Peat-laying rocks are alluvial and ancient alluvial deposits of light granulometric composition, which in turn are laid at different depths by carbonate loams and carbonates.[20] The area is correlated with lowland, an ancient channel of glacial flows. The predominant type of peat formation is lowland, but in the central, most deeply lying parts of the deposit there are mixed and top areas. The area of the peatland is about 6000 ha, the maximum capacity of the peat layer is 5.0 m, the average −1.71 m. Peat bog massif has been developed since 1949, now peat extraction is stopped. The control site on the peat bog massif was chosen in the ledum-moss pine forest on the light-drained undeveloped part of the peatland with the capacity of peat deposit about 1.5–2.0 m. Key areas were chosen in the territories with natural meadow and forest phytocenoses, which formed after the drainage and peat extraction, and in the territories involved in economic activity. The territories involved in economic activity are occupied with plantations of *Pinus silvestris* L. and *Picea abies* (L.) H. Karst. which ages varied from 9–11 to 35–40 years, or are used for sowing fodder crops. In these areas, the capacity of residual peat layer was from 1.3–1.5 to 0.5 m or less. Geographical sampling points have the following coordinates: control point 58°31′07.6″N, 48°59′04.0″E; developed areas—1. 58°30′09.5″N, 48°59′03.1″E; 2. 58°31′18.6″N, 49°00′47.3″E; 3. 58°30′40.2″N, 48°59′34.7″E.

The **Karinsky peatland** is located in Kirovo-Chepetsk district of Kirov region in the floodplain of the Cheptsa River. A dismembered ancient-dune sand relief characterizes the territory. By the nature of the deposit structure and peat formation conditions, bog-type peat deposits predominate. The area of the peatland is 8286 ha, the maximum capacity of the peat layer is 6.0 m, the average—2.21 m. Peat bog massif has been developed since 1943. On the peatland Karinsky there was not defined a control site due to the fact that the peatland is drained, and the territory is actively developed (a sufficiently large layer of peat has already been removed; peat production continues today). Peat deposit capacity varies from 0.8–1.5 m to 0.1–0.5 m in different areas. Key sites were laid in the territory where peat mining is currently under way, as well as in the territories with natural phytocenoses, which were formed as a result of the recovery after the drainage and extraction of peat. Geographical sampling points have the following coordinates: Developed areas—1. 58°35'13.6"N, 50°15'28.2"E; 2. 58°35'00.7"N, 50°13'55.8"E; 3. 58°34'54.8"N, 50°13'00.5"E; 4. 58°34'42.8"N, 50°15'01.0"E.

During three field seasons, four trial platforms were putted on each key site, where geobotanical description was carried out,[21] soil sections were made; soil and peat samples were taken by layers. Peat layers were separated visually by differences in the botanical composition of the column. According to conventional methods, the degree of peat decomposition and botanical composition were determined in the field.[22] In the following under laboratory conditions, the botanical composition was specified with more detailed study.[22,23] Analytical treatment of the collected material was carried out in the laboratory of the Department of Ecology and Zoology of the Vyatka State Agricultural Academy and Laboratory of Edaphic Plant Resistance of Federal Agricultural Research Center of the North-East named after N.V. Rudnitsky (Kirov). The content of the total and mobile forms of zinc, copper, lead, and cadmium was determined by inversion voltamperometry on an analyzer TA-4 (Tomsk Polytechnic Institute, Russian Federation) according to the manufacturer's methodological recommendations.

The tables show the average values of heavy metals content (Zenginsky peatland: control point—3 years × 5 repeats; for the rest—3 years × 3 plots × 5 repeats; Karinsky peatland—3 years × 4 sections × 5 repeats).

1.3   RESULTS AND DISCUSSION

1.3.1   CONTENT OF HEAVY METALS AND THEIR DISTRIBUTION ON A SOIL PROFILE OF CUTOVER PEAT BOGS

Obtained data have shown that content of the chemical elements under study does not exceed the average and background values characteristic to the different soil types of the region. From the literary sources and data of our previous studies, it is known that content of many chemical elements in peatland soils does not exceed 60 mg/kg (Table 1.1).

However, in the process of transformation of peatland soils toward the zone type of soil formation there is active mineralization of organic mass and involvement of preserved chemical elements in a biogeochemical cycle.

TABLE 1.1   Limits in Content of Investigated Elements in Soils of Kirov Region (mg/kg).

Heavy metal	Sod-podzolic soil[1]	Peat[2]		Average content in Earth crust[3]
		Bog-type peat	Transition-type peat	
Zn	30.00–70.00	8.50–56.10	3.60–30.60	70.00
Cu	45.00–50.00	2.00–6.00		60.00
Pb	5.00–43.00	–	–	14.00
Cd	0.70–1.03	–	–	0.15

[1]According to [24].
[2]According to [19].
[3]According to [25].

In order to assess whether peat is contaminated by a specific element, it is necessary to compare its content in peat with the average content in the Earth's crust (RI environmental risk index). According to Håkanson,[26] if the RI value is lower than one, it is possible to propose low degree of contamination; if the value is 1–3—moderate, 3–6—strong, and if it exceeded 6—very strong contamination. As can be seen from the data of table, on average for the Kirov region, the content of the test elements in peats corresponds to a low degree of contamination; although in sod-podzolic soil, which is the most common in the region, moderate lead contamination and severe/very severe cadmium contamination is noted.

At the control site (light-drained undeveloped part) of the peat bog massif Zenginsky, the lowest content of elements in the profile was observed compared to other surveyed territories (Table 1.2). Content of total zinc forms in different layers of soil profile varies from 0.16 to 2.60 mg/kg, mobile ones—from trace values to 1.60 mg/kg. The content of copper in peat varies from trace values to 0.90 mg/kg for total and to 0.12 mg/kg for mobile forms.

Increased content of both total and mobile compounds of both elements (Zn and Cu) in upper layers is noted, which coincides with literature data.[10,11,27] The increase in the content of these elements in the upper layers is due to the biogenic accumulation of these elements by plants. The concentration of zinc and copper in the underlying layers is decreased to trace values.

The content of total lead in the soils of the control site varies from 0.37 to 5.60 mg/kg, and of mobile forms from trace to 2.90 mg/kg (Table 1.2). The maximum content of the element in the soil profile is typical for the upper peat layers (0–20 cm).

TABLE 1.2 Heavy Metals' Average Content in Peat of Light-Dried Undeveloped (Control) Site of Zenginsky Peatland (mg/kg, Average Means of Metal Content (3 Years × 5 Repeats)).

Form of element	Soil layer, cm				
	0–14	15–29	30–59	60–89	90–105
		Zn			
Total	2.40	2.60	0.18	0.45	0.16
Mobile	1.60	0.17	0.01	0.34	0.02
Cu					
Total	0.90	0.68	0.28	0.04	0.00
Mobile	0.11	0.01	0.12	0.00	0.00
		Pb			
Total	5.60	3.60	0.42	0.37	1.80
Mobile	2.90	0.86	0.19	0.09	0.01
Cd					
Total	0.59	0.29	0.05	0.06	0.05
Mobile	0.04	0.02	0.01	0.02	0.00

Cadmium content in peat samples of the control site does not exceed background values. Low content of element in peat layers is noted (from trace values to 0.59 mg/kg total and to 0.038 mg/kg mobile).

Cadmium and lead are not vital elements for plants.[28] Despite their high phytotoxicity, some amounts of elements accumulate in phytomass. The accumulation of both lead and cadmium in the uppermost layers of peat may be due to both the biogenic accumulation of these elements and atmospheric anthropogenic pollution. According to Nikodemus et al.,[29] the decrease in the content of lead in the atmosphere in recent decades leads to an increase in the content of the element in the lower layers of the peat deposit. According to Veretennikova,[30] at present in peatlands remote from industrial sources of lead located in regions with low level of economic activity, lead concentration and its accumulation rate is mainly determined by the global level of atmospheric pollution typical for the beginning of the industrial era.

According to Gashkova and Ivanova[31] on the sites where peat extraction was carried out, plants accumulate cadmium ions most of all; the biological absorption coefficient of elements such as zinc, cadmium, lead, and copper is significantly higher in areas of high-disturbance peatlands than that of undisturbed areas.

In general, the low content of heavy metals in peat soils of the control site is due to peculiarities of peatland formation, as well as low content of elements in the sand rocks lining the peat deposit. For all four elements, there is accurately traced biogenic-accumulative accumulation in the upper layers of peat deposit. The input of elements with atmospheric flows at man-made atmospheric pollution also cannot be excluded. In this case, however, man-made contamination of peatlands is not considered as such. These territories are considered to be subject to minimal human economic activity.

The development of peat bog massif led to a change in content and redistribution of chemical elements in newly formed soil profiles. There are more sharp variations in the content of elements in depth in all investigated profiles (Table 1.3) compared to soils of undeveloped part of the massif (Table 1.2), which is consistent with the literary data concerning the analysis of heavy metals' content in disturbed peatlands.[32]

TABLE 1.3 Heavy Metals' Average Content in Peat of Developed (Cutover Peat Bog) Sites of Zenginsky Peatland (mg/kg, Average Means of Metal Content (3 years × 3 sites × 5 Repeats)).

Form of element	Soil layer, cm							
	0–15	16–34	35–57	58–102	103–124	125–129	*130–149	150–163
	Zn							
Total	9.3	41.0	29.0	113.0	35.0	97.0	40.0	14.0
Mobile	1.7	2.1	3.4	3.8	1.2	0.2	1.6	5.0
Cu								
Total	18.0	25.0	4.4	3.8	33.0	25.0	14.0	3.7
Mobile	0.2	0.7	0.0	0.0	0.2	1.0	0.1	0.0
	Pb							
Total	0.9	2.6	1.1	5.5	17.0	2.8	7.4	2.4
Mobile	0.1	0.8	0.5	1.0	1.0	0.4	0.8	2.0
Cd								
Total	0.5	2.4	0.2	2.5	8.6	0.6	1.1	0.7
Mobile	0.4	0.2	0.1	0.1	trace	trace	0.1	0.1

*Transition from peat deposit to mineral bedding rocks.

Content of the total copper forms varies from trace values up to 34.00 mg/kg; of Zinc in residual peat layers—from trace values to 126.00 mg/kg, and in the lower mineral part of soil profile—up to 40.00 mg/kg.

Increased copper and zinc concentrations (relative to the control site) are found in the upper horizons of some profiles in areas with well-developed plant cover. Both zinc and copper are biogenic elements actively absorbed and accumulated by plants.

The total lead content in the profiles varies from trace values up to 17.00 mg/kg, but the content of mobile forms of lead does not exceed 2.30 mg/kg.

The content of total cadmium compounds in the layers of the examined profiles on average does not exceed 1.00–1.20 mg/kg. At the same time, on some areas increased concentrations of element in separate layers of peat deposit are revealed, which reach 4.41–8.60 mg/kg. Content of mobile forms of cadmium in the profile of cutover peat bog soils is characterized as trace.

Unlike zinc and copper, cadmium and lead do not belong to biogenic elements. However, in some profiles it was noted an increase in the content of elements in the upper part of soil profile. On the one hand, this may be due to the biogenic accumulation of the element by living organisms and the active mineralization of the upper peat layers. Nevertheless, this may also be due to the presence of anthropogenic sources of cadmium input into the environment as a result of drainage and peat mining in the territory.

Variations in heavy metal content in the thickness of peat are obviously due to different reasons. One of the reasons is different botanical composition of peat layers.[33] Wood grass peat is known to concentrate molybdenum more than others do; wood-sedge peat—nickel and bismuth; wood-sphagnum—copper and zinc; and herbal—manganese and cobalt.[22,34] The scientific literature noted an increase in zinc, lead, and cadmium content in peat layers dominated by residues of *Scheuchzeria palustris* L. (Rannoch-rush or pod grass), *Menyanthes trifoliata* L. (bogbean or buckbean), *Carex lasiocarpa* Ehrh. (woollyfruit sedge or slender sedge), *Eriophorum* sp. (cottongrass), hardwood, sphagnum, and hypnum mosses are higher than peat of other botanical composition.[35]

In some cases, fluctuations in the content of the element are due to change of peat deposit with mineral underlying rocks. Transition (contact) horizons are compacted and have impurities of silted and clay particles in peat composition. On the boundary between peat thickness and mineral layers there is a sharp change of soil conditions.[34] Transition horizons tend to delay moisture along with chemical elements seeping from the upper layers with water. As a result, there may be an increase in the content of elements in such horizons.

The residual peat layers of the **Karinsky** peatland are also character-ized by a low content of total and mobile forms of heavy metals. The content of total zinc in peat varies within 4.20–62.00 mg/kg, the content of mobile compounds—from trace amounts to 1.46 mg/kg. The content of total copper varies within 3.70–22.0 mg/kg and the content of mobile metal forms vary from trace values to 0.69 mg/kg (Table 1.4).

TABLE 1.4　Heavy Metals' Average Content in Profile of Karinsky Peatland (mg/kg, Average Means of Metal Content (3 Years × 4 Sites × 5 Repeats)).

Form of element	Soil layer, cm								
	0–9	10–26	27–57	58–76	77–92	93–120	121–137	*138–145	146–155
					Zn				
Total	0.1	4.5	4.6	0.4	3.0	1.3	5.4	0.1	0.6
Mobile	0.0	Trace	0.4	0.0	0.0	Trace	Trace	0.0	0.4
					Cu				
Total	1.0	2.5	2.8	3.6	1.1	0.8	1.2	3.0	5.4
Mobile	0.1	0.0	0.1	0.0	0.0	0.0	trace	0.0	Trace
					Pb				
Total	0.1	0.5	0.7	0.2	0.1	Trace	1.1	1.4	2.5
Mobile	0.0	Trace	0.1	Trace	Trace	Trace	Trace	0.7	0.4
					Cd				
Total	Trace	0.1	0.1	Trace	Trace	Trace	0.1	0.3	1.2
Mobile	0.0	Trace	Trace	0.0	Trace	Trace	Trace	0.0	Trace

*Transition from peat deposit to mineral bedding rocks.

The lead content of peat and soils of this peat massif is within the background values typical of the Kirov region soils (Table 1.4).

Content of total forms of element varies from trace amounts up to 4.10 mg/kg; mobile forms are in trace concentrations. The content of both total and mobile forms of cadmium in peat and soils of peat massif Karinsky is low and varies from trace amounts up to 1.90 mg/kg for total forms. Low content of element in peat is caused by its low content in soil-forming rocks. In this area, they are represented by ancient alluvial sands, which are characterized by low cadmium content. Some key sites show an increase in lead and cadmium content at the top of soil profile. The accumulation of elements in this part of profile is likely due to active mineralization of top layers of peat deposit. As a result of peat destruction, preserved chemical elements are released from semi-decomposed plant mass. In this case, man-made pollution is not considered as such—the territory of the peat massif is subject to minimal impact from human production activities.

Study of botanical composition of peat of the Karinsky peatland showed a relatively poor species diversity of peat-forming plants.[35] Lower layers of the deposit in the peat are dominated by the remains of herbs—bogbean (*M. trifoliata* L.), different types of sedge (*Carex* sp.), the remains of horsetail (*Equisetums* sp.), cattail (*Typha* sp.), and common reed (*Phragmites* sp.), which in the upper layers are replaced with remains of wood plants—downy birch (*Betula pubescens* Ehrh), Scots pine (*Pinus sylvestris* L.), and Norway spruce (*Picea abies* (L.) H. Karst.). It is obvious that botanical composition of peat significantly affects the content of chemical elements. As a result of the studies carried out, the accumulation of the investigated elements in peat with the dominance of wood residues—*B. pubescens* Ehrh., *P. sylvestris* L., and *P. abies* (L.) H. Karst. and the residues of *Typha* sp. and *Phragmites* sp. In some sites increased zinc content in peat up to 27.00–46.00 mg/kg is noted. Lead content reaches 3.50–3.90 mg/kg. The content of cadmium in peat, formed from the remains of herbs (cattail and common reed), and wood (downy birch, Scots pine and Norway spruce), increases up to 0.45–1.50 mg/kg.

1.3.2 TRANSFORMATION OF PLANT COMMUNITIES OF PEATLAND BIOGEOCENOSES AFTER CUTOVER OF PEAT BOG

As noted above, in the development of peatlands there is also a transformation of plant communities in peatland biogeocenoses. Typical plant representatives of transition-type bogs and fens are species of the genus *Drosera* (sundew), various species of *Carex* (sedge), *Ledum palustre* L. (marsh Labrador tea), *Chamaedaphne calyculata* L. (leatherleaf), and a number of other species, as well as developed moos cover. At the same time, the species diversity of higher plants of peatland phytocenoses is low.

The discovered set of plant species on the control light-drained undeveloped site of Zenginsky peat massif gives an approximate view of plant composition of the peatland phytocenoses before peat drainage and extraction. The dominating types on the control site are marsh Labrador tea, leatherleaf, and some berries—cowberry (*Vaccinium vitis-idaea* L.), blueberry (*Vaccinium uliginosum* L.), bilberry (*Vaccinium myrtillus* L.) (Fig. 1.1).

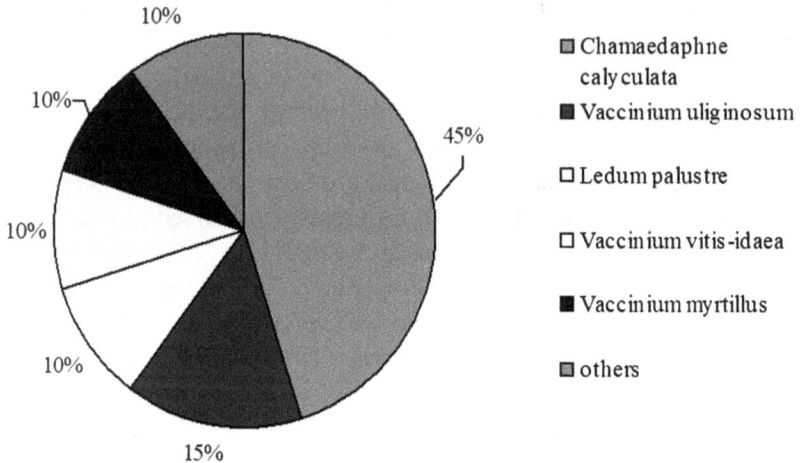

FIGURE 1.1 Species composition of the phytocenose of light-drained undeveloped (control) site of peat massif Zenginsky (projective cover of the species expressed as percentage).

Well-developed moos layer consists mainly of sphagnum mosses—*Shp. Magellanicum* Brid., *Shp. angustifolium* (Warnst.) C.E.O. Jensen, *Shp. fuscum* (Schimp.) H. Klinggr., and some others. *Pleurozium schreberi* (Brid.) Mitt. (red-stemmed feather moss) is also found. Typically, these species are characteristic of transition-type and bog peatlands.

After drying the peatland and peat mining, the plant cover changes radically. Generally, when the peat residue is drained and reduced, the participation rate of the peatland species in phytocenosis decreases. For example, with a residual peat layer capacity of 1.1 to 1.8 m at a number of key sites, the typical peatland species—*Drosera rotundifolia* L. (common sundew), *Trichophorum alpinum* (L.) Pers. (alpine bulrush), and *Parnassia palustris* L. (grass of Parnassus or bog star) are still dominant within phytocenoses. Nevertheless, at the same time there appear ruderal species, which on areas with peat capacity less than 1.0 m replace peatland species and turn to the category of dominant. Among them, there are *Antennaria dioica* (L.) Gaertn. (catsfoot), *Hieracium umbellatum* Michx. (Canadian hawkweed) and *Hieracium pilosella* Vaill. (mouse-ear-hawkweed), *Potentilla norvegica* L. (Norwegian cinquefoil) and *Potentilla goldbachii* Rupr. (European cinquefoil), and *Linaria vulgaris* Mill. (common toadflax) (Fig. 1.2). They were probably brought in by machinery when draining the peatland and mining peat or wild animals.

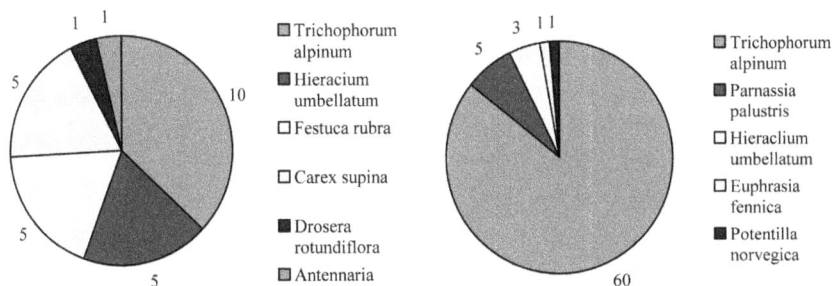

FIGURE 1.2 Species composition of the phytocenose of some areas with beginning stages of natural overgrowth on drained developed sites in peat massif Zenginsky (projective cover of the species expressed as percentage). The rest area is not covered with graminaceous plants.

At later stages of natural overgrowing of developed peatlands and at artificial reforestation at low capacity of peat layer (up to 50 cm) in phytocenosis there appear wood species, under the shelterwood of which a new plant community will be formed in the future. In natural restoration, the structure of the wood layer deciduous species are first dominated—*B. pubescens* (fluffy birch), *B. pendula* Roth., (silver birch), and their hybrid. In the following, coniferous species appear—*P. sylvestris* and *P. abies*. The herbal-shrub layer is dominated with light-demanding plant species—*Fragaria vesca* L. (alpine strawberry), *Galium mollugo* L. (hedge bedstraw), *Chelidonium majus* L. (greater celandine), *Urtica dioica* L. (stinging nettle), *Rubus idaeus* L. (red raspberry) (Fig. 1.3). These species are typical of mixed forests of southern taiga.

Pyrola rotundifolia L. (round-leaved wintergreen), *Majanthemum bifolium* (L.) F.W. Schmidt (false lily of the valley or May lily) and some other species characteristic of coniferous forests (Fig. 1.4) are growing at dominance of *P. sylvestris* and *P. abies* in the tree stands composition.

Subsequently, plant cover will affect the input and accumulation of chemical elements in the soil. In particular, in the upper horizons of soil profile, some metals will accumulate due to their biogenic accumulation by plants and plant residues.

Thus, after drying the peatland and extraction of peat, the cutover peat soils gradually approach the zonal soil types of the adjacent territories. The formation of a new soil profile is gradually proceeding, and the produced peat soils change in their characteristics in the direction of formation of a zone type of soil formation. On the peat massif Zenginsky, it is possible to trace the natural change of some plant communities to others. Temporary biogeocenoses, unable for a long time to be in a state of stable balance, are in succession, transforming into stable indigenous biogeocenoses. At the same time, according to the results obtained during the research, it is already possible to indicate some stabilization of natural processes and conditions in this territory.

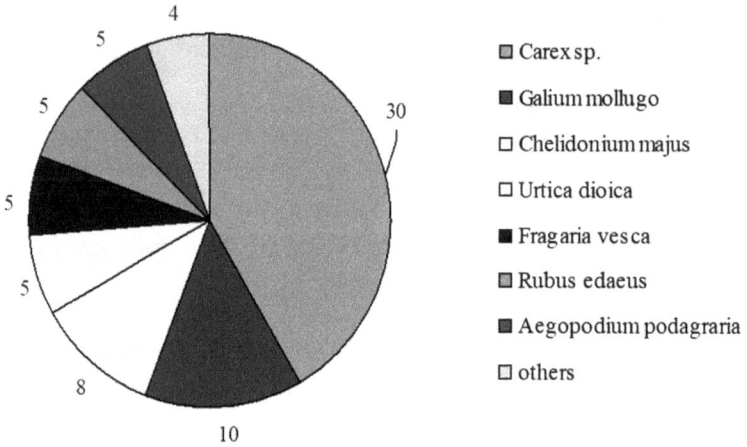

FIGURE 1.3 Species composition of the phytocenose of areas with latest stages of natural overgrowth on drained developed sites in peat massif Zenginsky (projective cover of the species expressed as percentage). The rest area is not covered with graminaceous plants.

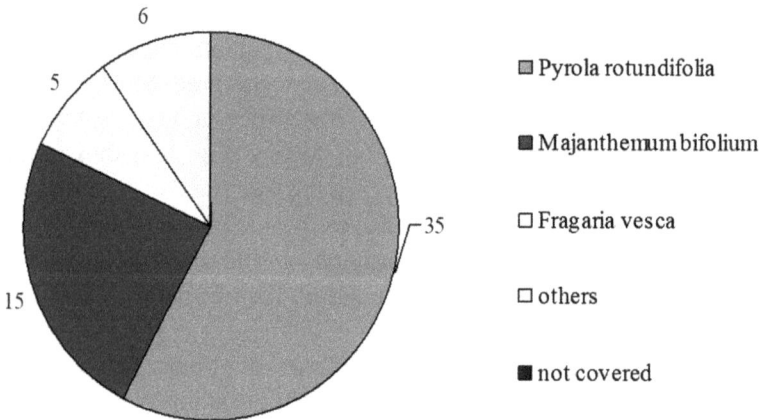

FIGURE 1.4 Species composition of the phytocenose with natural overgrowth on drained developed sites in peat massif Zenginsky at dominance of *P. sylvestris* and *P. abies* in the tree stands (projective cover of the species expressed as percentage). The rest area is covered with woody plants.

1.3.3 HEAVY METALS (Zn, Cu, Pb, AND Cd) IN SOME WILD PLANTS OF DRAINED PEATLAND BIOGEOCENOSES

The assessment of heavy metals in plants growing in peatlands, including anthropogenic transformed, is also of considerable interest. Soil is an important component in the cycle of elements and the flow of substances in any ecosystem. A significant proportion of chemical elements entering the soil, including heavy metals, subsequently enter plants. Studying the content of elements in the soil-plant system allows to assess the nature and ways of their migration, accumulation in individual organs of plants, to identify species features of plants in accumulation of certain chemical elements.

In order to study the content of heavy metals in wild plants, the following most common species were chosen on the peat-massif Zenginsky: *Betula pubescens*, *Salix caprea* L. (goat willow), *Rubus idaeus*, *Rosa majalis* Herrm. (cinnamon rose), and *Pyrola rotundifolia*.

The maximum allowance concentrations (MAC) of heavy metals in wild plants, as in peat, are not currently developed. There is also little evidence of heavy metals in plants growing on peat soil. In order to estimate the content of heavy metals in the vegetative mass of the studied plants, the standards of MAC for zinc, copper, lead, and cadmium in plant products (vegetables) were used in accordance with,[36] as well as literature information on the average content of heavy metals in the vegetative mass of wild plants (Table 1.5), including our previous data on wild plants growing on automorphic soils (Table 1.6).

TABLE 1.5 Content of Heavy Metals (Zn, Cu, Pb, and Cd) in Vegetables and Wild Plants (mg/kg of Dry Matter).

Metal	MAC in plant products (vegetables)[1]	Average content of heavy metals in vegetative mass of wild plants[2]
Zn	80.00	15.00–150.00
Cu	40.00	2.00–14.00
Pb	2.00	0.10–5.00
Cd	1.20	0.05–0.50

[1]According to [36].
[2]According to [28; 37].

TABLE 1.6 Average Content of Heavy Metals (Zn, Cu, Pb, and Cd) in Wild Plants Growing on Automorphic Soils of Unpolluted Areas in Kirov Region (mg/kg of Dry Matter).

Metal	Red raspberry (Rubus idaeus)	Downy birch (Betula pubescens)	Goat willow (Salix caprea)	Round-leaved wintergreen (Pyrola rotundifolia)	Cinnamon rose (Rosa majalis)
Zn	16.00	33.00	31.20	27.00	10.20
Cu	7.10	12.00	13.70	18.60	8.00
Pb	0.60	0.70	0.80	0.80	0.78
Cd	0.30	0.42	0.60	0.40	0.20

In the course of the conducted studies, the average content of zinc, copper, lead, and cadmium in the vegetative mass of the tested plants was determined (Table 1.7). Based on results of analysis of the data obtained during the study, it can be concluded that the content of heavy metals in the vegetative mass of the studied plant species does not exceed the values of the MAC of these elements in plant products. The obtained data also do not exceed the average values of element content in plants obtained in previous studies, including on automorphic soils of background territories of the Kirov region.

TABLE 1.7 Average Content of Heavy Metals (Zn, Cu, Pb, and Cd) in Vegetative Mass of Investigated Wild Plants (mg/kg of Dry Matter).

Metal	Red raspberry (Rubus idaeus)	Downy birch (Betula pubescens)	Goat willow (Salix caprea)	Round-leaved wintergreen (Pyrola rotundifolia)	Cinnamon rose (Rosa majalis)
Zn	11.55 ± 0.49	29.06 ± 0.60	29.28 ± 0.91	12.21 ± 0.45	8.91 ± 0.19
Cu	6.58 ± 0.16	$10.38 \pm 0,38$	12.63 ± 0.36	6.89 ± 0.22	7.03 ± 0.19
Pb	0.52 ± 0.01	0.55 ± 0.01	0.67 ± 0.02	0.35 ± 0.02	0.21 ± 0.01
Cd	0.26 ± 0.01	0.32 ± 0.01	0.51 ± 0.02	0.15 ± 0.01	0.05 ± 0.01

Copper and zinc belong to the group of biophilic elements necessary for normal plant life at certain concentrations. Plants accumulate them in relatively large amounts compared to other elements.

Among the studied plant species, the lowest copper content was found in *Rubus idaeus*, ranging from 5.90 to 7.80 mg/kg of dry matter. Plants growing on automorphic soils of the region also accumulate less copper than other studied species—on average 7.10 mg/kg of dry matter

(Table 1.5). Content of copper in the vegetative mass of *Pyrola rotundifolia* was 5.78–8.10 mg/kg, in the vegetative mass of *Rosa majalis*—6.04–8.10 mg/kg (on average 6.89 mg/kg and 7.03 mg/kg, respectively). In vegetative mass of *Betula pubescens* it ranges from 8.39 to 13.32 mg/kg dry matter. The highest content of copper was found in the vegetative mass of *Salix caprea*—10.65–15.10 mg/kg (on average 12.63 mg/kg) of dry substance. On automorphic soils of the region, *Pyrola rotundifolia* accumulates the most copper, averaging 18.60 mg/kg dry matter. *Salix caprea* plants accumulate slightly less amount of the element—13.70 mg/kg (Table 1.5).

The lowest zinc content among the studied species was found in the vegetative mass of *Rosa majalis*—8.04–9.76 mg/kg (average 8.91 mg/kg) of dry matter. On automorphic soils of Kirov region, plants of this species also accumulate less zinc than other studied species (average 10.20 mg/kg dry matter). The content of zinc in the vegetative mass of *Rubus idaeus* varies from 8.98 to 15.00 mg/kg, in the vegetative mass of *Pyrola rotundifolia*—from 10.34 to 14.50 mg/kg, and in the vegetative mass of *Betula pubescens*—from 26.41 to 33.50 mg/kg of dry matter. The highest zinc content was found in *Salix caprea* leaves—26.08–33.50 mg/kg (average 29.06 mg/kg dry matter). On automorphic soils of the region, the element is actively accumulated in *Betula pubescens* and *Salix caprea* leaves (on average 33.00 mg/kg and 31.20 mg/kg dry matter, respectively).

Cadmium and lead are toxicant elements whose role in plant metabolism has not been fully determined. Plants growing on non-contaminated soils accumulate these elements in relatively low amounts. Among the studied plant species, the lowest cadmium content is observed in the vegetative mass of *Rosa majalis*; it ranges from 0.02 to 0.08 mg/kg (on average 0.05 mg/kg) of dry matter. Plants of this species growing on automorphic soils of the Kirov region also accumulate less amount of cadmium compared to other studied species (on average 0.20 mg/kg dry matter). However, the content of the element in plants growing on peat soils is several times lower than in plants growing on automorphic soils. The content of cadmium in the vegetative mass of *Pyrola rotundifolia* plants varies from 0.11 to 0.23 mg/kg (on average 0.15 mg/kg dry matter). The content of the element in leaves of *Rubus idaeus* and *Betula pubescens* differs slightly and is 0.20–0.37 mg/kg and 0.26–0.38 mg/kg, respectively. Among the studied plant species, the highest cadmium content was found in *Salix caprea* leaves—0.48–0.61 mg/kg (average 0.51 mg/kg dry matter). On automorphic soils, *Salix caprea* plants also actively accumulate cadmium compared to other test plant species (average 0.60 mg/kg dry matter).

The lowest lead content among the studied wild plant species was found in vegetative mass of *Rosa majalis* plants—0.16–0.27 mg/kg dry matter. On automorphic soils lead content in vegetative mass of the studied plant species differs slightly; it is slightly lower in *Rubus idaeus* plants (on average than 0.60 mg/kg of solid). In *Pyrola rotundifolia* plants, lead content ranges from 0.27 to 0.40 mg/kg (average 0.35 mg/kg dry matter), in *Rubus idaeus* and *Betula pubescens* leaves 0.44 to 0.58 mg/kg and 0.45 to 0.62 mg/kg dry matter, respectively. The highest lead content among the test plant species was found in *Salix caprea* leaves and ranged from 0.58 to 0.79 mg/kg (average 0.67 mg/kg dry matter). On automorphic soils, slightly increased element content is observed in *Salix caprea* and *Pyrola rotundifolia* plants (average 0.80 mg/kg dry matter).

Thus, studied plant species growing on dried peat soils accumulate minor amounts of zinc, copper, lead, and cadmium in their vegetative mass. The values of element content obtained during the studies do not exceed the standards of MAC and are close to the average values of zinc, copper, lead, and cadmium content in plants growing on automorphic soils of the Kirov region.

Therefore, among the studied plant species *Rosa majalis* was found the least to accumulate zinc, lead, and cadmium. *Betula pubescens* and *Salix caprea* plants accumulate heavy metals more actively than others do. Well-developed root system of these species allows them to "extract" elements from a greater depth, from a larger volume of soil. *Salix caprea* leaves revealed the highest content of all four studied elements.

1.4   CONCLUSIONS

Thus, the conducted studies revealed that content of lead, cadmium, zinc, and copper in the peat layers of Karinsky and Zenginsky peat massifs in central part of the Kirov region is rather low and does not exceed the MAC and background content in the soils of the region.

Profiles of developed peatlands are characterized by significant fluctuations of heavy metals content both in different profiles and in different horizons of the same profile. These facts indicate the restructuring of soil processes and regimes taking place in developed and drained soils.

Drainage of the peatlands and their further economic development contribute to the transformation of these natural complexes. Economic use leads to active mineralization of the upper part of the remaining peat column. As a result of these processes, the stability of the biogeochemical cycles of the individual elements is impaired. At the same time, the results obtained allow to consider these peatlands as indicators of the state of the natural environment. Good safety of peat deposits makes it possible to study processes of accumulation and migration of chemical elements.

By the activity of heavy metals accumulating in the vegetative mass, the test plant species can be arranged as follows:

- Cu: *Rubus idaeus* < *Pyrola rotundifolia* < *Rosa majalis* < *Betula pubescens* < *Salix caprea*;
- Zn: *Rosa majalis* < *Rubus idaeus* < *Pyrola rotundifolia* < *Betula pubescens* < *Salix caprea*;
- Cd: *Rosa majalis* < *Pyrola rotundifolia* < *Rubus idaeus* < *Betula pubescens* < *Salix caprea*;
- Pb: *Rosa majalis* < *Pyrola rotudifolia* < *Rubus idaeus* < *Betula pubescens* < *Salix caprea*.

KEYWORDS

- **peat bog**
- **fen**
- **transition-type bog**
- **bog reclamation**
- **zinc**
- **copper**
- **lead**
- **cadmium**

REFERENCES

1. *The Canadian Wetland Classification System*, 2nd ed.; Warner, B. G., Rubec, C. D. A., Eds.; Wetlands Research Centre, University of Waterloo: Waterloo, Ontario, 1997; p 68.
2. *Our Earth's Changing Land: An Encyclopedia of Land-Use and Land-Cover Change*; Geist, H., Ed.; Greenwood Press: Westport, Connecticut, 2006; Vol. 2, p 463.
3. Vitt, D. H. Peatlands. *Encyclopedia of Ecology*, 2nd ed.; 2013; Vol. 2, pp 557–566. https://doi.org/10.1016/B978-0-12-409548-9.00741-7
4. Craft. C. Peatlands. *Creating and Restoring Wetlands. From Theory to Practice*; Elsevier: Amsterdam, Boston, Heidelberg, London, New York, Oxford, Paris, San Diego, San Francisco, Singapore, Sydney, Tokyo, 2016; pp 161–192. https://doi.org/10.1016/B978-0-12-407232-9.00007-5
5. Bourgeau-Chavez, L. L.; Endres, S. L.; Graham, J. A.; Hribljan, J. A.; Chimner, R. A.; Lillieskov, E. A.; Battaglia, M. J. Mapping Peatlands in Boreal and Tropical Ecoregions. *Compr. Remote Sens.* **2018**, *6*, 24–44. https://doi.org/10.1016/B978-0-12-409548-9.10544-5
6. Piavchenko, N. I. *Forest Peatland Science (Main Problems)*; Russian Academy of Science Publishing House: Moscow, 1963; p 193 (in Russian).
7. Novikov, S. M.; Usova, L. I. New Data on Swamp Area and Peat Reserves on the Territory of Russia. *Dynamics of Swamp Ecosystems of Northern Eurasia in Golocene*. Petrozavodsk: Karelian Scientific Center of RAS, 2000; pp 52–55 (in Russian).
8. Beilman, D. W.; MacDonald, G. M.; Smith, L. C.; Reimer, P. J. Carbon Accumulation in Peatlands of West Siberia Over the Last 2000 Years. *Glob. Biogeochem. Cycles* **2009**, *23*(1), GB1012. DOI: 10.1029/2007GB003112.
9. Mezhibor, A.; Arbuzov, S.; Rikhvanov, L.; Gauthier-Lafaye, F. History of the Pollution in Tomsk Region (Siberia, Russia) According to the Study of High-Moor Peat Formations. *Int. J. Geosci.* **2011**, *2*, 493–501. DOI:10.4236/ijg.2011.24052.
10. Borgulat, J.; Mętrak, M.; Staszewski, T.; Wiłkomirski, B.; Suska-Malawska, M. Heavy Metals Accumulation in Soil and Plants of Polish Peat Bogs. *Pol. J. Environ. Stud.* **2018**, *27*(2), 537–544. DOI: 10.15244/pjoes/75823.
11. Krumins, J.; Robalds, A. Biosorption of Metallic Elements onto Fen Peat. *Environ. Clim. Technol.* **2014**, *14*, 12–17. DOI: 10.1515/rtuect-2014-0008.
12. Forel, B.; Monna, F.; Petit, C.; Bruguier, O.; Losno, R.; Fluck, P.; Begeot, C.; Richard, H.; Bichet, V.; Chateau, C. Historical Mining and Smelting in the Vosges Mountains (France) Recorded in Two Ombrotrophic Peat Bogs. *J. Geochem. Explor* **2010**, *107*, 9–20. https://doi.org/10.1016/j.gexplo.2010.05.004
13. Serebrennikova, O. V.; Strelnikova, E. B.; Preis, Yu. I.; Averina, N. G.; Kozel, N. V.; Bambalov, N. N.; Rakovich, V. A. The Composition of Peat Extracts from Drained and Natural Raised Bogs of Belarus and Western Siberia. Bulletin of the Tomsk Polytechnic University. *Chem. Chem. Technol.* **2014**, *325*(3), 31–45 (in Russian).
14. Kõlli, R.; Asi, E.; Apuhtin, V.; Kauer, K.; Szajdak, L. W. Chemical Properties of Surface Peat on Forest Land in Estonia. *Mires Peat* **2010**, *6*. Article 06. http://www.mires-and-peat.net/
15. Givelet, N.; Le Roux, G.; Cheburkin, A.; Chen, B.; Frank, J.; Goodsite, M.; Kempter, H.; Krachler, M.; Noernberg, T.; Rausch, N.; Rheinberger, S.; Roos-Barraclough, F.;

Sapkota, A.; Scholz, C.; Shotyk, W. Suggested Protocol for Collecting, Handling and Preparing Peat Cores and Peat Samples for Physical, Chemical, Mineralogical and Isotopic Analyses. *J. Environ. Monit.* **2004,** *6,* 481–492. DOI: 10.1039/b401601g

16. Ulanov, A. N. Agrochemical Properties of Peat Deposits in Kirov Region. *Ecosystem Development and Environmental Management on Peat Soils*; Vyatka Publishing House: Kirov, 2003; pp 48–60 (in Russian).

17. Timofeev, A. F.; Komarova, L. A. *Complex use of Lands After Peat Extraction*; Center of Scientific Information: Kirov, 1973, p 36 (in Russian).

18. Ulanov, A. N. *Peat and Cutover Soils of South Taiga of Euro-North-East of Russia*; Vyatka Publishing House: Kirov, 2005; p 319 (in Russian).

19. Kosolapov, V. M.; Ulanov, A. N.; Zhuravleva, E. L.; Shel'menkina, Kh. Kh.; Mokrushina, O. G.; Smirnova, A. V.; Kosolapova, V. G.; Kovshova, V. N.; Pomaskina, Yu. V. *Perennial Cultural Pasture on Drained Peat Soils*; OOO VESI: Kirov, 2015, p 124 (in Russian).

20. Zverkov, Yu. V. *Second Life of Peat Bogs*; Volga-Vyatka Publishing House: Kirov, 1982; p 80 (in Russian).

21. Andreeva, E. N.; Bakkal, I. Yu.; Gorshkov, V. V.; Lyangusova, Sh. V. *Methods of Studying of Forests Communities.* Chemistry Research Institute of SPB State University Publishing House: Saint Petersburg, 2002; p 240 (in Russian).

22. Kulikova, G. G. *A Brief Guide to the Botanical Analysis of Peat.* Publishing House of the Moscow University: Moscow, 1974; p 95 (in Russian).

23. Dombrovskaya, A. V.; Koreneva, M. M.; Tyuremnov, S. N. *The Atlas of the Vegetative Remains Met in Peat*; State Energetic Publishing House: Moscow-Leningrad, 1959; p 228 (in Russian).

24. Shikhova, L. N.; Egoshina, T. L. *Heavy Metals in Soils and Plants of a Taiga Zone of the Northeast of the European Russia.* Agricultural Research Institute of the Northeast: Kirov, 2004; p 264 (in Russian).

25. *CRC Handbook of Chemistry and Physics*, 88th ed.; [Editor-in-Chief: David R. Lide (National Institute of Standards and Technology)]; CRC Press/Taylor & Francis Group: Boca Raton, 2007; p 2640.

26. Håkanson, L. An Ecological Risk Index for Aquatic Pollution Control: A Sedimentological Approach. *Water Res.* **1980,** *14,* 975–1001.

27. Syrovetnik, K.; Neretnieks, I.; Malmström, M. E. Accumulation of Heavy Metals in the Oostriku Peat Bog, Estonia: Determination of Binding Processes by Means of Sequential Leaching. *Environ. Pollut.* **2007,** *147*(1), 291–300. DOI: 10.1016/j.envpol.2005.10.048.

28. Kabata-Pendias, A. *Trace Elements in Soil and Plants*, 4th ed.; CRC Press: Boca Raton, 2010; p 548.

29. Nikodemus, O.; Brumelis, G.; Tabors, G.; Lapina, L.; Pope, S. Monitoring of Air Pollution in Latvia Between 1990 and 2000 Using Moss. *J. Atmospheric Chem.* **2004,** *49,* 521–531. https://doi.org/10.1007/s10874-004-1263-2

30. Veretennikova, E. E. Lead in the Natural Peat Cores of Ridge-Hollow Complex in the Taiga Zone of West Siberia. *Ecol. Eng.* **2015,** *80,* 100–107. http://dx.doi.org/10.1016/j.ecoleng.2015.02.001

31. Gashkova, L. P.; Ivanova, E. S. Accumulation the Heavy Metals in Plants Dominants of Anthropogenically Damaged Areas of Wetlands at the Territory of Tomsk Oblast. *News Samara Sci. Center Rus. Acad. Sci.* **2014,** *16*(1/3), 732–735 (in Russian).

32. Fiałkiewicz-Kozieł, B.; Smieja-Król, B.; Palowski, B. Heavy Metal Accumulation in Two Peat Bogs from Southern Poland. *Studia Quaternaria* **2011**, *28*, 17–24.

33. Rydin, H.; Jeglum, J. K. *The Biology of Peatland*. In *Biology of Habitats*, 2nd ed.; Oxford University Press: Oxford, 2013; p 382.

34. Tyuremnov, S. N. *Peat Fields*; Nedra ("Bowels" in Russian): Moscow, 1976; p 488 (in Russian).

35. Zubkova, K. A.; Gonina, E. S.; Shikhova, L. N.; Lisitsyn, E. M. The Analysis of Climatic Stages of Formation of Swamps on Botanical Structure. *Her. Orenburg State Univ.* **2016**, *5*(193), 57–64 (in Russian).

36. *Sanitary and Epidemiological Rules and Regulations SanPin 2.3.2 1078-01 "Hygienic Requirements for Food Safety and Nutritional Value"*; Moscow, 2002; p 269 (in Russian).

37. *Heavy Metals in System Soil-Plant-Fertilizer;* Ovcharenko, M. M., Ed.; Moscow: Proletarsky Svetoch ("Proletarian cresset" in Russian), 1997; p 290 (in Russian).

CHAPTER 2

Paleopalinology Studies of Conditions of Peatlands Formation on European Northeast of Russia

KSENYA A. ZUBKOVA[1*], LYUDMILA N. SHIKHOVA[2], and
EUGENE M. LISITSYN[2]

[1]*Vyatka State Agricultural Academy, 133 Oktyabrsky Av.,
Kirov 610017, Russia*

[2]*N.V. Rudnitsky Federal Agricultural Research Center of the North-East,
166a Lenin St., Kirov 610017, Russia*

Corresponding author. E-mail: ks.zubkowa@gmail.com

ABSTRACT

Peat deposits in three peatlands located in Kirov region in European Russia were analyzed for botanical and spores-pollen composition. Despite some differences in the history of formation of peat thickness in all three objects, quite accurate regularities are traced. The lowermost layers of peat with a thickness about 20–60 cm adjoining on the mineral horizons are characterized by existence and prevalence of green mosses. Such vegetative groups could be created in cool and damp conditions of the Preboreal period (about 8–9 thousand years ago) in territories of runoff of glacial waters. Above it, there are layers with prevalence of sedge. Thickness of sedge layers of peat on different objects varies from 40 to 80 cm. Groups (formation) with domination of sedge are characteristic of territories of periodic superficial flooding in period called boreal (about 7–8 thousand years ago). Maximum number of the wood remains is observed in the layers lying above. A large number of woods remains can be considered as sign of the "boundary" horizon, which existence in many peat bogs of the European Russia demonstrates wide circulation of forest vegetation during the Atlantic period (5–7 thousand years ago). At this time, the climate was warmer than modern one and forest vegetation had wider circulation. The peat layers stored presumably in the Subboreal period remained only on the undeveloped site of a peatland "Zenginsky". The specific composition of peat-forming plants demonstrates that the peat thickness was torn off from the mineral horizon and developed on transitional type. Thus, studying of botanical structure demonstrates alternation of different climatic eras. In each stage, there is a formation of the phytocenoses, which are most adapted for these climatic factors, that is visually reflected in the peat (organic) chronicle of bogs. Application of botanical and paleopalinologic analyses of peat allows to characterize vegetative groups at different stages of peatland formation. Such analyses make it possible to restore climatic and hydrological conditions of formation of phytocenoses.

2.1   INTRODUCTION

The current state of ecosystems is not only related to existing environmental conditions. Knowledge of the long-term ecosystems' evolution provides useful information to understand present ecosystems and predict their future changes.[1] Palaeoenvironmental research uses indicators from environmental archives in order to reconstruct how ecosystems have changed through time. Bogs are the unique natural ecosystems storing history of development of phytocenoses and climate in the territory of their location. Peat is easily datable material so it is relatively simple to chronologically frame the detected environmental changes. In peat deposits pollen, seeds, spores, and other parts of plants well remain. When investigate peat layers, it is possible to study history of an Earth's vegetation cover. Bogs are a component of natural complexes of a moderate belt of forest zone. They occupied about 10% of a northeast part of the European Russia, but until now, they are poorly studied.

In the last 50–70 years in the territory of the European Russia intensive activity on drainage of peat bogs and peat extraction were conducted. As a result, bog ecosystems disappear, and alongside with them the history of their formation and development which is stored in peat deposits. The semi-dissolute vegetative remains composing peat mass are witnesses of characteristics of conditions in which these phytocenoses were formed.

The vegetative remains serve for the stratigraphic partition of sedimentary deposits and restoration of paleogeographic and paleoclimatic conditions of time of accumulation of the studied matter.[2,3] Different calcareous or siliceous biological remains can be affected by dissolution, but organic-walled pollen and sporous are composed of resistant organic matter and are generally well preserved in peat deposits. Additionally, plant species have its own environmental preferences and their distribution in peat must correlate well with environmental conditions such as temperature, humidity, nutrient levels, and productivity.[4] Until middle of 1970s, pollen analysis was mainly a stratigraphic tool used for dating and stratigraphic correlation but it nowadays proved to be useful in botanical, biogeographical, and ecological research; palynology has developed into a fundamental tool to unravel the ecological and environmental trends and changes.[5,6]

According to Ref. [7] whereas 2422 peat geochemistry papers and 1355 peat palynology papers have been published (indexing in Web of Science database) since 1900, only 247 papers using both geochemical and palynological analyses to peat records were published during the same period of time. Rather more researches of dinocyst or ostracods assemblages were done.[8,9] Studying of botanical and sporous-and-pollen compositions of peat as well as stages of bog formation is necessary for the best understanding of peat-forming processes and preservation of bogs. Therefore, the main purpose of the given work is selection of succession stages of formation of peat deposits and studying the history of bogs' formation. Such study will provide useful information about the Holocene (the last ~11,600 years) environmental changes of a poorly known subarctic region.

2.2 MATERIALS AND METHODOLOGY

Objects of a research are peatland massifs of the central part of the Kirov region: "Karinsky," "Zenginsky," and "Gadovsky" with a different thickness of peat layer which remained after development.

The Kirov region is in northeast part of the European Russia. The territory of the area is located in a taiga zone and has sufficient moistening. Nearly 2 million hectares of the territory of the Kirov region is occupied with swampy and bogged soils. Among bog massifs, minerotrophic peatlands (also named fens) who receive their water inputs from both rain and ground water, and transitional-type peatlands generally prevail.

Considerable extent of the region from the North on the South and from the West on the East defines significant differences of a soil and vegetative cover. Therefore, bogs differ under conditions of origin and development. The surface relief and the nature of geological deposits exerted impact on processes of bog formation. The humidity of climate, smoothness of a relief, and water resistance of clay soil caused high degree of bogginess of a northern part of the region. Generally, bog massifs are associated with river valleys and hollows of runoff of glacial waters.[10]

During Holocene era, territory of the Kirov region was attacked periodically with glaciers from the North. Glaciations formed the landscapes promoting a formation of bogs. Therefore, in northern districts of the region boggy territories meet more often, than in southern one. To the south from border of glaciation bogs meet more seldom and they have smaller size (no more than 1000 ha).[10]

Unfortunately, data on botanical composition of peat-forming plants are not enough. Therefore, it is impossible to compare botanical composition of peat of the Kirov region with data from other regions. There is practically absent information in scientific articles on this matter. Separate data meet in Refs. [10–13].

The peatland "Zenginsky" locates in Orichevsky district of the Kirov region on the first fluvial terrace above floodplain of the Vyatka River. The total area of the massif is 6000 ha, the initial maximum thickness of peat layer—5 m. Ancient alluvial deposits presented by greenish-gray fine-grained sands serve as the underlying rock; these deposits in turn spread by silty loams and sandy loams.[10] Peatland "Zenginsky" belongs to transitional-type bogs, which are characterized by presence of signs both of ombrotrophic peatlands and of minerotrophic peatlands, namely existence of a large number of green mosses, some herbs, birches, and pines. Now peatland is drained and the top layers of peat (about 3 m) are cutover.

Peatland "Gadovsky" is located on the south-west from the Kirov city on a left-bank fluvial terrace above floodplain of the Bystritsa River that flows into the Vyatka River. The terrace is raised over floodplain level by 6–10 m. The total area of the peatland is 3895 ha. The peat deposit is directly spread with upper-quarternary alluvial and alluvial-diluvia middle- and fine-grained sands, sandy loams, more rare—sandy loams containing inclusions of carbonate rocks, and closer to bottom—with pebble and gravel. On the most part of the "Gadovsky" peatland, commercial production of peat was conducted in 1933–1970. Now the main part of peat thickness is cutover. For studying a residual part of massif with a peat thickness about 1.5 m was taken.

Peatland "Karinsky" is located to the east from the Kirovo-Chepetsk city. It is on the first fluvial terrace above floodplain of the Cheptsa River. The underlying rocks are ancient-alluvial sands. The area of a peatland is 8286 ha, the initial maximum thickness of a peat deposit—6 m. Now this massif is drained and the top layer of peat is cutover.

In the territory of a Peatland "Gadovsky" five trial areas, five soil profiles were put; in the territory of the "Zenginsky" peatland—8 trial areas and 8 soil profiles; and in the territory of a peatland "Karinsky"— three trial areas and three soil profiles. In total, 16 trial areas and 16 soil profiles were put. More than hundred soil samples were analyzed.

On trial areas, the geobotanical description is carried out. From each profile, peat samples according to layers were collected. Layers were selected on morphological features: extents of peat decomposition, botanical structure (field definition), color, peat density. In the selected samples of peat, the botanical structure of the vegetative remains was determined by Kulikova's technique[14] by means of determinants.[15,16] Regarding tests of peat for sporous and pollen structure was carried out by.[17]

All soil profiles were excavated down to the mineral underlying horizons. Thus, early stages of bog formation were available to studying.

Stratigraphic schemes of distribution of the vegetative remains on depth were constructed for each soil profile on the basis of the obtained data.

To reveal regularities and stages of formation of all massifs, the general stratigraphic schemes of the remained peat deposit were constructed based on all obtained data.

2.3 RESULTS AND DISCUSSION

The main bog massifs of the Kirov region are located in watercourse of an ancient runoff of glacial waters. Bogs of the central part of the Kirov region were formed in an extra-glacial zone after the beginning of thawing of the last Valdai glacier (70–15 thousand years ago). These bogs have similar history of formation; however, there are differences in formation stages linked with features of the underlying rocks and with situation in a relief.

The peat deposit is formed during several hundred or even thousands of years. There are no concrete data on age of bogs of this zone. In literature, there are some data on age of bog systems in the European part of Russia. Proceeding from these data, the age of bogs with rather thick peat deposit fluctuates within 5–12 thousand years that is their origination happened during a postglacial era.[18]

The peat massif is morphologically nonuniform and differs on a set of the peat-forming plants composing it. The number of peat layers of different botanical structure varies even within one bog massif. It is linked with roughness of a surface relief, on which a bog was formed, with roughness of a relief of a vegetative cover and other factors. The bogs are characterized by rather plain surface only at later stages of development when bog vegetation is closed completely and closes all roughness of a relief. Therefore, the amount of peat layers of different botanical structure, and peat thickness in different points of the bog differ considerably.

Studying the bog massif, it is very difficult to synchronize and generalize data on different profiles. A certain "marker" horizon, which has to be present at all profiles and have accurate dating, is necessary. According to Ref [19], for bogs of the European part of Russia such horizon is layers of wood and moss peat.

In the profiles of all studied peat massifs such horizons are present. Their thickness and depth vary, but allow to carry out a temporary binding of thickness. Owing to presence of these horizons, we tried to reduce all profiles studied on layers in the uniform stratigraphic scheme of the remained peat thickness for each peatland.

2.4   PEATLAND "GADOVSKY"

For a peatland "Gadovsky" the residual thickness of peat, available to studying, is about 150 cm. Three main stages of formation that are conditionally corresponded to the Preboreal, boreal and Atlantic periods (Fig. 2.1) are surely selected.

Within this peatland, sites with undisturbed peat layers did not remain; therefore, layers more young than the Atlantic period are absent. Stages of formations, in turn, are possible to divide into several various horizons, differing in set of peat-forming plants. The thickness of peat created presumably during **the Preboreal** period is nonuniform and consists of several layers. Change of vegetative formations from below up justified change of climatic factors. The remains of the plants living now in tundra and forest-tundra communities put the lowermost layers. It is remains of a marsh horsetail (*Equizetum palustris* L.), Rannoch-rush (*Scheuchzeria palustris* L.), cotton grass (*Eriophorum* L.); the insignificant amount of the remains of wood of a fir-tree (*Picea* Mill.) and its pollen meets. Such structure of phytocenosis can point to forest-tundra character of a landscape, with high humidity and a lack of heat. In direction to the upper bound of a layer, the specific structure of the vegetative remains extends. There are more thermophilic types. The Preboreal period was obviously nonuniform on climatic parameters. It is demonstrated with interlayers of the plants not characteristic of a layer in general.

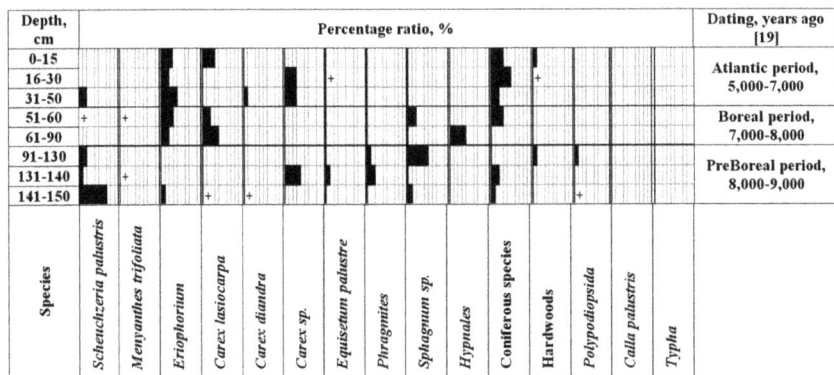

Depth, cm	Percentage ratio, %	Dating, years ago [19]
0-15		
16-30		Atlantic period, 5,000-7,000
31-50		
51-60		Boreal period, 7,000-8,000
61-90		
91-130		
131-140		PreBoreal period, 8,000-9,000
141-150		

Species (columns): *Scheuchzeria palustris*, *Menyanthes trifoliata*, *Eriophorium*, *Carex lasiocarpa*, *Carex diandra*, *Carex sp.*, *Equisetum palustre*, *Phragmites*, *Sphagnum sp.*, *Hypnales*, Coniferous species, Hardwoods, *Polypodiopsida*, *Calla palustris*, *Typha*

FIGURE 2.1 Stratigraphic scheme of botanical structure of peat profile of a peatland "Gadovsky".

In the layer located above (130–140 cm), the remains of the following plants are added to above-mentioned plants: cattail (*Typha* L.), sedge (*Carex* L.), Scots pine (*Pinus sylvestris* L.), march trefoil (*Menyanthes trifoliata* L.).

In the layer 90–130 cm there are remains of such plant species as a Rannoch-rush, common reed grass (*Phragmites australis* (Cav.) Trin. ex Steud.). Up to a half of volume of this layer is put by peat mosses (*Sphagnum* L.). There are remains of the hardwood tree species presented generally by a birch (*Betula* L.). In this layer, pollen of a pine (*Pinus* L.) and a larch (*Larix* Mill.) is found. In a grassy cover, sedge prevailed, but also ferns (*Polypodiophyta;* Cronquist, Takht. and W.Zimm.) is met. Formation of these peat layers can be referred to second stage of Preboreal period of the Holocene (about 8000 years ago) in which warming of climate was observed, tundra vegetation receded on the North. During this period in the southern part of the Kirov region, a role of the forests amplified.

Warming of climate and rather high-water content of the territory during the Preboreal period promoted that within a forest zone of the East European part of Russia formation of peatlands became possible. During this period, that is, about 9000 years ago actually accumulation of peat began.[19]

Peat layers at a depth of 50–90 cm are created presumably during **the boreal** period of the Holocene (7–8 thousand years ago). There are remains of the following plant species: cotton grass, bulrushes (*Trichophorum* Pers.), woollyfruit sedge (*Carex lasiocarpa* Ehrh.), male fern (*Dryopteris filix-mas* (L.) Schott), leafy mosses, (*Hypnales;* Buck and Vitt, 1986) and peat mosses. Among the wood remains the pine prevails, and the fir-tree is met. Such set of plants is characteristic of the boggy, slightly flooded territories.

The structure of peat-forming plants allows to assume that during formation of a deposit, the territory was covered with the light-coniferous forest in which gaps cotton-grass, bulrushes, and sedge is developed. Ferns grew under a canopy of the forest. Peat was moderately flooded. Given phytocenosis corresponds to the boreal period of the Holocene, which is characterized by expansion of pine formations due to reduction of a fir-tree.

Botanical composition of a phytocenosis corresponds to climatic conditions of the beginning of the boreal period of the Holocene (dry and cool). During this period, processes of bogging are slowed down; light-coniferous forests begin to develop.[20] In this layer, the greatest number of larch pollen was revealed. A larch is sun-loving plant species, undemanding to soils, growing in cold and moderate areas. Now in forest phytocenoses of this region, the larch is absent.

Peat layers 0–50 cm were formed presumably during **the Atlantic** period of the Holocene (5–7 thousand years ago). From bottom to top, the specific structure of plants increases. A large number of species of sedge, including lesser panicled sedge (*Carex diandra* Schrank) is met. In peat samples, a significant amount of a Rannoch-rush appears. Among tree species, a large number of pines remain and its pollen is revealed.

At a depth of 2–30 cm, the remains of broad-leaved species of plants, such as basswood (*Tilia* L.), alder (*Alnus* L.) are added to above-mentioned species. The horsetail marsh is met. Now the basswood does not meet in phytocenosis of this territory. Also, in this layer spore of peat mosses, swamp saw grass (*Cladium mariscus* (L.) Pohl), and adder's-tongue ferns (*Ophioglossum* L.) were found. Emergence of broad-leaved tree species can indicate considerable warming of climate which corresponds to the Atlantic period of the Holocene (6–4.5 thousand years ago), to a so-called climatic optimum of the Holocene. The warm and humid climate resulted in the maximum biological efficiency of bogs' vegetation. For example, during this period speed of peat accumulation increased sharply in bogs of Volga Hills.[20]

The Atlantic stage is marked by the maximum development of light-coniferous forests[10] that also is confirmed by existence of a significant amount (up to 50%) of remains and spores of a Scots pine in peat samples. Possibly, broad-leaved trees were additive to the dominating pine formation. Under forest canopy sedge, horsetails, and various herbs grew.

Now in the territory of partially cutover peatland, the meadow and mixed forest types of phytocenosis develops. Different types of herbs, sedge, small fir-trees, pines, European white birch (*Betula pubescens* Ehrh.), a goat willow (*Salix caprea* L.), and an aspen (*Populus tremula* L.) grow.

2.5 PEATLAND "ZENGINSKY"

Thickness of peat layer on a peatland "Zenginsky" also considerably decreased as a result of a peat cutover. On the most part of the investigated territory, the thickness of a residual peat layer does not exceed 1–1.5 m. Only in a northeast part of the massif, there was a small site unaffected with peat excavation that allowed to study later stages of a deposit formation.

On the stratigraphic scheme of peat profile of this massif, it is possible to select four main eras of formation with characteristic vegetative associations (Fig. 2.2).

The lowermost layers at a depth of 260–270 cm, which are directly adjoining on the mineral underlying horizons, consist of the remains of sedge, horsetails, and Rannoch-rush. Existence of the remains of these types in the lower layer on contact with the mineral horizon demonstrates strong moistening of the territory and rich mineral nutrition during formation of a layer.

As a result of analysis, spores of *Hypnales* mosses (undulate atrichum moss *Atrichum undulatum* (Hedw.) P. Beauv.; cord moss *Funaria hygrometrica* Hedw.; common haircap *Polytrichum commyne* Hedw.), and pollen of larch, cypress (*Cupressus* Tourn.), violet marsh (*Viola palustris* L.) were found in the mineral and near-mineral horizons. The found spores and pollen could be brought on the territory of a peatland by the flowing waters from watersheds or from adjacent territories. It is an initial stage of bog formation, which possibly belongs to the 7th millennium BC.

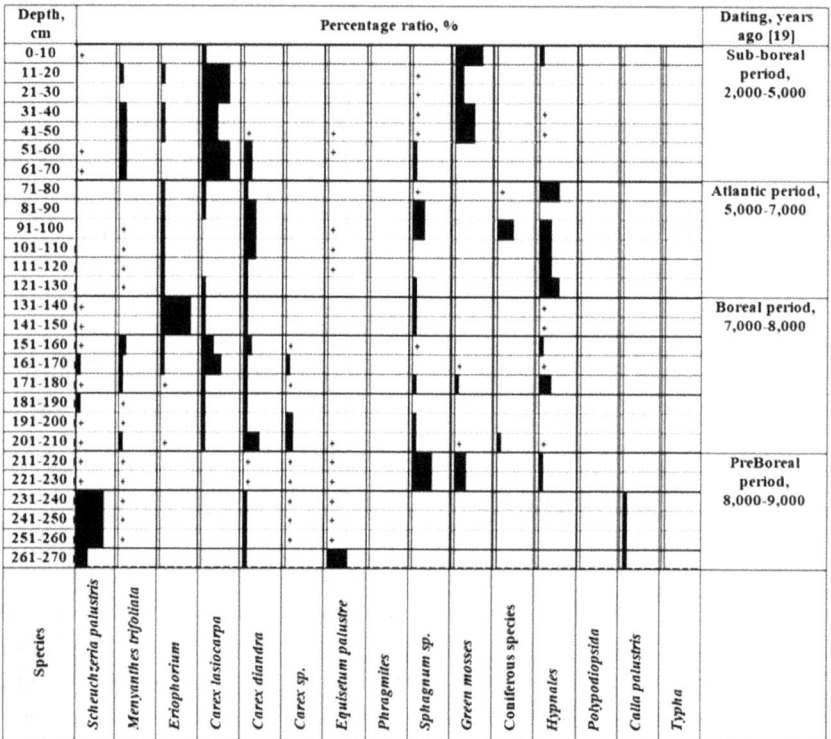

FIGURE 2.2   Stratigraphic scheme of botanical structure of peat profile of a peatland "Zenginsky".

Overlying layers approximately up to the depth of 210 cm were formed obviously during **the Preboreal** period of the Holocene.[19] In the lower part of a layer at a depth of 230–260 cm increase in content of the remains of Rannoch-rush, appearance of Rannoch-rush, and reduction in amount of the horsetail remains is noted. Formation of the lowermost layers of a deposit happened, obviously, in hard-watering conditions, on shallow water, during removal from the course of a water stream.

At a depth of 210–230 cm peat moss and *Hypnales* mosses appear with additives of tree species. Such phytocenosis forms the peat detaining water and increasing the thickness of a peat. For the first time in these peat layers, the remains meet of wood plants, namely an alder. It is the main forest-forming species in the conditions of minerotrophic peatlands and in floodplains of the rivers. The specific structure of the main peat-forming plants gives the chance to claim that the studied territory during formation of this layer was characterized by stagnant moistening with enough mineral salts. Existence of these three stages in which presence of hygrophilous species is noted, justified long-time flood of this territory.

At a depth of 130–210 cm, there are peat layers created presumably during **the boreal** period of the Holocene.

At a depth of 180–210 cm, the remains of wood plants disappear. Woody-moss peat passes into grassy-sedge one. A set of peat-forming plants changes sharply. Sedge (woollyfruit sedge, lesser panicled sedge, turfy sedge *Carex cespitosa* L.) and Rannoch-rush prevail. Change of vegetative groups from forest on marsh-and-meadow allows to assume that during formation of this peat layer, the climate became more severe and water less available. Perhaps, periodically there were droughty periods. Data of researches of change of vegetative communities in glacial and interglacial eras justified to it.[21]

Above on a profile (depth of 130–180 cm), the content of sedge remains in peat increases. There appear black alder (*Alnus glutinosa* (L.) Gaertn.), cotton grass, *Sphagnum* and *Hypnales* mosses, Rannoch-rush, alpine bulrush (*Trichophorum alpinum* (L.) Pers.). The wood-grassy-sedge vegetative group can point both to moistening of the territory, and to its siccation. Existence of the remains of the plants differ in their hydrological requirements can demonstrate frequent change of climatic conditions.

A large number of the remains of cotton grass and alpine bulrush (up to 80%) characterize the peat layer at a depth of 130–150 cm. At the same time, the wood remains almost completely disappear. Disappearance of wood vegetation and emergence of cotton grass and alpine bulrush can indicate inaccessibility of water or mineral nutrients. Most likely, the root system of plants at that time did not reach the mineral horizon, and plants lacked nutrition. Perhaps, during this period water was in enough, but is physiologically inaccessible for the majority of plants. Sharp climate change was the characteristic phenomenon of the Holocene. After change of climatic conditions vegetative communities changed. From the North of the area forest-tundra periodically came which at warming was replaced by a taiga and, at some periods, by mixed forests.[21] Change of climate influenced specific structure of peat-forming plants and the speed of formation of a peat deposit.

Based on a set of peat-forming plants in overlying layers of peat it can be assumed that they were formed during colder era. Approximately, at this depth it is possible to draw line of the Atlantic and boreal periods.

In general, the peat layer lying at a depth from 130 to 210 cm is distinguished with high content of sedge. In this layer, the part of wood vegetation (mainly deciduous species) raises. Heyday of forest and grassy vegetation happened during the boreal period of the Holocene, about 7–8 thousand years ago.[19] According to many authors, the climate of a boreal era was characterized by frequent change of the cold and warm periods at different variations of humidity.

The peat layer from 70 to 130 cm depth was created, in our opinion, during **the Atlantic** period of the Holocene. Large number of remains as deciduous (a birch, an alder), and coniferous (a pine, a fir-tree) species are found in peat. The content of the remains of cotton grass and alpine bulrush decreases, and *Sphagnum* mosses appear. The amount of fragments of a lesser panicled sedge increases. Increase in a specific variety of the formed community is observed. Complex botanical structure of peat can justify frequent change of factors of a bog formation during this period and formation of some different types of phytocenosis, including phytocenosis with tree species.

A large number of wood remains can be considered as indicator of the boundary horizon which existence in many peat bogs of the European Russia demonstrates wide distribution of forest vegetation during the Atlantic period. The Atlantic period (7000–5000 years ago)—time of the maximum distribution of the forests.[19] Many authors identify the Atlantic period with a climatic optimum of the Holocene. In the Atlantic Time, the climate was warmer than modern. The arrangement of climatic zones of this period was similar to modern. The difference consisted in wider distribution of a forest zone in the northern direction and almost total disappearance of tundra vegetation from the European part of Eurasia.[22]

The topmost layers of peat (depth of 10–70 cm) are characterized by sharp reduction of content of the wood remains, existence of a large number of fragments of sedge, and green mosses. Reduction of the remains of wood plants in these peat layers indicates considerable thickness of peat and a full separation of the horizon from a mineral substratum. The result of it is peat depletion by mineral elements. Trees disappeared, which roots are not able to reach the horizons rich in mineral substances. They were succeeding with less demanding species, for example, woollyfruit sedge, forming lasiocarpic layers of peat. The bog from minerotrophic type turns in transitional type one that affects specific structure of peat-forming plants. Existence of the remains of Rannoch-rush in a layer demonstrates remoistening and excess moistening of peat thickness. Obviously, the periods of excess and insufficient moistening alternated. Formation of this peat layer took place possibly **in the subboreal** period (2000–5000 years ago) to which overlapping of the boundary horizon with *Sphagnum* layer was often corresponded.[19]

Unfortunately, a considerable part of a peat deposit is developed and cutover. Therefore, it is difficult to judge botanical structure of the top layers of peat of an initial deposit. To gain an approximate impression about what type of bog represented the excavated peat layer and what plants it was created, it is possible having studied the small undeveloped site of the peatland.

In the top peat layers of this site, the remains of the following plant species were found: white birch, black alder, gray alder, aspen, Scots pine, fir-tree, bog cranberry (*Oxycoccus microcarpus* Turcz. ex Rupr.), sweet gale (*Myrica gale* L.), cotton grass, Rannoch-rush, a horsetail, reed, cattail, Rannoch-rush, woollyfruit sedge, lesser panicled sedge, turfy sedge, fine bogmoss (*Sphagnum angustifolium* (C.E.O. Jensen ex Russow) C.E.O. Jensen), Magellanic bogmoss (*Sphagnum magellanicum* Brid.), tomentypnum moss (*Tomentypnum nitens* ((Hedw.) Loeske, 1911), and hooked scorpion-moss (*Scorpidium scorpioides* (Hedw.) Limpr.). The specific structure of peat-forming plants demonstrates that the peat layer was torn off from the mineral horizon and developed on transitional type. Now, the given site of the bog develops on ombrotrophic peatlands type. It is confirmed by existence in the top peat layers and on a surface of peatland of such plant species as sweet gale, bog cranberry, cotton grass, woollyfruit sedge, Magellanic bogmoss. A specific variety of vegetative community decreased. Wood-moss-subshrub phytocenosis now develops on a deposit surface; its top layer is presented by Scots pine and birch, a subshrub layer—sweet gale, bog cranberry. The lower layer is presented by *Sphagnum* and *Hypnales* mosses.

Peatlands "Zenginsky" and "Gadovsky" are located very close territorially; however, there are also similarities and distinctions in the history of their development. At all stages of bog formation specific structure of peat-forming plants of peatland "Zenginsky" is more various, than of peatland "Gadovsky". Perhaps, it is linked with different geological conditions of bog formation and according to their different geochemical structure. In the same time in a peatland "Gadovsky," the remains meet of plants which practically do not meet in peatland "Zenginsky." These are, for example, such species as reed and cattail.

Both considered peatland massifs were formed by bogging of dry land. Data of sporous-and-pollen analysis indicate that at the time of formation of the peatland "Zenginsky" in the territory adjoining to it, forest phytocenosis existed with the increased soil humidity. Existence of spores of green mosses, including common haircap moss (*Polytrichum commune* Hedw.), cord moss, and common smoothcap moss (*Atrichum undulatum* (Hedw.) P. Beauv.) in near-mineral and mineral horizons justified to it. The found types of leafy mosses grow mainly in the woods and perhaps were brought on peatland by the flowing waters from adjacent territories or watersheds. Sporous-and-pollen analysis of the near-mineral horizons of the peatland "Gadovsky" finds spores of a fir-tree; botanical analysis—the remains of Rannoch-rush and marsh horsetail. These results can indicate that the forest community with elements of bog community could grow in the territory of the peatland at initial stages of its formation. The soil most likely was remoistened as a result of frequent flooding.

Thus, it is possible to make a conclusion that peatlands "Gadovsky" and "Zenginsky" were formed by bogging of dry land; forest phytocenoses settled down in their territory. The territory of the peatland "Gadovsky" was humidified more considerably than "Zenginsky," but the period of flooding was shorter.

Both peatlands began the formation during the Preboreal period of the Holocene, about 8000–9000 years ago. Landscapes of the Kirov region in Preboreal time were presented by fir-tree and pine massifs with birch additives, including dwarf birch (*Betula nana* L.). Warming of climate in a late glacial and the preboreal period promoted peat accumulation.

In general, the obtained data of botanical and sporous-and-pollen analyses of peatlands "Zenginsky" and "Gadovsky" confirm distribution of forest phytocenoses in initial stage of bogs formation.

2.6 PEATLAND "KARINSKY"

Peatland "Karinsky" differs from peatlands "Gadovsky" and "Zenginsky" on geological conditions of formation. The peat layer is underlyed by ancient alluvial sands of poor mineralogical composition. Only small part of peat, with thickness no more than 100–150 cm was available to studying.

Three main stages of forming of a peatland "Karinsky" (Fig. 2.3) may be separated on the stratigraphic scheme of the rest part of peat deposit.

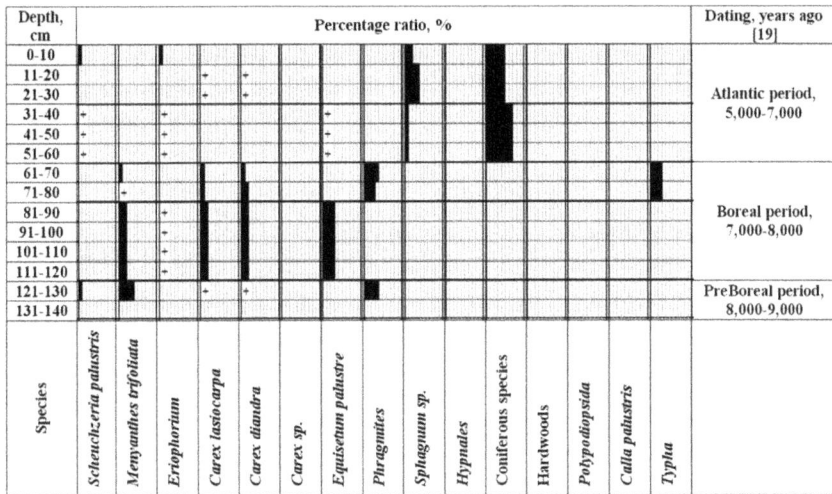

Depth, cm	Percentage ratio, %	Dating, years ago [19]
0-10		
11-20		
21-30		Atlantic period,
31-40		5,000-7,000
41-50		
51-60		
61-70		
71-80		
81-90		Boreal period,
91-100		7,000-8,000
101-110		
111-120		
121-130		PreBoreal period,
131-140		8,000-9,000

Species: *Scheuchzeria palustris*, *Menyanthes trifoliata*, *Eriophorium*, *Carex lasiocarpa*, *Carex diandra*, *Carex sp.*, *Equisetum palustre*, *Phragmites*, *Sphagnum sp.*, *Hypnales*, Coniferous species, Hardwoods, *Polypodiopsida*, *Calla palustris*, *Typha*

FIGURE 2.3 Stratigraphic scheme of botanical structure of peat layers in peatland "Karinsky".

The peat layers, which are directly contacting to the mineral underlying layers (depth 120–140 cm) are characterized by existence of the remains of a Rannoch-rush, march trefoil, and common reed grass. There are single remains of woollyfruit sedge and lesser panicled sedge.

Now these hygrophilous plants grow in floodplains of the rivers, on coast of reservoirs, on bogs and wet meadows. Therefore, the bog formation began in the flooded territory. Peat accumulation was promoted by long remoistening with slowly flowing waters. Presumably, these peat layers were formed at the end of **the Preboreal** period (about 8 thousand years ago).

Layers of peat **of boreal** age lie presumably at a depth from 60 to 120 cm.

In the lower part of layer (depth 80–120 cm), a large number of the remains of cotton grass, woollyfruit sedge, and lesser panicled sedge is revealed. The insignificant content of the remains of march trefoil is noted. The reed remains disappear completely. Such set of peat-forming plants can justify reduction of moisture content of the territory.

At a depth of 60–80 cm, there is a noticeable change of botanical structure of the vegetative remains. In peat, the content of sedge and march trefoil decreases, but reed and cattail appear. The remains of a cotton grass disappear completely. If to consider that reed and cattail are hygrophilous plants, which grow submerged in water, then it is possible to assume that for some reason the territory of a peatland "Karinsky" was waterlogged again. Perhaps, it demonstrates change of climatic conditions from severe on softer and damp. Interlayers of cattail-reed peat were revealed in this peat layer. It indicates that tangles of these plants at that moment were rather dense. Having created dense tangles, reed forced out other plant species.

The topmost residual layers of peat (at a depth of 10–60 cm) are created presumably during **the Atlantic** period of the Holocene. A large number of woods remain demonstrate it. Sometimes in profiles at this depth there are continuous heavy layers of completely wood peat with well remained fragments of trunks and branches. The lower part of this layer (depth of 30–60 cm) contains a large number of remains of coniferous trees, namely fir-trees and pines. Alongside with the wood remains at peat there are remains of *Sphagnum* mosses (*Sphagnum angustifolium* (Warnst.) C.E.O. Jensen; *S. palustre* L., *S. magellanicum* Brid.). There are also single remains of cotton grass, horsetail, and Rannoch-rush. Hydrophilic plants—reed and cattail—completely disappear. This vegetative community was obviously formed in the conditions of change of the hydrological mode, retreat of water, or its binding by low temperatures. Most likely, the climate during this period became colder and dry.

In the top part of the Atlantic layer (at a depth of 10–30 cm), the content increases of remains of *Sphagnum* mosses, especially a fine bogmoss, which is often found in the boggy woods. At this stage of formation of a deposit, tree species began to die-off. Only low-growing fir-trees and pines in depression state remained. It can justify both to severe environmental conditions, and to flooding of the territory.

Significant extent of the Kirov region from the North on the South and from the West on the East results in a variety of climatic, geological, orographical conditions. Therefore, the bogs, which are formed in this territory, differ in the properties, structure, and history of formation. Peatlands "Zenginsky," "Gadovsky," and "Karinsky" differ under the terms and stages of formation of a peat deposit, on botanical composition of peat and a specific variety.

At all stages of bog formation, specific structure of peat-forming plants of "Zenginsky" peatland is more various, than "Karinsky" peatland. Perhaps, it is linked with different geological conditions of the bog formation and according to their different geochemical composition. In peat thickness of the "Zenginsky" peatland there are plant species which practically do not meet in peat of the "Karinsky" massif. These are, for example, such species as prairie sphagnum (*S. palustre* L.), hooked scorpion-moss, sweet gale, turfy sedge. Peat of a peatland "Zenginsky" is rather richer with mineral components in comparison with peat of a peatland "Karinsky."[23] The increased content of mineral components in peat of a peatland "Zenginsky," in comparison with "Karinsky," promotes increase in a specific variety. Besides, differences in the hydrological mode affect difference in peat formation of the bogs.

Significant differences and in stages of formation of peat deposits are observed. The difference in vegetative groups is traced on botanical structure of the lower layers of peat (Figs. 2.1 and 2.3). On the "Karinsky" peatland there are considerable deposits of the remains of reed and march trefoil in the lower peat layers. On the "Zenginsky" peatland, the lower layer of peat is formed by a horsetail, Rannoch-rush, and lesser panicled sedge. On this basis it is possible to draw a conclusion that the "Zenginsky" peat massif began its formation on the re-humidified substratum, but "Karinsky"—was completely flooded or was formed by overgrowing of a superficial reservoir.

2.7 CONCLUSIONS

Despite some differences in the history of formation of peat thickness in all three objects quite accurate regularities are traced. The lowermost layers of peat with a thickness about 20–60 cm adjoining on the mineral horizons are characterized by existence and prevalence of green mosses. Generally, it is hooked scorpion-moss. Slightly above the remains of sphagnums are added to it. Such vegetative groups could be created in cool and damp conditions of the Preboreal period (about 8–9 thousand years ago) in territories of runoff of glacial waters.

Above it there are layers with prevalence of sedge. Thickness of sedge layers of peat on different objects varies from 40 to 80 cm. Groups (formation) with domination of sedge are characteristic of territories of periodic superficial flooding. In literature,[19] this period is called boreal (about 7–8 thousand years ago) and is characterized by some decrease in moistening. At the end of this period in peat layers there are remains of the *Sphagnum* mosses and some type of characteristic for tundra formations (cotton grass, dwarf birch) that testifies to a cold snap at the end of this era.

In peat layers of presumably Preboreal and boreal periods, the remains of tree species, both deciduous and coniferous also meet in a low amount. However, the maximum number of the wood remains is observed in the layers lying above. Sometimes it is the layers presented by completely semi-decayed wood. A large number of wood remains can be considered as sign of the "boundary" horizon, which existence in many peat bogs of the European Russia demonstrates wide circulation of forest vegetation during the Atlantic period (5–7 thousand years ago). The Atlantic period is identified with a Holocene optimum. At this time, the climate was warmer than modern one and forest vegetation had wider circulation.

The peat layers stored presumably in the subboreal period remained only on the undeveloped site of a peatland "Zenginsky" (2000–5000 years ago).

In the top peat layers of this site, the remains of the following plant species were found: white birch, black alder, an aspen, Scots pine, fir-tree, bog cranberry, sweet gale, cotton grass, Rannoch-rush, horsetail, reed, cattail, march trefoil, woollyfruit sedge, turfy sedge, lesser panicled sedge, fine bogmoss, a Magellanic bogmoss, tomentypnum moss, and hooked scorpion-moss. The specific composition of peat-forming plants demonstrates that the peat thickness was torn off from the mineral horizon and developed on transitional type. Now the given site of the bog develops on transitional type. It is confirmed by existence in the top peat layers and on a surface of a peatland of such plant species as sweet gale, bog cranberry, cotton grass, woollyfruit sedge, and Magellanic bogmoss. A specific variety of vegetative community decreased.

On a deposit surface, wood-moss-subshrub phytocenosis now develops which top layer is presented by a Scots pine and white birch, a subshrub layer—sweet gale and bog cranberry. The lower layer is presented by *Sphagnum* and *Hypnales* mosses.

Thus, studying of botanical structure demonstrates alternation of different climatic eras. In each stage, there is a formation of the phytoce-noses, which are most adapted for these climatic factors, that is visually reflected in the peat (organic) chronicle of bogs.

Application of botanical and paleopalinologic analyses of peat allows to characterize vegetative groups at different stages of peatland formation. Considering biotic preferences of different plant species, it is possible to restore to a certain degree climatic and hydrological conditions of forma-tion of phytocenoses with their participation. Such researches will allow to concretize the climatic periods of postglacial time in the territory of the Kirov region.

KEYWORDS

- **bog**
- **Boreal period**
- **Atlantic period**
- **Holocene**
- **peat**
- **botanical composition**
- **paleopalinologic analysis**
- **stratigraphic scheme**

REFERENCES

1. Thuiller, W.; Lavorel, S.; Araujo, M. B.; Sykes, M. T.; Prentice, I.C. Climate Change Threats to Plant Diversity in Europe. *PNA* **2005,** *102,* 8245–8250.
2. Zhao, X.; Dupont, L.; Schefuß, E.; Bouimetarhan, I.; Wefer, G. Palynological Evidence for Holocene Climatic and Oceanographic Changes off Western South Africa. *Quat. Sci. Rev.* **2017,** *165,* 88–101.
3. Yao, Q.; Liu, K-b. Dynamics of Marsh-Mangrove Ecotone Since the Mid-Holocene: A Palynological Study of Mangrove Encroachment and Sea Level Rise in the Shark River Estuary, Florida. *PLoS ONE* **2017,** *12*(3), e0173670.

4. Zonneveld, K. A. F.; Pospelova, V. A Determination Key for Modern Dinoflagellate Cysts. *Palynology* **2015**, *39*, 387–409.

5. Rull, V.; Montoya, E.; Giesecke, T.; Morris, J. L. Editorial: Palynology and Vegetation History. *Front. Earth Sci.* **2018**, *6*, 186.

6. Carter, V. A.; Chiverrell, R. C.; Clear, J. L.; Kuosmanen, N.; Moravcová, A.; Svoboda, M.; Svobodová-Svitavská, H.; van Leeuwen, J. F. N.; van der Knaap, W. O.; Kuneš, P. Quantitative Palynology Informing Conservation Ecology in the Bohemian/Bavarian Forests of Central Europe. *Front. Plant Sci.* **2018**, *8*, 2268.

7. Sánchez, N. S. Late-Holocene Environments Reconstructed from Peatlands: Linking Geochemistry and Palynology. Ph.D. Thesis, Universidade de Santiago de Compostela 2016; p 212.

8. Ledu, D.; Rochon, A.; Vernal, A.; St-Onge, G. Palynological Evidence of Holocene Climate Change in the Eastern Arctic: A Possible Shift in the Arctic Oscillation at the Millennial Time Scale. *Can. J. Earth Sc.* **2008**, *45*, 1363–1375.

9. Khazin, L. B.; Khazina, I. V.; Krivonogov, S. K.; Kuzmin, Ya. V.; Prokopenko, A. A.; Yi, S.; Burr, G. S. Holocene Climate Change in Southern West Siberia Based on Ostracod Analysis. *Rus. Geol. Geophys.* **2016**, *57*, 574–585.

10. Ulanov, A. I. *Boggy and Cutover Soils of South Taiga of Euro-North-East Russia*. Kirov: Vyatka Publ., 2005; p 320 (in Russian).

11. Turygina, O. V. Paleobioklimatic Factors of Environment of the Valley of Central Yenisei in the Holocene (on the Example of Inundated Ecosystems). *Her. Krasnoyarsk State Agric. Univ.* **2017**, *8*, 124–128 (in Russian).

12. Prokushkin, A. S.; Karpenko, L. V.; Tokareva, I. V.; Korets, M. A.; Pokrovskii O. S. Carbon and Nitrogen in the Bogs of the Northern Part of the Sym-Dubches Interfluve. *Geogr. Nat. Res* **2017**, *2*, 114–123 (in Russian).

13. Inisheva, L. I.; Maslov, S. G.; Dementyeva, T. V.; Porokhina, E. V.; Dyrin, V. A. Estimation of Organic Matter of the West-Siberian Peats. *Proc. Komi Sci. Centre Ural Div. Rus. Acad. Sci.* **2017**, *1*(29). 36-43 (in Russian).

14. Kulikova, G. G. *Brief Guidance on Botanical Analysis of Peat*; Moskow State Univ. Publ.: Moscow, 1974; p 96 (in Russian).

15. Dombrovskaya, A. V.; Koreneva, M. M.; Tuyremnov, S. N. *Atlas of Plant Residuals Met in Peat*; 'Gosenergoizdat' Publ.: Moscow, 1959; p 223 (in Russian).

16. Kats, N. Ya.; Kats, S. V.; Skobeeva, E. I. *Atlas of Plant Residuals in Peat*; 'Nedra' ('Mineral Resources' in Rus.) Publ.: Moscow, 1977; p 372 (in Russian).

17. Pokrovskaya, I. M. Methods of Field and Laboratory Works. In *Methods of Paleopalynologic Investigations and Morphology of Some Fossil Spores, Pollen and Other Plant Microfossils*; Pokrovskaya, I. M., Ed.: 'Nedra' ('Mineral Resources' in Rus.) Publ. Leningrad branch: Leningrad, 1966; pp 29–60 (in Russian).

18. Ulanov, A. N.; Zhuravleva, E. L. Bogs. *Encyclopedia of Vyatka Land. V.7. Nature*; Regional Writers' Organization Publ.: Kirov, 1997; pp 223–233 (in Russian).

19. Khotinsky, N. A. Holocene of North Eurasia; "Nauka" ("Science" in Rus.) Publ.: Moscow, 1977; p 198 (in Russian).

20. Tyuremnov, S. N. *Peat Fields*; "Nedra" ("Mineral Resources" in Rus.) Publ.: Moscow, 1976; p 488 (in Russian).

21. Kolchanov, V. I.; Zhukova, I. A.; Pakhomov, M. M.; Prokashev, A. M. Geological Past. In *Encyclopedia of Vyatka Land. V.7. Nature;* Regional Writers' Organization Publ.: Kirov, 1997; pp 58–80 (in Russian).

22. *Paleoclimate and Paleo-Landscapes of Extratropical territories of Northern Hemisphere. Late Pleistocene–Holocene*; Velichko, A. A., Ed.; GEOS Publ.: Moscow, 2009; p 120 (in Russian).

23. Shikhova, L. N.; Gonina, E. S.; Ulanov, A. N. Content of Biogenic Elements (Zinc and Copper) in Soil Component of Bog Biogeocenoses (on Example of Peatbog Zenginsky in Kirov Region). *Agric. Sci. Euro-North-East* **2016,** *2*, 41–47 (in Russian).

CHAPTER 3

Use of *Artemia salina* Biomarkers for the Evaluation of the Effect of Static Magnetic Fields

IRINA I. RUDNEVA and VALENTIN G. SHAIDA[*]

A.O. Kovalevsky Institute of the Southern Seas Research Russian Academy of Sciences, Nahimov av., 2, 299011 Sevastopol, Crimea

[*]Corresponding author. E-mail: svg-41@mail.ru

ABSTRACT

The purpose of the present study was to determine the response of *Artemia salina* cysts and to evaluate oxidative stress level in hatching nauplia treated with static magnetic 50 mT field (SMF), taking into account of north (N) and south (S) poles. Four groups of *Artemia* cysts were investigated: (1) control group without treatment; (2) intact cysts, incubated in marine water, which was treated during 24 h with SMF; (3) dry cysts were irradiated with SMF during 24 h and incubated in marine water for 48 h; (4) cysts in marine water were exposed to SMF during the entire hatching period (48 h). Hatching rate of the cysts and chemiluminescence (ChL) level of their extracts were measured. No significant differences in hatching percentage of cysts incubated in treated magnetic water and dry cysts, pretreated with SMF were observed. Hatching percentage of the developmental cysts irradiated near S pole was higher, than those in the control, while ChL response of the nauplia extracts in various experimental groups was not uniform. Direct and indirect treatment of SMF modifies *Artemia* early development. The separate pole of SMF plays a role in the response of *Artemia* hatching process and ChL level in nauplia.

3.1 INTRODUCTION

Magnetic fields (MFs) as ecological factor play an important role in the evolution of living organisms. Biological effects of MFs were demonstrated on plants, vertebrate, and invertebrate animals and in man. The interest to biological effects of MF has been intensified at present because various MF products have been developed and used in human activity (therapy, agriculture, aquaculture, engineering, etc.). MFs were used successfully as presowing seed treatments to increase vigor, seeding growth, and yield.[1-4] Biological effects of MFs depend on the responses of living organisms, and they connect with the intensity of static magnetic field (SMF), which ranges: low intensity in the microtesla (μT), including geomagnetic field (25–65 μT), moderate SMF, which intensity is estimated as millitesla (mT), and high intensity SMF in the, and ultrahigh intensity which intensity is more than 5 T.[5,6] Various behavioral, developmental, physiological, and biochemical responses on MF treatment were documented in insects.[7,8] Because in the development of new technologies people use a potential exposure to high intensities of static MFs, the effects of MF on living organisms have been the object of numerous investigators. However, they are widely varied and sometimes the results are contradicted because they depended on experimental duration (MF strength, time of exposure, tools of measurements, etc.). Additionally, not much information is available on the mechanism occurring in magnetically treated organisms, and there has not been a unique theory. Several mechanisms of the SMF impact on living systems were proposed. The increase of free radical generation and oxidative stress caused SMF is one of the possible mechanism underling the effects of MF.[5,9,10] Oxidative stress in SMF-treated organisms were observed in the rats.[11] The authors used the biomarkers namely nitric oxide (NO), malondialdehyde (MDA), and advanced oxidation protein products (AOPP), and a decrease in superoxide dismutase (SOD), glutathione (GSH), and glycation end products (AGEs) for the evaluation of the biological response on SMF impact, and they demonstrated that the response to SMFs was time-dependent. Clinical studies have suggested that magnetic stimulation could accelerate the healing process and provide an effective complementary therapy for different pathologies.[12] It may also induce changes in living systems, at the organism, tissue, cellular, membrane, and subcellular levels. Therefore, study of SMF effects on living organisms is very important both for theoretical, practical, and especially therapy purposes for improving the productive characteristic of plants seeds, animal's eggs hatching, and for human health problems.

The brine shrimp *Artemia* is widely distributed in hypersaline water bodies namely salt lakes and lagoons all over the world, including the coastal part of Black Sea.[13] *Artemia* contains many kinds of essential compounds (amino- and fatty acids, hormones, vitamins, carotenoids, etc.), which play a role in food quality of the brine shrimp and its use in aquaculture. Hatching characteristics and nutritional value of *Artemia* cysts in various biotops are differed each from other, and they depend on specificity of climate, geographical position of water bodies, mineral and microalgae composition, season, year, etc.[14,15] Artemia is a good tool in ecotoxicological studies, and it is used as test organism in the evaluation of heavy metal, organic, and many other kinds of toxic materials.[16,17] However, during the period of storage cysts quality and hatching percentage tend to decrease. Several procedures help to increase hatching characteristics such as decapsulation, stimulation with chemicals, freezing, etc.[18] MFs also change *Artemia* cysts hatching rate as we described previously.[19] We also studied the effects of the gradient of SMF, considering of southern and northern poles, on the hatching rate of *Artemia salina* cysts, and the uniform trend of the hatching rate was shown.[20]

However, current information of MFs effects on *Artemia* cysts hatching and its possible application in bioindication and aquaculture purposes is very limited. Taking into account that in the development of new technologies people widely use a potential exposure to high intensities of SMFs, the biological effects of MF studies are very important. Therefore, the goal of the present study was to determine the response of *Artemia* cysts hatching process and to evaluate oxidative stress values in the hatching nauplia treated with SMF.

3.2 MATERIALS AND METHODOLOGY

3.2.1 MATERIALS

Artemia cysts were collected in the coast of Sasyk-Sivash lake (Crimea, Russia) in December, 2016. Cysts were washed by salt water and then by fresh water for viable cysts separation according the standard method.[21] The viable cysts were collected and dried at 28°C for further studies. Dry cysts were incubated under the standard conditions for determination of their hatching characteristics.[21,22]

3.2.2   EXPERIMENTAL DESIGN

The scheme of experimental design for determination of the MF effects on hatching percentage of *Artemia* cysts is present in Figure 3.1. Experimental platform was designed and fabricated utilizing two rectangular (80 × 57 × 20 mm, mass 380 g) ferrite static magnets (50 mT). Glass tubes with saline water, dry cysts, and incubated cysts were placed near the north (N) and south (S) poles of the magnets. Duration of experiment includes four groups (Table 3.1):

1) Control group without treatment;
2) Sterile marine water (salinity 35‰) in glass tubes, treated with SMF 50 mT near north (NW) and south (SW) poles of the magnet during 24 h, after treatment intact dry cysts were incubated in the water according the standard procedure during 48 h;
3) Dry cysts in glass tubes were irradiated with MF (near north (NC) and south (SC) poles, respectively) during 24 h, then they were incubated in marine sterile water for 48 h according the standard procedure;
4) Cysts in glass tubes in sterile marine water salinity 35‰ were exposed to SMF near N (NT) and S (ST) poles during the entire hatching period (48 h). Hatching percentage of the examined samples was measured as described by Refs [21, 22].

3.2.3   CHEMILUMINESCENCE ASSAY

Hatching nauplia were contained, washed with distilled water, homogenized in glass homogenizer with cool 0.85 NaCl solution and centrifuged 15 min at 5000 g. Protein concentration in nauplia extracts were determined using the standard kit. Chemiluminescence (ChL) parameters of nauplia extracts were assayed on the basis on stimulation of 3% H_2O_2 and $FeSO_4$ used Chemiluminometer 2011 (LKB, Sweden). The values of ChL estimated as the ratio of the intensity of ChL maximum to protein concentration in the sample.[23]

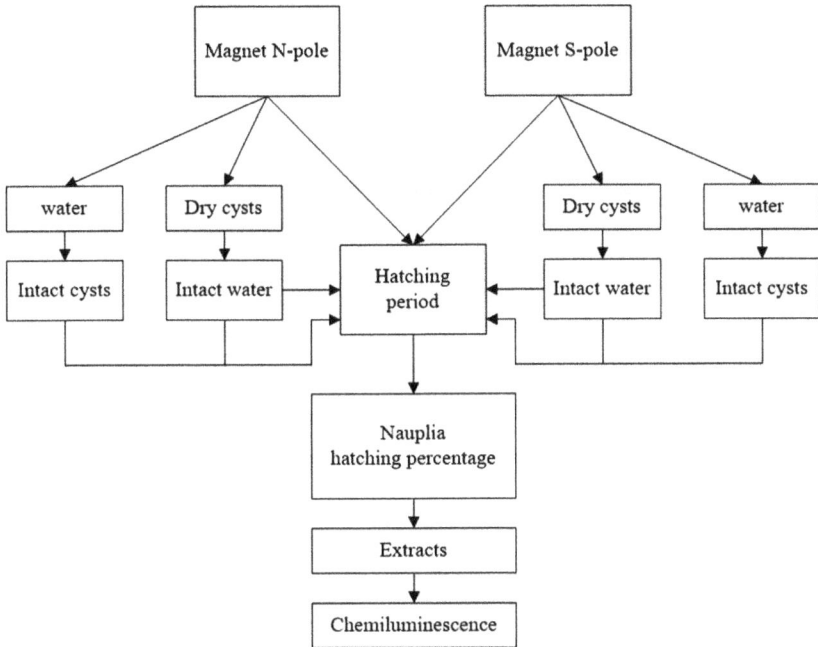

FIGURE 3.1 Scheme of the experimental design.

TABLE 3.1 Experimental Groups and Experimental Protocol of SMF Effect Study on *Artemia salina* Cysts.

Groups	Treatment	Title of group	Number of determinations
Group 1	Without treatment	Control	15
Group 2	Water	NW + SW	13 + 18
Group 3	Dry cysts	NC + SC	23 + 18
Group 4	Hatching process	NT + ST	6 + 6

3.2.4 STATISTICAL ANALYSIS

All analysis were performed in three replicates. The results from different treatments were compared using ANOVA. Statistical differences were calculated using Mann–Whitney tests. All numerical data are given as means ± SE.[24] The significance level was $p<0.05$. Statistical correlations between hatching percentage and ChL parameters in experimental groups were calculated by the least-squares method using the computer program CURVFIT (Version 2.10-L).

3.3   RESULTS

3.3.1   ARTEMIA HATCHING RATE

No significant differences of hatching percentage of *Artemia* cysts incubated in pretreated water (groups NW and SW) as compared with control were observed (Fig. 3.2).

However, we found insignificant increase of hatching percentage of cysts in SW-group as compared with the control and NW-group. The values of percentage of *Artemia* nauplia hatching in pretreated dry cysts did not observe significant differences also (groups NC and SC) (Table 3.2). We could mark, that the hatching percentage of the cysts in SC-group tended to decrease as compared with the control, while hatching value of cysts in NC-group tended to increase. In ST-group, the hatching percentage of the cysts was significantly higher ($p<0.05$) as compared with control and NT-group.

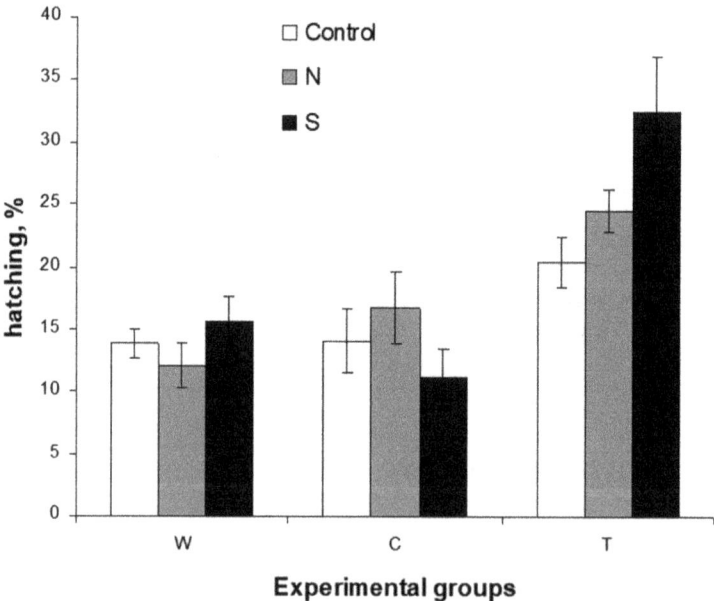

FIGURE 3.2   Hatching percentage of *Artemia* cysts in different experimental groups treated with static magnetic field 50 mT. W—nauplia hatching in water pretreated with static magnet 50 mT during 24 h., C—nauplia hatching in cysts pretreated with static magnet 50 mT during 24 h, T—nauplia hatching treated with static magnet 50 mT. N—north pole, S—south pole.

3.3.2 CHEMILUMINESCENCE LEVEL

ChL level parameters of hatching nauplia are present in Figure 3.3. ChL of nauplia extracts in groups NW, SW, NC, and SC was significantly ($p<0.01$) lower as compared with control. ChL of nauplia in group NW was significantly lower ($p<0.01$) than in group SW, while in NC- and SC-groups the values were the similar. The ChL value in the group NT was significantly ($p<0.01$) lower as compared with the control, while in group ST it was greater ($p<0.01$) than in the control and in group NT.

The correlation between hatching percentage and ChL values was r = 0.47 and regression was as following

$$Y = -0.22 + 0.04X$$

Therefore, the results obtained showed the differences of hatching percentage and ChL values of *Artemia* cysts and nauplia induced by SMF 50 mT in different experimental groups. The obtained results demonstrated that the influence of MF of moderate intensity was not uniform and depended on the experimental group treatment conditions.

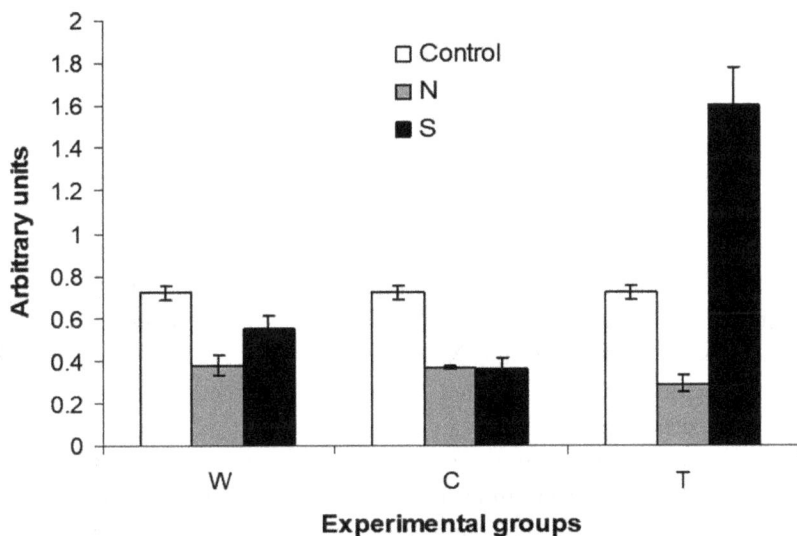

FIGURE 3.3 Chemiluminescence values of *Artemia* hatching nauplia extracts of different experimental groups treated with static magnetic field 50 mT. For other explanations see Figure 3.1.

3.4   DISCUSSION

The biological effects of various ecological factors and their complex play an important role in the processes of reproduction and development of living organisms. For instance, different aquatic animals are able to perceive geomagnetic field and use it for navigation purposes, the others are able to generate the electromagnetic field for the starvation purposes.[25] However, both, static and alternating artificial fields might disturb this ability.[26,27] Experimental studies into effects of MFs on aquatic organisms are still very scanty. Sometimes, it is very hard to determine the evolutionary outcome of environmental conditions in organism's interaction and adaptation to them. Several studies have been carried out to evaluate the effects of MFs on the living organisms in various levels of their biological organization (from molecular to the organism) in experimental conditions, and the authors postulated that they depend on the duration of exposure, namely intensity of magnetic induction, light, temperature, humidity, etc.[4,7,28,29] Time of exposure also plays an important role in response of living systems on MF irradiation in different levels of their biological organization.[30] The authors postulated that EMF is considered much more hazardous for living organisms, than static MF,[31,32] because the effect of EMF on mutagenicity is estimated to be 200 times higher than that of static MF.[33] The information is very limited to the effects of static MF on early life stages of aquatic organisms, especially marine species.[34–38]

Our results have shown that the response of *Artemia* cysts on SMF irradiation 50 mT depended on the duration of exposure of the samples. We found no significant effect on hatching process of *Artemia* cysts incubated in pretreated magnetic water (groups NW and SW). However, we could mark the insignificant increase of hatching percentage of cysts incubated in water pretreated near S pole (SW-group) as compared with the values of the control and NW-group, which could be explained the stimulating effect of the treated water. The changes of magnet-treated water properties may stimulate cysts hydration and oxygen consumption, induction of metabolic processes in developing embryos, and growth hatching percentage. SMF modifies water molecular structure, pH, electric charge, polarization, dzeta-potential, energy, etc. Chemical reactions are increased in magnet-treated water, and heavy metals are precipitated in it.[39,40] Additionally, reactive oxygen species (ROS) generation is one of the possible mechanisms underlying the effects of SMFs which modify the processes in the environment and in the living organisms.[3,10,41] At the other hand, signals of the magnetic particles directly influence on the cells and living functions of the organisms[42] (Table 3.2).

In our study, no significant differences were shown between the hatching percentage of cysts in the control group and in dry samples treated with SMF (groups NC and SC). However, hatching percentage of the cysts exposed near S-pole (SC-group) tended to decrease as compared with the control, while hatching value of cysts irradiated near N-pole (NC-group) tended to increase. Therefore, we could propose the different effects of N- and S-poles on pretreated dry cysts which are visualized during hatching process.

At least, hatching percentage was higher in SMF-irradiated hatching conditions (groups NT and ST), especially near S-pole. We could suggest that SMF stimulates hatching process of *Artemia* cysts particular near S-pole. As we see, the effect of SMF in groups ST and NT may reflect the complex of each individual interaction (water and dry cysts) exposed with SMF or the combination of these interactions.

Stimulating effects of SMFs on development processes of various living organisms were documented in several publications. SMF exposure increased the germination of low viability seeds and improved their quality and growth.[3] In the MF ranged from 9.4 and 14.1 T, the hatching delay of the mosquito eggs increased nonlinearly with the intensity of the SMF and biological effects of MFs could be reversible or partially reversible.[7] The papers report the influence of MF on the early development stages of aquatic organisms namely fish embryos[52] and larvae.[53] Our results agree with the data obtained by Pan,[54] who studied the effect of different SMF intensity on fresh insect eggs *Heliothis virescens* (tobacco bugworm). The results indicated that SMF up to 0.16 T had no detectable biological effects on the hatching of insect eggs, while at 7 T MF the hatching was delayed, the hatching rate was slower, and the hatching efficiency was lower. The effects of MFs on early life stages of aquatic organisms include changes in the permeability of fish egg shells,[55] behavior of melanophores of embryos,[56] changing in respiration and metabolic rate of embryos and larvae[57] and can affect sperm motility.[58]

In our study, we observed several different trends in the hatching percentage of cyst incubated near S- and N-pole of the static magnet. The researchers also detected different biological effects in *Drosophila mela-nogaster* after exposure to strong static magnet 2.4 T near north and south

TABLE 3.2 Biological Effects of Magnetic Fields on Aquatic Organisms.

Species	MF intensity	Effects	References
Freshwater crab *Barytelphusa cunicularis*	Low-frequency electromagnetic fields	Total aggregation between 60 and 90 min. Feeding rate was found to be higher in LF-EMF-induced crabs; however, in eyestalk ablated crabs, the feeding was voracious and nonselective.	[43]
Crustacean (*Crangon crangon, Rhithropanopeus harrisii and Saduria entomon*); mussel (*Mytilus edulis*) and the flounder (*Planthichthys flesus*)	3.7 mT	Changes of survival rate and fitness.	[35]
Chum salmon (*Oncorhynchus keta*	2.5, 5, 7.5 mT	Changes in migratory behavior.	[44]
Zebrafish (*Danio rerio*)	50 Hz	Significant change in cortisol, glucose, 17β-estradiol (E2), and 17-α hydroxy progesterone (17-OHP) levels by enhancing the intensity and time of exposure to SMF.	[45]
Zebrafish (*Danio rerio*)	50 Hz 1 mT electromagnetic	Delays the hatching period.	[46]
Rainbow trout (*Oncorhynchus mykiss*) at early stages, common ragworm *Hediste diversicolor* and the Baltic clam *Limecola balthica*	Low-frequency EMF of different induction values field	Induction of formation of micronuclei (MN), nuclear buds (NB), nuclear buds on filament cells (NBf) and cells with blebbed nuclei (BL) were assessed as genotoxicity endpoints, and 8-shaped nuclei, fragmented (Fr), apoptotic (Ap), and binucleated (BN) cells as cytotoxicity endpoints.	[47]
Embryos of sea urchin *Strongylocentrotus purpuratus*	50 Hz magnetic fields	Disturbs the mitotic cycle.	[48]
Blue mussel *Mytilus galloprovincialis*	Pulsing magnetic fields	Activation of MAP kinases and the expression of heat shock proteins.	[49]
Freshwater brook trout *Salvelinus fontinalis*	Electromagnetic field emissions	Increase of a pineal melatonin level.	[50]
Edible crab (*Cancer pagurus*)		Changes in behavior.	[51]

poles.[8] They proposed that different poles of static magnet may modify the process underlying development and thereby to achieve the effects on different life stages.

SMF irradiation could be potential stress factor for the organisms because it generates free radicals and stimulates oxidative stress, synthesis of heat shock proteins and induction of antioxidant enzyme activities.[8-10,30] ChL method was used to measure the production of ROS generated by MF in three experimental groups. Examinations of ChL parameters of hatching nauplia responses to SMF also suggest a dependence on experimental variables. Our results demonstrated the significant decrease ($p<0.01$) of ChL parameters in *Artemia* nauplia hatching in pretreated water (NW- and SW-groups) and in dry cysts (NC- and SC-groups) as compared to the control. The increase of ROS concentration in magnetic-treated water[3,41] may modify prooxidant/antioxidant balance of the developing embryos during the process of their hydration and elevate antioxidant properties of the hatching nauplia, which correlated with the decrease of ChL values of the extracts related to control. At the case of dry cysts, we could suggest that SMF changes the metabolic processes or/and biomolecules of the developing embryos and modifies their antioxidant/prooxidant balance tended to increase the antioxidant status.

The applied MF modified the ChL parameters of the nauplia hatching from the cysts, which were exposed to SMF during the entire hatching period. The ChL values in group NT was significantly lower, while in ST-group it was twofold greater than in the control and above threefold higher than in NT-group. Researchers also documented different biological effects of N- and S-poles of static magnet on antioxidant enzyme activities of *Drosophila subobscura*.[8] They observed that the total GSH content was significantly decreased, while catalase activity was increased in exposed groups, and the response of SOD activity was not uniform. The different effects of S- and N-poles of static magnet do not have any explanations and need further investigations.

Therefore, from our findings we could propose that SMF modifies the responses of *Artemia* developing embryos in different mechanisms, which attributed with the changes in ROS production. MFs stimulate the increase of ROS generation both in the environment and in the organism and damage prooxidant/antioxidant balance.[59-61] In both NW- and SW-groups, SMF generates ROS in the water and changes its properties which were postulated by several researchers.[41,42] At the beginning of the development, cysts are hydrated and they take up water from the environment. In groups NW and SW, the pretreated water molecules with changing properties may absorb in the cysts and modify the prooxidant/antioxidant balance in developing embryos. In addition, increase of free radicals concentration in magnetic-treated water soaked up into embryos could stimulate their antioxidant activities and quench ChL level, which was observed in both NW- and SW-groups.

In the case of groups NC and SC, we could propose another mechanism of ChL inhibition in hatching nauplia. Polarization by MF may play an important role in the modification of the embryo status of the shell and its fitness components, which could influence on the further development and hatching process.[7] However, our results demonstrated insignificant fluctuations in hatching percentage, while the ChL level was significantly lower in the nauplia obtained from pretreated cysts. It revealed the decrease of the intensity of the level of ROS or (and) the increase of antioxidant defense in the hatching nauplia.

In groups NT and ST, we have found the significant differences between the effects of separate magnetic poles. In group NT, the ChL level decreased, while in group ST it was significantly increased. The mechanism could explain the synergistic effect of treated water and treated cysts in the case of N-pole and antagonistic effect in the case of S-pole which was stimulated ROS production at the present experimental conditions. However, this triggering mechanism of ROS stimulation in nauplia hatching process exposed SMF is unknown and need further investigations.

3.5 CONCLUSIONS

Therefore, the experimental results show that SMF may affect metabolism of the living systems through different mechanisms. Our study demonstrated that 50 mT SMF has apparent biological effects on *Artemia salina* cysts hatching percentage and ChL level of nauplia extracts. Direct and indirect mechanisms of MFs effects on *A. salina* early development were observed. Indirect mechanism modified hatching process via magnetic water properties changes. Direct mechanism affected cysts and entire hatching process which changed under SMF exposure. SMF irradiation modifies prooxidant/antioxidant status of the developing brine shrimp embryos. The separate pole of the magnet also plays a role in the response of *A. salina* hatching process and ChL level of nauplia. Further investigations are needed to explain the effects of separate magnetic poles of SMF on the brine shrimp.

ACKNOWLEDGMENTS

This work was conducted in accordance with the governmental assignment. The functional, metabolic, and toxicological aspects of existence of aquatic organisms and their populations in biotopes with different physicochemical regimes (no. AAAA-A18-118021490093-4) to A.O. Kovalevsky Institute of Marine Biological Research.

We thank the Head of Physical Department of V. Vernadsky Federal University Prof. V.N. Dobzhansky and Dr. A.I. Gorbovanov for the help of SMF dosimetry.

KEYWORDS

- **magnetic fields**
- **oxidative stress**
- **brine shrimp**
- **chemiluminescence**
- **nauplia**

REFERENCES

1. Moon, J. D.; Chung, H. S. Acceleration of Germination of Tomato Seed by Applying an Electric and Magnetic Field. *J. Electro Stat.* **2000,** *48*(2), 103–114.
2. Chen, Y. P.; He, J. M. Magnetic Field can Alleviate Toxicologicla Effect Induced by Cadmium in Mungbean Seedlings. *Ecotoxicology* **2011,** *20*(4), 760–769.
3. Shine, M. B.; Guruprasad, K. N.; Anand, A. Effect of Stationary Magnetic Field Strengths of 150 and 200mT on Reactive Oxygen Species Production in Soybean. *Bioelectromagnetics* **2012,** *33*(5), 428–437.
4. Halgamuge, M. N. Weak Radiofrequency Radiation Exposure from Mobile Phone Radiation on Plants. *Electomagn. Biol. Med.* **2017,** *36*(2), 213–235.
5. Okano, H. Effects of Static Magnetic Fields in Biology: Role of Free Radicals. *Front. Biosci.* **2008,** *13*, 6106–6125.

6. Bejaoui, M.; Khalloufi, N.; Touaylia, S. Effect of Static Magnetic Field on Terrestrial Isopods (Isopoda: Oniscidea). *Crustacean Biol.* **2019**, *1*, 1–5.

7. Pan, H.; Liu, X. Apparent Biological Effect of Strong Magnetic Field on Mosquito Egg Hatching. *Bioelectromagnetics* **2004**, *25*(2), 84–91.

8. Todorovich, D.; Peric-Mataruga, V.; Mircic. D.; Ristic-Djurovic, J.; Prolic, Z.; Petkovie, D.; Savie, T. Estimation of Changes in Fitness Components and Antioxidant Defense of Drosophila Subobscura (Insecta, Diptera) After Exposure to 2.4 T Strong Magnetic Field. *Environ. Sci. Pollut. Res.* **2015**, *27*(7), 5305–5314.

9. Chekhun, V. F.; Demash, D. V.; Naleskina, L. A. Evaluation of Biological Effects and Possible Mechanisms of Action of Static Magnetic Fields. *J. Physiol.* **2012**, *58*(3), 85–94 (*in Ukrainian*).

10. Jouni, F. L.; Abdolmaleki, P.; Ghanati, F. Oxidative Stress in Broad Bean (*Vicia faba* L.) Induced by Static Magnetic Field Under Natural Radioactivity. *Mutat. Res.* **2012**, *741*, 112–116.

11. Coballase-Urrutia, E.; Navarro, L.; Ortiz, J. T.; Verdugo-Díaz, L.; Gallardo, J. M., Hernández, M. E.; F. Estrada-Rojo F. Static Magnetic Fields Modulate the Response of Different Oxidative Stress Markers in a Restraint Stress Model Animal. *Hindawi BioMed. Res. Int.* **2018**, *9*, Article ID 3960408.

12. Kiss, B.; Laszlo J. F.; Szalai, A.; Porszasz, R. Analysis of the Effect of Locally Applied Inhomogeneous Static Magnetic Field Exposure on Mouse Ear Edema - A Double Blind Study. *PLoS ONE* **2015**, *10*(2), Article ID e0118089.

13. Gulov, O. A. Ecocyd of Crimean Salt Lakes. In *Theoretical and Practical Approach of Inland Water Bodies Remedies*; RAS: Saint-Petersburg, 2007; 60–78 (*in Russian*).

14. Sorgeloos, P.; Lavens, P.; Leger, P.; Tackaert, W.; Versichele, D. *Manual for the Culture and use of Brine Shrimp Artemia in Aquaculture*; State University of Ghent Press: Belgium, 1986; p 319.

15. Rudneva, I. I. *Artemia: Perspectives of Application in Common Wealth*; Naukova Dumka: Kiev, 1991; 142 (*in Russian*).

16. Rotini, A.; Rotini, L.; Manfra, S.; Canepa, A.; Tornambè, L.; Migliore, A. *Artemia* Hatching Assay be a (Sensitive) Alternative Tool to Acute Toxicity Test?! *Bull. Environ. Contam. Toxicol.* **2015**, *95*, 745–751.

17. Neumeyer, C. H.; Gerlach, J. L.; Ruggiero, K. M.; Covi, J. A. A Novel Model of Early Development in the Brine Shrimp, *Artemia franciscana*, and its Use in Assessing the Effects of Environmental Variables on Development, Emergence, and Hatching. *J. Morphol.* **2014**, *276*, 342–360.

18. Lavens, P.; Tackaert, W.; Sorgeloos, P. International Study on Artemia. XLI. Influence of Culture Conditions and Specific Diapause Deactivation methods on the Hatchability of Artemia Cysts Produced in a Standard Culture System. *Marine Ecol. Prog. Series* **1986**, *31*, 197–203.

19. Shckorbatov, Yu.; Rudneva, I.; Pasiuga, V.; Grabina, V.; Kolchigin, N.; Ivanchenko, D.; Kazanskiy, O.; Shaida, V.; Dumin, O. Electromagnetic Field Effects on *Artemia* Hatching and Chromatin State. *Cent. Eur. J. Biol.* **2010**, *5*(6), 785–790.

20. Shaida, V. G.; Rudneva, I. I. Hatching Rate of Artemia Cysts Under the Impact of Constant Magnetic Field Gradient. *Pollution of Marine Environment: Ecological Monitoring, Bioassay, Standardization*: Proceedings of Russian Scientific Conference

with the International Participation Devoted to 125th Anniversary of Professor V.A. Vodyanitsky, (Sevastopol, May 28–June 1); Colorit: Sevastopol, 2018; 301–305.

21. Van Stappen, G. *FAO Fisheries The Technical Paper № 36*. Use of Cysts, 1996; 102–123.

22. El-Magsodi, M. O.; El-Ghebli, H. M.; Hamza, M.; Van Stappen, G.; Sorgeloos, P. Characterization of Libyan Artemia from Abu Kammash Sabkha. *Libyan J. Mar. Sci.* **2005**, *10*, 19–29.

23. Vladimirov, J. A. Activated Chemiluminescence and Bioluminescence as a Tool in Medical – Biological Investigations. *Soros Educ. J.* **2001**, *7*, 16–20 (*in Russian*).

24. Halafyan, A. A. Analysis of Variance. In *Statistica 6. Data Statistic Processing*; Binom: Moscow; 2008; 133–152 (in Russian).

25. Wiltschko, W.; Wiltschko, R. Magnetic Orientation and Magneto Reception in Birds and Other Animals. *J. Comp. Physiol. A Neuroethol. Sens. Neural. Behav. Physiol.* **2005**, *191*, 675–693.

26. Gill, A. B. Offshore Renewable Energy - Ecological Implications of Generating Electricity in the Coastal Zone. *J. Appl. Ichthyol.* **2005**, *42*, 605–615.

27. Gill, A. B.; Gloyne-Philips, I.; Kimber, J., Sigray, P. Marine Renewable Energy, Electromagnetic (EM) Fields and EM-sensitive Animals; In *Marine Renewable Energy Technology and Environmental Interactions*; 2014; 61–79.

28. Ohkubo, Ch.; Okano, H. Clinical Aspects of Static Magnetic Field Effects on Circulatory System. *Environmentalist* **2011**, *31*(2), 97–106.

29. Loghmannia, J.; Heidari, D.; Rpzati, S. A.; Kazemi, S. The Physiological Responses of the Caspian Kutum (*Rutilus frisii kutum*) Fry to the Static Magnetic Fields with Different Intensities During Acute and Subacute Exposures. *Ecotoxicol. Environ. Safety* **2015**, *111*, 215–219.

30. Laramee, C. B.; Frisch, P.; McLeod, K.; Li, G. C. Evaluation of Heat Shock Gene Expression from Static Magnetic Field Exposure in Vitro. *Bioelectromagnetics* **2014**, *35*(6), 406–413.

31. Panagopoulos, D. J.; Karabarbounis, A.; Margaritis, L. H. Mechanism for Action of Electromagnetic Fields on Cells. *Biochem. Biophys. Res. Commun.* **2002**, *298*, 95–102.

32. Directive, 2013. 2013/35/EU of the European Parliament and of the Council of 26 June. On the Minimum Health and Safety Requirements Regarding the Exposure of Workers to the Risks Arising From Physical Agents (electromagnetic fields) and Repealing Directive 2004/40/EC.

33. Suzuki, Y.; Toyama, Y.; Miyakoshi, Y.; Ikehata, M.; Yoshioka, H.; Shimizu, H. Effect of Static Magnetic Field on the Induction of Micronuclei by Some Mutagens. *Environ. Health. Prev. Med.* **2006**, *11*, 228–232.

34. Sadowski, M.; Winnicki, A.; Formicki, K.; Sobocinski, A.; Tanski, A. The Effect of Magnetic Field on Permeability of Egg Shells of Salmonid Fishes. *Acta Ichtyol. Pisc.* **2007**, *37*,129–135.

35. Bochert, R.; Zettler, M. L. Long-Term Exposure of Several Marine Benthic Animals to Static Magnetic Fields. *Bioelectromagnetics* **2004**, *25*, 498–502.

36. Loghmannia, J.; Heidari, B.; Rozati, S. A.; Kazemi, S. The Physiological Responses of the Caspian Kutum (*Rutilus frisii kutum*) Fry to the Static Magnetic Fields with Different Intensities During Acute and Subacute Exposures. *Ecotoxicol. Environ. Saf.* **2015**, *111*, 215–219.

37. Fey, D. P.; Greszkiewicz, M.; Otremba, Z.; Andrulewicz, E. Effect of Static Magnetic Field on the Hatching Success, Growth, Mortality, and Yolk-Sac Absorption of Larval Northern Pike *Esox lucius*. *Sci. Total Environ.* **2019a,** *647,* 1239–1244.
38. Fey, D. P.; Jakubowska M.; Greszkiewicz, M.; Andrulewicz, E.; Otremba, Z.; Urban-Malinga, B. *Aqautic Toxicol.* **2019b,** *209,* 150–158.
39. Alkhazan, M. M. K.; Saddiq, A. A. N. The Effect of Magnetic Field on the Physical, Chemical and Microbiological Properties of the Lake Water in Saudi Arabia. *Evolut. Biol. Res.* **2010,** *2*(1), 7–14.
40. Zeilinski, M.; Debowski, M.; Krzmieniewski, M.; Dudek, M.; Grala, A. Effect of Constant Magnetic Field (CMF) with Various Values of Magnetic Induction on Effectiveness of Dairy Wastewater Treatment Under Anaerobic Conditions. *Pol. J. Environ. Stud.* **2014,** *23*(1), 255–261.
41. Lesser, M. P. Oxidative Stress in Marine Environments: Biochemistry and Physiological Ecology. *Ann. Rev. Physiol.* **2006,** *68*(1), 253–278.
42. Smirnov, J. V. Bio Magnetic Hydrology. The Effect of a Specially Modified Electromagnetic Field on the Molecular Structure of Liquid Water; *Global Quantec Incorporation of the U.S.A.* 2003; pp 122–125.
43. Rosaria, J. C.; Martin, E. R. Behavioral Change in Freshwater Crab, *Barylephusa cunicularis* After Exposure to Low Frequency Electromagnetic Fields. *World J. Fish Marine Sci.* **2010,** *2,* 487–494.
44. Yano, A.; Baba N.; K. Nagasawa, K.; Ogura, M.; Sato, A.; Sakaki, Y.; Shimizu, Y. Effect of Modified Magnetic Field on the Ocean Migration of Maturingchum Salmon, *Oncorhynchus keta. Marine Biol.* **1997,** *129,* 523.
45. Sedigh, E.; Heidari, B.; Roozati, A.; Valipour, A. The Effect of Different Intensities of Static Magnetic Field on Stress and Selected Reproductive Indices of the Zebrafish (*Danio rerio*) During Acute and Subacute Exposure. *Bull. Environ. Contam. Toxicol.* **2019,** *102*(2), 204–209.
46. Skauli, K. S.; Reitan, J. B.; Walther, B. T. Hatching in Zebrafish (Danio rerio) Embryos Exposed to a 50 Hz Magnetic Field. *Bioelectromagnetics* **2000,** *21,* 407–410.
47. Stankevičiūtė, M.; Jakubowska, M.; Janina Pažusienė, Makarasa, T.; Otrembac, Z.; Barbara Urban-Malinga, B.; Fey, D. P.; Greszkiewicz, M.; Sauliutė, G.; Baršienė, J.; Eugeniusz Andrulewicz, E. Genotoxic and Cytotoxic Effects of 50 Hz 1 mT Electromagnetic Field On Larval Rainbow Trout (*Oncorhynchus mykiss*), Baltic Clam (*Limecola balthica*) and Common Ragworm (*Hediste diversicolor*). *Aquatic Toxicol.* **2019,** *208,* 109–117.
48. Levin, M.; Ernst, S. Applied AC and DC Magnetic Fields Cause Alterations in the Mitotic Cycle of Early Sea Urchin Embryos. *Bioelectromagnetics* **1995,** *16,* 231–240.
49. Malagoli, D.; Lusvardi, M.; Gobba, F.; Ottaviani, E. 50 Hz Magnetic Fields Activate Mussel Immunocyte p38 MAP Kinase and Induce HSP70 and 90. *Comp. Biochem. Physiol.* **2004,** *137,* 75–79.
50. Lerchl, A.; Zachmann, A.; Ali, M. A.; Reiter, R. J. The Effects of Pulsing Magnetic Fields on Pineal Melatonin Synthesis in a Teleost Fish (Brook Trout, *Salvelinus fontinalis*). *Neurosci. Lett.* **1998,** *256,* 171–173.
51. Scott, K.; Harsanyi, P.; Lyndon, A. R. Understanding the Effects of Electromagnetic Field Emissions from Marine Renewable Energy Devices (MREDs) on the Commercially Important Edible Crab, *Cancer pagurus* (L.). *Mar. Pollut. Bull.* **2018,** *131,* 580–588.

52. Tański, A.; Formicki, K.; Śmietana, P.; Sadowski; M., Winnicki, A. Sheltering Behaviour of Spiny Cheek Crayfish (*Orconectes limosus*) in the Presence of an Artificial Magnetic Field. *Bull. Fr. Peche Piscicult.* **2005,** *376–377,* 787–793.

53. O'Connor, J.; Muheim, R. Pre-Settlement Coral-Reef Fish Larvae Respond to Magnetic Field Changes During the Day. *J. Exp. Biol.* **2017,** *220,* 2874–2877.

54. Pan, H. The Effect of a 7 T Magnetic Field on the Egg Hatching of *Heliothis virescens*. *Magn. Reson. Imaging* **1996,** *14*(6), 673–677.

55. Sadowski, M.; Winnicki, A.; Formicki; K., Sobociński, A.; Tański, A. The Effect of Magnetic Field on Permeability of Egg Shells of Salmonid Fishes. *Acta Ichtyol. Pisc.* **2007,** *37*(2), 129–135.

56. Brysiewicz, A.; Formicki, K.; Tański, A.; Wesołowski, P. Magnetic Field Effect on Melanophores of the European whitefish *Coregonus lavaretus* (Linnaeus, 1758) and Vendace *Coregonus albula* (Linnaeus, 1758) (Salmonidae) During Early Embryo-genesis. *Eur. Zool. J.* **2017,** *84*(1), 49–60.

57. Winnicki, A., Korzelecka-Orkisz, A., Sobociński, A., Tański, A., Formicki, K. Effects of the Magnetic Field on Different Forms of Embryonic Locomotor Activity of Northern Pike, *Esox lucius* L. *Acta Ichtyol. Pisc.* **2004,** *34,* 193–203.

58. Formicki, K., Szulc, J., Tański, A., Korzelecka-Orkisz, A., Witkowski, A., Kwiat-kowski, P. The Effect of Static Magnetic Field on Danube huchen, *Hucho hucho* (L.) Sperm Motility Parameters. *Arch. Pol. Fish.* **2013,** *21*(3), 189–197.

59. Rosen, A. D. Mechanism of Action of Moderate-Intensity Static Magnetic Fields on Biological Systems. *Cell Biochem. Biophys.* **2003,** *39*(2), 163–173.

60. Rosen, A. D. Studies on the Effect of Static Magnetic Fields on Biological Systems. *PIERS Online* **2010,** *6*(2), 133–136.

61. Miyakoshi, J. Effects of Static Magnetic Fields at the Cellular Level. *Prog. Biophys. Mol. Biol.* **2005,** *87*(2–3), 212–223.

PART II

Genetic Diversity: Experimental Induction and Evaluation of Ecosystems in the Modern Natural Environments

CHAPTER 4

Theoretical and Applied Aspects of Mutations Induction for Improving Agricultural Plants

NINA A. BOME[1*], LARISSA I. WEISFELD[2*], NATALIA N. KOLOKOLOVA[1], NIKOLAY V. TETYANNIKOV[3], MARAL U. UTEBAYEV[1,4], and ALEXANDER Y. BOME[5]

[1] *Tyumen State University, Institute Biology, Volodarsky St., 6, Tyumen 625003, Russia*

[2] *Emanuel Institute of Biochemical Physics, Russian Academy of Sciences, Kosygin St., 4, Moscow 119334, Russia*

[3] *All-Russian Horticultural Institute for Breeding, Agrotechnology and Nursery, Zagoryevskaya St., 4, Moscow 115598, Russia*

[4] *A.I. Barayev Research and Production Centre of Grain Farming, Barayev St., 12, Shortandy-1, 021601, Kazakhstan*

[5] *Exeter Produce and Storage ltd, 149A Thames Rd. W, N0M 1S3, Exeter, ON, Canada*

Corresponding author. E-mail: bomena@mail.ru or liv11@yandex.ru

ABSTRACT

Abiotic and biotic environmental stresses can directly or indirectly affect the morphophysiological and biochemical status of a plant, changing its metabolism, growth, and development. Changing climatic conditions present new requirements for the parameters of the designed agroecosystems. Agrocenosis with a variety of varietal composition have an advantage due to a more complete use of environmental resources and greater resistance to adverse environmental factors.

Therefore, the use of various breeding and genetic methods to increase the resistance of plants to stresses (heat, drought, frost, colds, salinization, diseases, pests) is one of the priority national economic problems. The solution to this problem, which does not require tremendous anthropogenic costs at all stages of agricultural development, is the creation and use of new stress-resistant cultivars.

The driving force behind plant diversity is the process of mutation. Induction of artificial mutations significantly increases the diversity of crops, increases the possibilities of growing crops in ecologically diverse regions, and the spread of crops with important agronomic traits. Chemical mutagenesis has advantages over radiation in terms of the quantity and quality of plant traits which can be obtained for further selection.

The article presents up-to-date data on methods for obtaining and efficient use of induced mutations to create new cultivars or source material for selection.

Here are the results obtained on the basis of the association of scientists of the Tyumen State University, Emanuel Institute of Biochemical Physics of the Russian Academy of Sciences (Russia), Scientific and Production Center for Grain Management named after A.I. Barayev (Kazakhstan) and School of Biological Sciences: College of Science and Technology, Flinders University (Australia).

4.1 INTRODUCTION

Due to the loss of plant resources over the past 50 years[1,2] and the recognition by world science of the urgent need to take measures in support of biodiversity for all stakeholders, the United Nations adopted the "Strategic Plan: Conservation and Sustainable Use of Biodiversity 2011–2220."[3]

The famous Russian scientist Nikolai I. Vavilov was the first in the world to raise the problem of the need to preserve and increase the diversity of cultivated plants, created genetic banks, and organized expeditions to collect plant seeds from exotic regions of the world.[4–6]

Currently, researchers are attracted by wild relatives of cultivated plants from the collections of genetic banks, since they contain unused genetic variations.[5,6]

A potential source of new variations may be wild-growing forms of flax (*Linum bienne* Mill.), which is the progenitor of cultivated flax (*Linum usitatissimum* L.).[7] Wild forms of barley *Hordeum spontaneum* can serve as a source of genes for many desirable traits can be a potential source of new variations.[8,9]

4.2 USING MUTATION TO OBTAIN AGRICULTURAL USEFULL PLANTS

4.2.1 DISCOVERY OF CHEMICAL MUTAGENESIS

The phenomenon of chemical mutagenesis was discovered in Russia by the famous scientist, geneticist Joseph Abramovich Rapoport (1912–1990). He set himself the task of causing changes in genes that are inherited, that is, mutations. Initially, he conducted a series of experiments on homomorphism in dipterans. Modifier substances influenced the phenotype but did not lead to the inheritance of new traits. At the next stage, Rapoport tested substances that can interact with proteins (following N. K. Koltsov's hypothesis that the chromosome is made of proteins). Chemical compounds were discovered that induce gene mutations in *Drosophila* that are inherited. Carbonyl compounds, formaldehyde, and its derivatives, ethylene imine, and many others caused mutations that were inherited in subsequent generations in the mutagen treatment of sperm cells in *Drosophila*. Thus, the phenomenon of chemical mutagenesis was discovered. However, Rapoport's participation in the Second World War delayed publications. The first publication of Rapoport on the discovery of the phenomenon of chemical mutagenesis in experiments on *Drosophila* took place only in 1946 "Carbonyl compounds and the chemical mechanism of mutations."[10]

In this and subsequent works on larvae, adult insects and plants, he experienced a large number of chemical mutagens and received a high level of hereditary changes.

The main works of Rapoport, which consistently highlight the stages of development and the scientific justification of chemical mutagenesis, one of the largest scientific discoveries of the 20th century, are published in the book "The Discovery of Chemical Mutagenesis. Selected Works."[9]

In the article "Chemical mutagens in the reproduction and protection of nature,"[9, pp 266–267] Rapoport writes: "Among the useful services that modern genetics can provide to plant breeding in agriculture, one of the prominent places belongs to chemical mutagenesis, that is, the induction of hereditary changes using chemicals...."

With the introduction of chemical mutagenesis, chemically induced mutations are useful and desirable for selection with a frequency of 10–30%. Moreover, the absolute frequency ranges from 30 to 100% and higher. The high frequency of useful mutations, in particular the appearance of rare and completely new ones among them, the ability to repeat the procedure at intervals of two generations without a significant accumulation of harmful mutations contributed to the fact that breeding is increasingly starting to use chemical mutagenesis.

It was written in… 1970. Now mutational breeding is recognized as a valuable method, has spread throughout the world, and is used in breeding everywhere.

In the book "Microgenetics,"[10] Rapoport divided mutagens into groups: strong mutagenic stimulus (pp 82–83), moderate and weak (pp 86–88).

The history of the discovery and development of chemical mutagenesis by Rapoport is described in detail in the book of O. G. Stroeva.[11, pp 84–109]

"We are not given to foresee, how our word will come back"… (F. I. Tyuchev), but Rapoport knew, how our word would respond, and devoted himself to this matter. At the Institute of Chemical Physics of the Academy of Sciences of the USSR, with the support of the director and Nobel Prize laureate N. N. Semenov, Rapoport organized a center of chemical mutagenesis for scientists and for the introduction of the method in agronomy. In the USSR and in some other republics, Rapoport widely disseminated a scientific study of chemical mutagenesis. He was the coordinator of the work on the application of chemical mutagenesis in agriculture. Annually, at the Institute of Chemical Physics of RAS, the open scientific seminars carried out the study of chemical mutagens in experiments and in agricultural applications. In parallel, Rapoport published a series of collections at the Nauka (Sciences0 Publishing House with the experimental using chemical mutagens and para-aminobenzoic acid (PABA). Each book was preceded by a theoretical article by Joseph. By 1991 (the year Rapoport passed away), 380 cultivars on a mutant basis were created on the basis of chemical mutagenesis, 166 of them were State registered.

4.2.2 CHEMICAL MUTAGENESIS: DEVELOPMENT TODAY

Genetic diversity created through experimental mutagenesis is regarded as a valuable resource for agricultural plants,[12–14] as evidenced by the number of mutant cultivars created (3222) and their successful cultivation in countries of the world (China, Japan, India, Germany, the United States, Russia, etc.).[15–17]

Induced mutagenesis makes it possible to create hitherto unknown alleles.[12–14] The main successes have been achieved in changing such important characteristics as the duration of the growing season, resistance to diseases, tolerance to abiotic and biotic stresses, and productivity. Induced mutations will be even more in demand for creating cultivars of crops with such properties as the content and quality of oil, starch, protein; deeper root system, providing resistance to drought; resistance to soil salinization, diseases.[18]

A mutant cultivar of barley "Diamant" provided additional income in the brewing industry. More than 150 cultivars of barley in several countries in Europe, North America, and Asia were obtained by hybridization with the participation of the "Diamant."[19]

The economic significance of some mutant cultivars is described. For example, from the inedible flax *Linum usitatissimum* L. by the induced mutagenesis was obtained the "Linole" (the trade name for the edible flax culture). "Linole" has significant economic potential. Its introduction into production was facilitated by the Organization of Scientific and Industrial Research of the Australian Commonwealth.[20,21] "Linole" is now manufactured in Australia, Canada, the United Kingdom (UK), and the United States. The oil from the wild type of flax rancid pretty quickly, while Linole oil is comparable to sunflower and rapeseed oils.[22]

As a result of self-pollination of the mutant forms, a line was created with a linolenic acid content of less than 2% compared to 49% for wild-type parents ("McGregor").[23]

This confirms the theoretical and practical significance of research on the creation of induced plant mutations in complex, often extreme conditions of Western Siberia.

Our studies on the mutagenesis of crops in the conditions of the North Trans-Urals were started in the 70s of the 20th century on cultivars and hybrids of spring soft wheat: at that time, they were relevant and were of theoretical and practical importance.

Experimental mutagenesis has been classified as a relatively new method. Much attention was paid to the emerging trend of sharing sources of mutational and combinational variability to create new cultivars. For the first time in the region, work was carried out to determine the effectiveness of radiation mutagenesis in 8 cultivars and 28 hybrids obtained as a result of crossbreeding according to an incomplete diallelic scheme. It turned out that interest in reproduction is represented by forms with a complex of improved traits obtained using gamma rays on interspecific hybrids.

Induced mutational variability in hybrids increases the livelihood of the appearance of forms with a transgressive expression of signs of productivity.

Since 1980, field and laboratory experiments have been performed to study the mutational variability of self- and cross-pollinated plant species induced by chemical mutagens and gamma radiation. Comparison of long-term data on various plant species made it possible to establish general patterns in their sensitivity (M_1) and mutability (M_2) when exposed to physical and chemical mutagens. In the sensitivity of cultures to mutagens, polymorphism was observed from pronounced resistance to its complete absence. Hybrid forms in most cases were characterized by higher resistance to mutagens. Hybridization, perhaps, creates a new type of organism, the offspring of which is not only more viable, but also has a special mechanism for the repair after mutagenic damage.[24]

The effectiveness of using chemical mutagens to create forms valuable for plants has been shown in various soil and climatic conditions.[25-33]

In the problem of artificial mutagenesis, much attention was paid to studying the influence of various systems on the restoration of genetic disorders. Genetically active substances, including PABA, can significantly affect the process of genetic recovery. Rapoport[10,34] first drew attention to the role of PABA in mutagenesis.

PABA as chemical compound is known since 1863 but its high biological activity in low concentrations was first discovered in 1939 with well-known geneticist J. A. Rapoport on *Drosophila*. He showed that the positive effect on living systems is based on the previously unknown phenomenon of its interaction with ferments. This interaction results in the restoration of the ferments activity, decreased in some cases at the genetic level (e.g., because of excess of recessive genes), or because of damaging environmental factors. In subsequent studies, the ranges of PABA suitable for different objects were determined. It was proved that PABA is a promoter of phenotypic activity and increases immunity; it has virucidal and antimicrobial action. There are data about PABA effect decreasing harmful mutagens action, PABA positively effect on all characters determining yield structure and increasing adaptive plant properties including the resistance to a series of diseases.[10,34]

Our studies have shown the possibility of using PABA to reduce disorders due to the combined action of this substance and chemical mutagens (NMU, NEM, DMS, DES, EI) on the seeds of certain plant species (amaranth, sainfoin, alfalfa, boneless bonfire6 clover, etc.).[35,36] The revealed properties of PABA can be used on other plant species that are inhibited by the influence of the mutagenic factor. For example, we observed a decrease in the seed germination field and the height of flax plants in the first mutant generation under the influence of nitrosomethylurea (NMU) and nitrosoguanidine (NG) in four cultivars ("Grant," "Laska," "Aramis," and "Rod-829") in the Republic of Belarus.[37]

New forms can be obtained using various methods (from simple selection of plants with desirable characteristics to more complex molecular levels).[40] Breeders use in vitro culture for rapid propagation, molecular methods for selection of the desired genotypes, and mutagens for increase variation, environmental factors for selections.[40,41]

Further development of mutational selection is associated with the use of well-known now chemical supermutagens (N-nitrosomethylurea, N-nitrosoethylurea, ethyleneimine), as well as the discovery of new highly active substances (chiral stereoisomers, alkylating agents, antibiotics).[25,42–44]

4.2.3 NEW MUTATION BREEDING METHODS

One of the tasks of mutational breeding is the selection from populations of plant forms with hereditarily altered traits. Point mutations induced by chemical mutagens can be detected using new technologies such as TILLING (local leounds targeting induced in genomes or targeting induced local damage in genomes).[45–48]

The TILLING reverse genetics method accompanies traditional mutagenesis and provides high throughput screening to detect mutant genotypes containing mutations in the gene of interest. When implementing that method, plants of the second mutant generation (M_2) are used as starting material, plants of the first generation (M_1) after self-pollination. First you need to determine the genotype of explant, structure, and concentration of mutagen, and population size. For example, in the studies of V. Talamè[46] it was shown that in order to obtain information on the toxicity and/or lethality of sodium azide (NaN_3) and assessment about its mutational activity, a series of experiments was required for studying the ability of seeds to germinate and the variability of some morphometric parameters plants at different concentrations of the mutagen solution. Moreover, due to the low population density of some species, it is necessary to scan populations from several thousand plants in order to induce new alleles. For example, in the studies of Chantreau,[47] there are no split studies performed on flax; mutant populations from 4894 M_2 families were used as the TILLING platform. For example, in the studies of Chantreau,[47] there are no split studies performed on flax; mutant populations from 4894 M_2 families were used as the TILLING platform. The TILLING population can be used for screening for features of interest. The method is characterized by high throughput, low cost, and is applicable to most organisms. Large-scale TILLING methods have transmitted thousands of induced mutations to the international scientific community.[18]

A modern understanding of plant breeding and genetic improvement involves the integration of traditional and molecular methods. The effectiveness of molecular genetic analysis depends on the properties of the original platform (mutant population), which determine the frequency of mutations, their diversity and quality. Considering that recombination variability can increase under changing environmental conditions (which is typical for the sharply continental climate of Western Siberia), one can expect an increase in the likelihood appearance of plant forms with transgressive expression of characters in generations.

In partnership with the Emanuel Institute of Biochemical Physics RAS conducted experiments using a new chemical mutagen: phosphemid (from the group of phosphazines). The full name of phosphemid is di-(ethyleneimide)-pyrimidyl-2-amidophosphoric acid. The synthesis of phosphemid was carried out in the laboratory of physical and chemical methods of analysis M.V. Lomonosov Moscow State University by Professor E. V. Babaev.

Initially, phosphemid was studied on a model object *Crepis capillaris* (L.) Wallr.[49–51] It showed that in phosphemid at concentrations $2 \cdot 10^{-3}$ M and $1 \cdot 10^{-2}$ M there is a statistically significant suppression of mitotic activity, as well as it has a high level of chromosome rearrangement. It was noted that at higher concentrations, also chromosome fragmentation occurred. This shows that phosphemid is able to affect chromosomes and, accordingly, at lower concentrations, induce mutations.

Currently, research is being conducted at Tyumen State University (Institute of Biology) as part of the scientific project "Study of the genetic resources of cultivated plants and the formation of a collection fund for the conditions of the Northern Trans-Urals." A gene pool of seeds of cultivated plants has been created, numbering more than 2500 samples of cereals, legumes, and oilseeds from 60 countries, hybrid and mutant forms have been obtained. Methodological approaches have been developed for a comprehensive analysis of the adaptive properties of plants in simulated laboratory conditions and in a field experiment.[38,39]

In two cultivars of spring soft wheat ("Cara" and "Scant 3") and the hybrid form F4 ("Cara" x "Scant 3"), the efficacy of the chemical mutagen phosphemid at concentrations of 0.002 and 0.01% (exposure time 3 h) was studied. In the first generation (M_1), an inhibitory effect was observed with respect to field germination of seeds and morphometric parameters of seedlings (root and shoot lengths), which depended on the concentration of mutagen. Compared to the control, stimulation effects were observed in the Cara × Scent 3 hybrid combination in terms of seed germination energy in a laboratory experiment and biological resistance of plants in the field (by 14.0–80.0%, respectively).

Under the influence of mutagen, there was a significant increase in grain mass from 1 m² in the hybrid (by 16.0%) and a decrease in this indicator in the selection cultivars "Cara" and "Skent 3" (by 67.0 and 57.0%, respectively). The proportion of families with altered plants in the cultivars was 5.3% less than in the hybrid. The maximum number of changes in cultivars and hybrids was found in variants with a mutagen concentration of 0.01%.[44,52–54] The decision to include hybrid forms in experiments is based on the fact that, in hybrids, induced mutational variability increases the likelihood of forms with transgenic expression of valuable traits. The treatment of air-dried flax seeds in a solution of phosphemid (concentrations of 0.005%, 0.01%, and 0.1%) was studied. As a result of laboratory experiments, differences between the samples by the morphometric parameters of the seedlings and the growth rate of the aboveground and underground biomass were revealed. Field experiments revealed the effect of phosphemid on seed germination and plant survival during the growing season, and morphological characters were studied.

The decision was made to include hybrid forms in experiments based on the fact, that in hybrids, induced mutational variability increases the likelihood of occurrence of forms with transgenic expression of selection-valuable traits.

The technique of processing air-dried flax seeds in a solution of phosphemid (concentration of 0.005%, 0.01%, and 0.1%) was studied.

As a result of laboratory experiments, differences between the samples were revealed by the morphometric parameters of the seedlings and the growth rate of the aboveground and underground biomass. Field experiments revealed the effect of phosphemid on seed germination and plant survival during the growing season, and studied morphological characters. A series of laboratory experiments showed that the response of flax to seed treatment with phosphemid is significantly different from the reaction of cereals. Concentrations of phosphemid worked out by us earlier on grain crops turned out to be unsuitable for flax seeds. The spectrum of phenotypic changes was characterized by narrow limits at low concentrations of phosphemid solution, and high concentrations for the studied cultivars turned out to be lethal.[55]

In barley, analysis of the field germination of seeds and the survival of plants of the M_1 generation made it possible to determine the concentrations of phosphemid 0.002% and 0.01% as optimal for the growth and development of plants of sample C.I. 10995 (k-30630). A concentration of 0.01% was applied to the semilethal for samples Dz02-129 (k-22934) and "Zernogradsky 813" (k-30453), since seed germination when using it was below 50%. In the "Zernogradsky 813" in generations M_1 and M_2, the effect of plant growth stimulation was recorded in the variants of seed treatment with phosphemid at concentrations of 0.002% and 0.01%. The use of mutagen contributed to increasing the resistance of barley plants to lodging in M_1 (sample C.I. 10995, 6.0 points) and M_2 (sample Dz02-129, 6.2 points). A structural analysis of plant productivity elements in the M_1 generation showed that the highest sensitivity to the effect of the mutagenic factor on the grain weight on the spike and plant was demonstrated by the "Zernogradsky" grade 813 (k-30453) at two concentrations and the sample Dz02-129 (k-22934) in the concentration variant (0.002%). A stimulation effect was noted for a mass of 1000 grains in samples Dz02-129 (k-22934) and C.I. 10995 (k-30630) (at a concentration of phosphemid of 0.01%).[56]

4.2.4 INSTRUMENTAL CHLOROPHYLL ASSESSMENT METHODS

As one of the markers of the sensitivity of the genotype to mutagen in the first generation (M_1), we use the dynamics of accumulation and degradation of chlorophyll in leaf cells based on the indicators of the chlorophyll counter SPAD 502 (Minolta Camera Co, LTD, Tokyo, Japan).

A number of studies have shown the ability to quickly and accurately determine the chlorophyll content using an SPAD 502 optical counter (Minolta Camera Co, Ltd, Tokyo, Japan) in wheat leaves[57,58]; oats[61]; rice[62]; potatoes;[63] and soybeans.[64–66]

Differences in the amount of chlorophyll upon exposure to salinity,[67] contrast temperature,[68,69] and lack of moisture[70] in vitro were revealed.[71] The relationship between the chlorophyll content in leaf cells detected by the SPAD 502 counter and signs of productivity and quality was revealed in some plant species.[72] Thermal and fluorescence visualization of chlorophyll is considered as an opportunity to screen stress-resistant genotypes.[73] In Russia, studies using this device are fragmented.[74]

In our studies, this rapid diagnostic method was first tested on selected samples of flax and oil flax. In the laboratory experiment, measurements were carried out three times with an interval of 5 days, in the field seven times depending from phenological phases: full shoots, herringbone, rapid growth, budding, flowering, green ripeness, and early yellow ripeness.

Positive correlations were found between the results of chlorophyll measurements using SPAD 502 and traits characterizing the surface of plant assimilation (linear dimensions, area and number of leaves, plant height), as well as plant survival during the growing season. Samples with a relatively rapid increase in SPAD 502 in the first half of the growing season and their uniform decrease during the ripening period had a significant advantage in the yield.[75]

SPAD 502 readings were also successfully used to determine differences in the chlorophyll content in soybean leaves between variants with seeds inoculated by *Bradyrhizobium japonicum* seeds and control at three stages of phenological development when grown in Russia and Germany.[65]

The results became the basis for the use of the SPAD 502 chlorophyll counter in mutation induction at flax studies. The reaction to the treatment of seeds with the phosphemid mutagen according to the testimony of SPAD 502 in young plants (vegetative experience in simulated conditions) was confirmed in most cases on adult plants (field experiment in natural conditions). All samples responded to an increase in phosphemid concentration by a decrease in the chlorophyll content in the leaves. Variants with mutagenic treatment revealed changes in morphological characters (plant height, leaf blade size), structure, and color of inflorescences, physiological characteristics of growth and development.[55]

In our studies, this method of rapid diagnosis was first tested on collection samples of flax fiber and oil flax. In a laboratory experiment, measurements were made three times with an interval of 5 days, in the field seven times according to the time of the onset of phenological phases: full shoots, fir-tree, rapid growth, budding, flowering, green ripeness, early yellow ripeness. Positive correlations were found between the results of measurements using SPAD 502 and traits characterizing the surface of plant assimilation (linear sizes, area and number of leaves, plant height), as well as with plant survival during the growing season. Samples with a relatively rapid increase in SPAD 502 in the first half of the growing season and their uniform decrease during the ripening period had a significant advantage in the yield of trusts and seeds.[75]

SPAD 502 readings were also successfully used to determine differences in the chlorophyll content in soybean leaves between variants with inoculation of nitrogen-fixing bacteria *Bradyrhizobium japonicum* seeds and control at three stages of phenological development when grown in Russia and Germany.[65]

The results became the basis for the use of SPAD 502 in studies on the induction of mutations. The reaction to the treatment of seeds of crops with mutagen according to the testimony of SPAD 502 young plants: vegetative experience (growing experience in simulated conditions) in most cases was confirmed on adult plants (field experience in vivo). All samples responded to an increase in phosphemid concentration by a decrease in the chlorophyll content in the leaves. In variants with mutagenic treatment, changes in morphological characters (plant height, leaf blade size), structure, and color of inflorescences, physiological characteristics of growth and development were revealed.[55]

A decrease in the content of chlorophyll under the influence of a mutagenic factor in the leaves of three barley samples was noted in stem elongation phase, cereal milk ripeness phase, waxy and full ripeness of grain. In most cases, the accumulation of chlorophyll in the leaves of control and experimental plants was of a similar nature, while degradation was accelerated in variants with phosphemid.[56]

Of the biochemical parameters, we include the starch content in barley grain as informative, given that stress factors can disrupt its normal level. The literature describes the loss of starch in the leaves of *Hordeum vulgare* L. in response to abiotic stress.[76–78] The study of this indicator is carried out by us in partnership with scientists of the All-Russian

Research Institute of Starch Products V. Goldstein and L. A. Wasserman (IBCP RAS).[76]

Treatment with phosphemid did not lead to a loss in the mass fraction of starch by the dry matter content in M_2 grain in samples "Zernogradsky 813" and Dz02-129. A decrease in the indicator is observed in sample C.I. 10995 (at a mutagen concentration of 0.002%) as a result of an increase in bound starch in cellulose during grain processing. Phosphemid (0.002%) in samples Dz02-129 and "Zernogradsky 813" led to a decrease in the mass fraction of amylose in starch by 5.26.0% in sample C.I. 10995 did not have a negative impact. Starch remobilization under stress is known to be an alternative source of energy and carbon, which contributes to the production of high-quality seeds even in adverse conditions.[56]

The effectiveness of the mutagenic factor was determined by the sensitivity of barley and flax to phosphemid in M_1, the frequency and spectrum of mutations in M_2 and M_3, both in the laboratory and in the field. In the experimental work, one should take into account the genotype features (for example, differences in the set of chromosomes: in *Hordeum vulgare* 2n = 14; *Linum usitatissimum* 2n = 30), mutagen concentration, seed treatment technology).[79,80]

4.2.5 MOLECULAR METHODS FOR ASSESSING-INDUCED GENETIC DIVERSITY

The effectiveness of the new chemical mutagen phosphemid is confirmed by the isolation of new plant forms with valuable traits that can be used as starting material for priority breeding areas.

Assessment of intra- and interpopulation natural or induced genetic diversity is carried out using various methods. Recently, a wide variety of molecular DNA markers have been used. They have an advantage over classical phenotypic approaches, since they are stable and are found in all tissues, do not mix with environmental, pleiotropic and epistatic effects.[81–84]). There are more than 15 different types of markers used for molecular genetic analysis of the plant genome (RFLP-, CAPS-, STS-, SSR-, SNP-, RAPD-, SCAR-, AFLP-, SSCP-, ISSR-markers, etc.).[82,83]

Studies on the identification of alleles of genes that control the synthesis of protein—wheat gliadin, are carried out by us together with scientists from Kazakhstan and Australia. Based on the molecular genetic analysis of the allelic composition of the collection of spring soft wheat, a variety of genes of HMW subunits of high molecular weight gluteins was revealed and the most promising cultivars were proposed for use in the breeding process for selection for baking quality.[84] It has been shown that gliadin electrophoresis spectra do not change depending on environmental conditions and are useful for identifying wheat germplasm in addition to DNA markers.[85,86] The high frequency of gliadin alleles can be associated with the selection of genotypes with improved characters in Kazakhstan[87] and Northern Zauralie.[88] A study is being made of the accumulation and preservation of carotenoid pigments in wheat grain, which makes a significant contribution to solving the problem of obtaining food products with desired or functional properties.[89] Morphological markers based on available quantitative and qualitative traits that are important for plant status have not lost their significance. For example, in flax (*L. usitatissimum*), about 60 genes control the color of flowers and seeds.[90] According to the State variety testing of the Russian Federation, 60 cultivars of long flax and 33 cultivars of oil flax are allowed to be used.[91] Only in two cultivars of flax, the color of flowers and seeds does not differ from the wild type (blue and brown, respectively) more than half of the cultivars of oil flax are characterized by a changed color.[92] The advantage of these markers is that they allow us to evaluate genetic diversity in changing environmental conditions, which cannot be ignored in the analysis of genotypic variability. Undoubtedly, the collection of mutant lines obtained by the method of radiation mutagenesis with a description of the inheritance of seven flower and seed coloring genes and six chlorophyll deficiency genes[93–95] is of scientific value. Some results of the use of induced mutagenesis in the breeding process of flax in Belarus, Lithuania, Russia, and Ukraine are presented in a review article.[96]

It has been shown that mutational breeding allows you to create cultivars several times faster than conventional hybridization. The main methods and methods of mutagenic exposure to various crops (seed treatment, vegetative seedlings, or pollen), physical factors of mutagenic exposure (radiation, electromagnetic radiation, gamma rays, etc.) are presented; chemical (alkylating compounds, ethyleneimine and its derivatives, nitrogen base inhibitors), biological (viruses, exogenous DNA). The effectiveness of chemical, physical-induced mutagenesis in crop breeding in different countries of the world is described.

The effectiveness of the selection of valuable genotypes is determined by an understanding of the mechanisms of plant resistance to stress factors. To test genotypes for resistance to stress factors, various signs are used. It was found that drought screening using some technological parameters of seeds is very effective for experiments in laboratory conditions.[97,98] The samples of millet in the conditions of the Central Himalayan region revealed morphophysiological and biochemical characters responsible for resistance to abiotic stresses. It has been shown that plant height, activity of superoxide dismutase, and catalase, glutathione content, the number of days before flowering, the weight of 1000 seeds, and yield are characteristics suitable for screening large populations and selecting economically valuable forms.[99]

We consider the evaluation of valuable genotypes in simulated stressful laboratory conditions with further testing in extremely difficult environmental conditions as a tool for selecting forms with altered traits, the formation of trait collections of mutants for use in genetic research, traditional, and molecular selection. Methods for studying stress resistance (salinization, lack of moisture, low temperatures, plant diseases) have been developed,[100–103] and application of modern technologies for the study and conservation of genetic resources.[104]

The scope of the task is determined by: the relevance of solving the problem of food security; a comprehensive analysis of plant adaptation processes in simulated (laboratory) and natural (field) conditions using informative morphological, physiological, biochemical, and genetic criteria; the total effect of mutational and recombination variability in obtaining a wide range of original and valuable macro- and micro mutations; the significance of the results for the northern regions, characterized by complex (often extreme) soil and climatic conditions and limited resources of cultivated plant species.

The scientific results on the induction of mutations in agricultural plants have been tested and presented (1) at the International Forums "BIOTECH WORLD 2018, 2019. Biotechnology: Status and Development Prospects. Life Sciences", Moscow; (2) V International Scientific Conference "Genetics, Genomics, Bioinformatics and Plant Biotechnology"—PlantGen 2019. Novosibirsk, "Akademgorodok"; (3) 3rd Agriculture and Climate Change Conference "The research of adaptive ability of agricultural plants in extreme conditions of the Northern Trans-Urals area" Budapest, Hungary, 2019, as well as at Russian competitions, scientific and methodological seminars.

4.3 CONCLUSION

The scope of the task posed in this review is determined by the following reasons: (1) the relevance of the search for solutions to the food security problem; (2) by the need for a comprehensive analysis of plant adaptation processes in simulated (laboratory) and natural (field) conditions; (3) at the use of informative morphological, physiological, biochemical, and genetic criteria; (4) by an assessment of the overall effect of mutational and recombination variability upon receipt of a wide range of original and valuable macro- and micromutations; and (5) by the significance of the results for the northern regions, characterized by complex (often extreme) soil and climatic conditions and limited resources of cultivated plant species.

KEYWORDS

- **chemical mutagenesis**
- **mutational breeding**
- chlorophyll
- **tilling**
- **phosphemid**
- **SPAD 502**
- cultivar

REFERENCES

1. Ecosystems and Human Well-being: General Synthesis. http://www.millenniumassessment.org/en/Synthesis.aspx
2. Convention UN on Biological Diversity, 1993. https://www.un.org/ru/documents/decl_conv/conventions/biodiv.shtml
3. Strategic Plan for the Conservation and Sustainable Use of Biodiversity 2011–2220 https://www.cbd.int/undb/media/factsheets/undb-factsheets-ru-web.pdf
4. Vavilov, N. I. *Five Continents*; Nauka (The Science): Leningrad, 1987; p 213 (in Russian).
5. Vavilov, N. I. Selected Works. In *Origin and Geography of Cultivated Plants*; Centers of Origin of Cultivated Plants: Moscow—Leningrad, 1965; Vol. 5, pp 9–107 (in Russian).

6. Vavilov, N. I. World Resources of Varieties of Cereals, Grains, Legumes, Flax and Their use in Breeding. In *Agroecological Review of the Most Important Field Crops*; Publishing House of the Academy of Sciences of the USSR: Moscow—Leningrad; 1957, p 471 (in Russian).

7. Braulio, J.; Soto-Cerda, B. J.; Diederichsen, A.; Duguid, S.; Booker, H.; Rowland, G.; Cloutier, S. The Potential of Pale Flax as a Source of Useful Genetic Variation for Cultivated Flax Revealed Through Molecular Diversity and Association Analyses. *Mol. Breed.* **2014**, *34*(4), 2091–2107.

8. Pournosrat, R.; Kaya, S.; Shaaf, S.; Kilian, B.; Ozkan, H. Geographical and Environmental Determinants of the Genetic Structure of Wild Barley in Southeastern Anatolia. *PloS One* **2018**, *13*(2), P. e0192386.

9. Rapoport, J. A. *Selected Works. Discovery of Chemical Mutagenesis*; Nauka (The Science): Moscow, 1993; p 304 (in Russian).

10. Rapoport, J. A. *Microgenetics*; Nauka (The Science): Moscow, Reprint Edition (2010); p 534 (in Russian).

11. Stroyeva, O. G. Josef Abramovich Rapoport. 1912–1990; Nauka (The Science): Moscow, 2009; p 215 (in Russian).

12. Martín, B.; Ramiro, M.; Martínez-Zapater, J. M.; Alonso-Blanco, C. A High-Density Collection of EMS-Induced Mutations for TILLING in *Landsberg erecta* Genetic Background of Arabidopsis. *BMC Plant Biol.* **2009**, *9*(1), 147.

13. Wang, N.; Long, T.; Yao, W.; Xiong, L.; Zhang, Q.; Wu, C. Mutant Resources for the Functional Analysis of the Rice Genome. *Mol. Plant.* **2013**, *6*, 596–604.

14. Anh, T. T. T.; Khanh, T. D.; Dat, T. D.; Xuan, T. D. Identification of Phenotypic Variation and Genetic Diversity in Rice (*Oryza sativa* L.) Mutants. *Agriculture* **2018**, *8*, 30.

15. Suprasanna, P.; Mirajkar, S. J.; Patade, V. Y.; Jain, S. V. Induced Mutagenesis for Improving Plant Abiotic Stress Tolerance. In *Mutagenesis: Exploring Genetic Diversity of Crops*; Tomlekova, N. B., Kozgar, M. I., Wani, M. R., Eds.; Wageningen Academic Publishers, Netherlands, 2014; pp 349–378.

16. FAO/IAEA Mutant Variety Database, 2015. [Электронный ресурс]. Режим доступа: https://mvd.iaea.org/#!Search

17. FAO/IAEA Mutant Variety Database, 2018. [Электронный ресурс]. Режим доступа: https://mvd.iaea.org/#!Search

18. Raina, A.; Laskar R. A.; Khursheed A. R. Role of Mutation Breeding in Crop Improvement-Past. *Present Future Asian Res. J. Agric.* **2016**, *2*(2).

19. Bouma, J.; Ohnoutka, Z. Importance and Application of the Mutant 'Diamant' in Spring Barley Breeding. In *Plant Mutation Breeding Crop Improvement*; IAEA: Vienna, 1991; Vol. 1, p 127–133.

20. Askew, M. F. Novel Oil, Fibre and Protein Crops in United Kingdom—a Future Perspective. In *Brighton Crop Protection Conference Weeds*; 1993; Vol. 2, pp 653–662.

21. Larkin, P. J. Introduction. In "*Somaclonal Variation and Induced Mutations in Crop Improvement*"; Jain, S. M., Brar, D. S., Ahloowalia, B. S., Eds.; Kluwer Academic Publishers: Dordrecht, Holland, 1998; pp 3–13.

22. Rowland, G. G. An EMS-Induced Low-Linolenic-Acid Mutant in McGregor Flax (*Linum usitatissimum* L.). *Can. J. Plant Sci.* **1991**, *71*(2), 393–396.

23. Green, A. G.; Marshall, D. R. Isolation of Induced Mutants in Linseed (*Linum usitatissimum*) Having Reduced Linolenic Acid Content. *Euphytica* **1984**, *33*(2), 321–328.

24. Bome, N. A. Selection of Crops and Methods of Creation the Varieties for Extreme in Conditions of the Northern Trans-Urals. International Congress of Russian Bred and Grain. Abstract of Dis. SPb. 1996; p 46 (in Russian).
25. Morgun, V. V.; Katerinchuk, A. M.; Chugunkova T. V. The use of New Stereoisomers of Nitrosoalkylurea in the Selection of Winter Wheat. *Bulletin of the Samara Scientific Center of the Russian Academy of Sciences.* **2013,** *15*(3/5), 1666–1669.
26. Popolzukhina, N. A. On the Genetic Nature of Mutations in Spring Soft Wheat Plants. *Agric. Biol.* **2003,** *38*(3), 108–111.27 (in Russian).
27. Popolzukhina, N. A. *The Selection of Spring Soft Wheat in Western Siberia is Based on a Combination of Induced Mutagenesis and Hybridization*; Abstract of DSc: Tyumen, 2004; p 31 (in Russian).
28. Popolzukhina, N. A.; Popolzukhin, P. V.; Yakunin, R. A.; Suponin, M. S. Use of Mutational and Allocytoplasmic Variability in Breeding Spring Soft Wheat for Adaptability. In *Biotechnology: State and Development Prospects;* 2017; pp 97–100.
29. Krotova, L. A. Chemical Mutagens as a Factor in the Production of Various Mutations in Spring Common Wheat. *Bull. Altai State Univ.* **2009,** *9*, 12–15 (in Russian).
30. Krotova, L. A. Chemical Mutagenesis as a Method of Creating Starting Material for Breeding Soft Wheat. *Electron. Sci. Methodol. J. Omsk State Agrar. Univ.* **2015,** *2*(2), 1–5.
31. Eiges, N. S. The Historical Role of Joseph Abramovich Rapoport in Genetics. Continuation of Research Using the Method of Chemical Mutagenesis. *Vavilov J. Genet. Breed.* **2013,** *17*(1), 162–172 (in Russian).
32. Ushchapovsky, I. V; Lemesh, V. A; Bogdanova, M. V; Guzenko, E. V. Features of Selection and Prospects for the use of Molecular Genetic Methods in Genetic Selection Studies of Flax (*Linum usitatissimum* L.) (Review). *Agric. Biol.* **2016,** *51*(5), 602–616. DOI: 10.15389/agrobiology.2016.5.602eng.
33. Weisfeld, L. I.; Bome, N. A.; Korolev, K. P.; Tetyannikov, N. V.; Shiryaev, P. A. The use of the Chemical Mutagen Phosphemid for the Induction of Diversity in Plants. In *Genetics and Breeding in the Modern Agro-Complex.* Materials of all-Ukrainian Scientific-Practical Conference, June 26, 2019; Uman, 20119; pp 10–13 (in Russian).
34. Rapoport, J. A. Chemical Mutagens and *Para*-Aminobenzoic Acid in Increasing the Yield of Agricultural Plants. The Effect of PABA in Connection with the Genetic Structure. Editor-in-chief J. A. Rapoport; Nauka (The Science): Moscow, 1989; pp 3–39 (in Russian).
35. Bome, N. A; Bome, A. Ya.; Petrova, A. A. The Biological Activity of Para-Aminobenzoic Acid (PABA) at Some Types of Cultivated Plants. In *Bioantioxidant,* Proceedings of the IX International Conference. Section 5: Antioxidants in the Life of Organisms (Antimutagenesis); IBCP RAS, Russian Peoples' Friendship University: Moscow, 2015; pp 305–313 (in Russian).
36. Bekuzarova, S. A.; Weisfeld, L. I.; Belyaeva, V. A. Bioantioxidants, Application in Presowing Treatment of Legume Seeds. In *Bioantioxidant,* Proceedings of the IX International Conference. Section 5: Antioxidants in the Life of Organisms (Antimu-tagenesis); IBCP RAS, Russian Peoples' Friendship University: Moscow, 2015; pp 314–320 (in Russian).
37. Korolyov, K. P.; Bogdan, V. Z.; Bome, N. A.; Bogdan, T. M. Influence of Factors of Chemical Nature on Quantitative Characters of Flax Genotypes (*Linum*

usitatissimum L.) of the First Mutant Generation (M$_1$). Actual Questions of Biology, Selection, Technology of Cultivation and Recycling of Oilseeds and Other Industrial Crops. Collection of Materials of the 9th All-Russian Conference with International Participation of Young Scientists and Specialists. Krasnodar, February 21–22, 2017. All Russian Scientific Research Institute of Oil-Bearing Cultures Named V.S. Pustovoyt, 2017; pp 47–52.

38. Korolev, K. P.; Bome, N. A. Evolutions Adaptability of Flax (*Linum usitatissimum* L.) Genotypes on the North-Eastern Belarus. *Agric. Biol.* **2017**, *52*(3), 615–621. DOI: 10.15389/agrobiology.2017.3.615eng.

39. Korolev, K. P.; Bome, N. A. Use of Morphophysiologocal Markers in Intraspecific Polymorphism Analysis of Flax *Linum usitatissimum* L. *Agric. Biol.* **2018,** *53*(5), 927–937. DOI: 10.15389/agrobiology.2018.5.927eng.

40. Goyalyal, S.; Khan, S. A Comparative Study of Chromosomal Aberrations in *Vigna mungo* Induced by Ethylmethane Sulphonate and Hydrazine Hydrate. *Thai J. Agric. Sci.* **2009,** *42*(3), 177–182.

41. Ahloowalia, B. S.; Maluszynski, M. Induced Mutations–a New Paradigm in Plant Breeding. *Euphytica* **2001**, *118*(2), 167–173.

42. Shu, Q. Y., Forster, B. P., Nakagawa, H., Eds.; Plant Mutation Breeding and Biotechnology. Plant Breeding and Genetics Section, Joint FAO/IAEA Division of Nuclear Techniques in Food and Agriculture, International Atomic Energy Agency, Vienna, Austria, 2012. DOI: 10.1079/9781780640853.0000.

43. Meng, Q.; Redetzke, D. L.; Hackfeld, L. C.; Hodge, R. P.; Walker, D. M.; Walker, V. E. Mutagenicity of Stereochemical Configurations of 1,2-Epoxybutene and 1,2:3,4-Diepoxybutane in Human Lymphoblastoid Cells. *Chem.Biol. Interact.* **2007**, *166*(1–3), 207–218. DOI: 10.1016/j.cbi.2006.06.001.

44. Bome, N. A.; Weisfeld, L. I.; Babaev, E. V.; Bome, A. Ya.; Kolokolova, N. N. Influence of Phosphemid, a Chemical Mutagen, on *Triticum aestivum* L. *Agric. Biol.* **2017**, *52*,(3).

45. Till, B. J.; Cooper, J.; Tai, T. H.; Colowit, P.; Greene, E. A.; Henikoff, S.; Comai, L. Discovery of Chemically Induced Mutations in Rice by TILLING. *BMC Plant Biol.* **2007,** *7*(1), 19.

46. Talamè, V.; Bovina, R.; Sanguineti, M. C.; Tuberosa, R.; Lundqvist, U.; Salvi, S. TILLMore, a Resource for the Discovery of Chemically Induced Mutants in Barley. *Plant Biotechnol. J.* **2008**, *6*(5), 477–485.

47. Chantreau, M.; Grec, S.; Gutierrez, L.; Dalmais, M.; Pineau, C.; Demailly, H.; et al. PT-Flax (Phenotyping and TILLinG of Flax): Development of a Flax (*Linum usitatissimum* L.) Mutant Population and TILLinG Platform for Forward and Reverse Genetics. *BMC Plant Biol.* **2013**, *13*(1), 159.

48. Cooper, J. L.; Till, B. J.; Laport, R. G.; Darlow, M. C.; Kleffner, J. M.; et al. TILLING to Detect Induced Mutations in Soybean. *BMC Plant Biol.* **2008**, *8*, 1–9.

49. Weisfeld, L. I. About Cytogenetic Mechanism of Chemical Mutagenesis. In *Ecological Consequences of Increasing Crop Productivity. Plant Breeding and Biotic Diversity*; Opalko, A. I., Weisfeld, L. I., Bekusarova, S. A., Bome, N. I., Zaikov, G. E., Eds.; Apple Academic Press: Toronto, New Jersey, 2015; pp 259–269.

50. Weisfeld, L. I. Chromosomes and Mitotic Activity by Influence of Alkylating Agent Phosphemidum. *Polym. Res. J.* **2013**, *7*(1), 9–22.

51. Weisfeld, L. I. Induction of Chromosomal Rearrangements in Connection with the Possibilities of Studying "Mutational" Heterosis. Heterosis of Agricultural Plants. International Symposium. Materials of Reports and Messages; Moscow, 1977; pp 13–20 (in Russian).

52. Ripberger, E. I. Adaptive Potential of Inter-Varietal Hybrids of Soft Spring Wheat (*Triticum aestivum* L.) in Various Soil and Climatic Conditions. Abstract PhD. Tyumen, 2015; p 20 (in Russian).

53. Bome, N. A.; Weissfeld, L. I.; Ripberger, E. I.; Bome, A. Ya. New Chemical Mutagen to Create a Biotic Diversity of *Triticum aestivum* L. The Collection of Theses of the All-Russian Conference 50 Years of VOGIS: Successes and Prospects; Moscow, November 8–10, 2016; p 98. www.vogis.org (in Russian).

54. Bome, N. A.; Weissfeld, L. I.; Ripberger, E. I.; Arsentiev, S. V. The Sensitivity of Winter and Spring Forms of *Triticum aestivum* L. to the Effects of Phosphemid. Materials of the Congress "Biotechnology: State and Development Prospects" (March 17–20, 2015. Section "Agricultural Biotechnology"). Part 2. Moscow. "Expo-biochemistry", RCTU ithem. D. I. Mendeleev, 2015; pp 81–82.

55. Korolyov, K.; Bome, N.; Weisfeld, L. Study of Reaction of Flax (*Linum usitatissumum* L.) Genotypes on Mutagen Stressor in the Modeled and Natural Environmental Conditions. *Zemdirbyste Agric.* **2019,** *1*, 29–36. DOI: 10.13080/z-a.2019.106.004.

56. Tetyannikov, N. V. Ecological and Biological Features of the Intraspecific Diversity of *Hordeum vulgare* L. and its use to Create New Forms. Abstract PhD. Moscow, 2019; p 22 (in Russian).

57. Pour-Aboughadareh, A.; Ahmadi, J.; Mehrabi, A. A.; Etminan, A.; Moghaddam, M.; Siddique, K. H. Physiological Responses to Drought Stress in Wild Relatives of Wheat: Implications for Wheat Improvement. *Acta Physiol. Plant.* **2017,** *39*(4), 106.

58. Ogbonnaya, F.; Rasheed, A.; Okechukwu, E. Genome-Wide Association Study of Agronomic and Physiological Traits in Spring Wheat Evaluated in a Range of Heat Prone Environments. *Theor. Appl. Genet.* **2017,** *130*(9), 1819–1835. DOI: 10.1007/s00122-017-2927-z.

59. Buschmann, C.; Konanz, S.; Zhou, M.; Lenk, S.; Kocsányi, L.; Zhou, M.; Barócsi, A. Excitation Kinetics of Chlorophyll Fluorescence During Light-Induced Greening and Establishment of Photosynthetic Activity of Barley Seedlings. *Photosynthetica* **2013,** *51*(2), 221–230. DOI: 10.1007/s11099-013-0017-2.

60. Honsdorf, N.; March, T.; Hecht, A.; Eglinton, J. Evaluation of Juvenile Drought Stress Tolerance and Genotyping by Sequencing with Wild Barley Introgression Lines. *Mol. Breed* **2014,** *34*(3), 1475–1495. DOI: org/10.1007/s11032-014-013120.

61. Sánchez-Martín, J.; Mur, L. A. J.; Rubiales, D. Targeting Sources of Drought Tolerance Within an *Avena* spp. Collection Through Multivariate Approaches. *Planta* **2012,** *236*(5), 1529–1545. DOI: 10.1007/s00425-012-1709-8.

62. Aghaee, A.; Moradi, F.; Zare-Maivan, H.; Zarinkamar, F.; Irandoost, H.; Sharifi, P. Physiological Responses of Two Rice (*Oryza sativa* L.) Genotypes to Chilling Stress at Seedling Stage. *Afr. J. Biotechnol.* **2011,** *10*(39), 7617–7621.

63. Gianquinto, G.; Goffart, J.; Olivier, M. The use of Hand-Held Chlorophyll Meters as a Tool to Assess the Nitrogen Status and to Guide Nitrogen Fertilization of Potato Crop. *Potato Res.* **2004,** *47*, 35. DOI: org/10.1007/BF02731970.

64. Thompson, J.; Schweitzer, L.; Nelson, R. Association of Specific Leaf Weight, an Estimate of Chlorophyll, and Chlorophyll Concentration with Apparent Photosynthesis in Soybean. *Photosynth. Res.* **1996,** *49*, 1–10.

65. Kühling, I.; Hüsing, B.; Trautz, D.; Bome, N. Soybeans in High Latitudes: Effects of *Bradyrhizobium* Inoculation in Northwest Germany and Southern West Siberia. *Organ. Agric* **2018,** *8*(2), 159–171. DOI: 10.1007/s13165-017-0181-y.

66. Gratani, L.; Crescente, M.; Fabrini, G.; Varone, L. Growth Pattern of *Bidens cernua* L. Relationships Between Relative Growth Rate and its Physiological and Morphological Components. *Photosynthetica* **2008,** *46*(2), 179–184. DOI: org/10.1007/s11099-008-0029-5.

67. Kiani-Pouya, A.; Rasouli, F. The Potential of Leaf Chlorophyll Content to Screen Bread-Wheat Genotypes in Saline Condition. *Photosynthetica* **2014,** *52*(2), 288–300. DOI: 10.1007/s11099-014-0033-x.

68. Figueiredo, N.; Carranca, C.; Trindade, H.; Pereira, J.; Goufo, P.; Coutinho, J.; de Varennes, A. Elevated Carbon Dioxide and Temperature Effects on Rice Yield, Leaf Greenness, and Phenological Stages Duration. *Paddy Water Environ.* **2015,** *13*(4), 313–324.

69. Hund, A.; Frascarol, E.; Leipner, J.; Jompuk, C. Cold Tolerance of the Photosynthetic Apparatus: Pleiotropic Relationship Between Photosynthetic Performance and Specific Leaf Area of Maize Seedlings. *Mol. Breed.* **2005,** *16*(4), 321–331. DOI: 10.1007/s11032-005-1642-7.

70. Caser, M.; Lovisolo, C.; Scariot, V. The Influence of Water Stress on Growth, Ecophysiology and Ornamental Quality of Potted Primula vulgaris 'Heidy' Plants. New Insights to Increase Water use Efficiency in Plant Production. *Plant Growth Regul.* **2017,** *83*(3), 361–373. DOI: 10.1007/s10725-017-0301-4.

71. Erturk, U.; Sivritepe, N.; Yerlikaya, C.; Bor, M.; Ozdemir, F.; Turkan, I. Responses of the Cherry Rootstock to Salinity in vitro. *Biol. Plant.* **2007,** *51*(3), 597–600. DOI: 10.1007/s10535-007-0132-7.

72. Uddling, J.; Gelang-Alfredson, J.; Piikki, K.; Pleijel, H. Evaluating the Relationship Between Leaf Chlorophyll Concentration and SPAD-502 Chlorophyll Meter Readings. *Photosynth. Res.* **2007,** *91*, 37–46.

73. Chaerle, L.; Leinonen, I.; Jones, H. G.; Van Der Straeten, D. Monitoring and Screening Plant Populations with Combined Thermal and Chlorophyll Fluorescence Imaging. *J. Exp. Bot.* **2007,** *58*(4), 773–784.

74. Kapotis, G.; Zervoudakis, G.; Veltsistas, T. Comparison of Chlorophyll Meter Readings with Leaf Chlorophyll Concentration in *Amaranthus vlitus*: Correlation with Physiological Processes. *Rus. J. Plant Physiol.* **2003,** *50*(3), 395–397. DOI: 10.1023/A:1023886623645eng.

75. Korolev, K. P.; Bome, N. A. Use of Morphological Markers in Intraspecific Polymorphism Analysis of Flax *Linum usitatissimum* L. **2018,** *53*(5) 927–937. DOI: 10.15389/agrobiology.2018.5.927eng.

76. Villadsen, D.; Rung, J. H.; Nielsen, T. H. Osmotic Stress Changes Carbohydrate Partitioning and Fructose-2,6-Bisphosphate Metabolism in Barley Leaves. *Funct. Plant Biol.* **2005,** *32*(11), 1033–1043.

77. Damour, G.; Vandame, M.; Urban, L. Long-Term Drought Modifies the Fundamental Relationships Between Light Exposure, Leaf Nitrogen Content and Photosynthetic

Capacity in Leaves of the Lychee Tree (*Litchi chinensis*). *J. Plant Physiol.* **2008**, *165*, 1370–1378.

78. Bome, H. A.; Korolev, K. P.; Tetyannikov, N. V.; Kolokolova, H. H.; Bome, A. Ya.; Weisfeld, L. I.; Wasserman, L. A.; Goldstein, V. G. Materials of the Study of this Indicator is Carried Out by us in Partnership with Scientists of the All-Russian Research Institute of Starch Products the All-Russian Scientific-Practical Conference with International Participation "Modern Approaches and Methods in Plant Protection," November 12–14, 2018; Ural Federal University: Yekaterinburg, 2018; pp 105–107 (in Russian).

79. Bomé, N. A.; Korolev, K. P.; Tetyannikov, N. V.; Weisfeld, L. I.; Kolokolova, N. N. Mutagenic Effect of Phosphemid for Induction of Mutation of *Hordeum vulgare* L. and *Linum ustitatissimum* L. Plant Genetics, Genomics, Bioinformatics, and Biotechnology (PlantGen2019) Abstracts. Institute of Cytology and Genetics. Siberian Branch of the Russian Academy of Sciences, 2019a, 49. DOI: 10.18699/ PlantGen2019-031.

80. Bomé, N. A.; Korolev, K. P.; Tetyannikov, N. V.; Weisfeld, L. I.; Kolokolova, N. N. *Current Challenges in Plant Genetics, Bioinformatics, and Biotechnology Federal Research Center Institute of Cytology and Genetics, Siberian Branch of the Russian Academy of Sciences (Novosibirsk),* Proceedings of the Fifth International Scientific Conference PlantGen2019, Novosibirsk, June 24–29, 2019, 2019b. DOI: 10.18699/ PlantGen2019-031.

81. Chesnokov, Yu. V. Molecular Genetic Markers and their use in Preselection Studies, *SPb. APhI*, 2013; p 116.

82. Khlestkina, E. K. Molecular Markers in Genetic Research and in Breeding. *Vavilov J. Genet. Breed.* **2013**, *17*(4/2), 1044–1053.

83. de Vienne, D. *Molecular Markers in Plant Genetics and Biotechnology*; Sci. Publ. Inc.: Enfield, Plymouth, 2003; p 248.

84. Zaitseva, O. I.; Burakova, A. A.; Babkenov, A. T.; Babkenova, S. A.; Utebayev, M. U.; Lemesh, V. A. Allelic Variation of High-Molecular-Weight Glutenin Genes in Bread Wheat. *Cytol. Genet.* **2017**. DOI: 10.3103/s009545271706012.

85. Utebayev, M.; Dashkevich, S.; Kunanbayev, K.; Bome, N.; Sharipova, B.; Shavrukov, Y. Genetic Polymorphism of Glutenin Subunits with High Molecular Weight and their Role in Grain and Dough Qualities of Spring Bread Wheat (*Triticum aestivum* L.) from Northern Kazakhstan. *Acta Physiol. Plant.* **2019**, *41*(5), 71. DOI: 10.1007/ s11738-019-2862-5.

86. Utebayev, M.; Dashkevich, S.; Bome, N. A.; Bulatova, K.; Shavrukov, Yu. Genetic Diversity of Gliadin-Coding Alleles in Bread Wheat (*Triticum aestivum* L.) from Northern Kazakhstan. *Peer J.* **2019**, *2019*(7). DOI: 10.7717/peeri.7082.

87. Utebayev, M.; Dashkevich, S.; Babkenov, A.; Shtefan, G.; Fahrudenova, I.; Bayahmetova, S.; Sharipova, B.; Kaskarbayev, Z.; Shavrukov, Yu. Application of Gliadin Polymorphism for Pedigree Analysis in Common Wheat (*Triticum aestivum* L.) from Northern Kazakhstan. *Acta Physiol. Plant.* **2016**, *38*(4), 204. DOI: 10.1007/ s11738-016-2209-4.

88. Utebaev, M. U.; Bome, N. A.; Dashkevich, S. M.; Fasylova, D. D. Intra-Varietal Polymorphism of the Variety of Spring Common Wheat (*Triticum aestivum* L.). Tyumen 29 in the Conditions of the Northern Trans-Urals. Biotechnology in

Crop Production, Animal Husbandry and Agricultural Microbiology. Collection of Abstracts of the 19th All-Russian Conference of Young Scientists; All-Russian Research Institute of Agricultural Biotechnology: Moscow, 2019; pp 90–91.

89. Utebaev, M. U.; Dashkevich, S. M.; Babkenov, A. T.; Bome, N. A. Accumulation of Carotenoid Pigments in Spring Soft Wheat (*Triticum aestivum* L.) in North Kazakhstan. In *Gene Pool and Plant Breeding*, Proceedings of the IV International Scientific and Practical Conference, Novosibirsk, Russia, April 4–6, 2018; pp 363–366.

90. Keijzer, P.; Metz, P. Breeding of Flax for Fibre Production in Western Europe. *Biol. Process. Flax* **1993,** 33–66.

91. Plant Cultivars Included in the State Register of Breeding Achievements Approved for Use of the Variety of Flax Oilseed Cultivar http://reestr.gossort.com/reestr/culture/117

92. Porokhovinova, E. A. Genetic Collection of Flax (*Linum usitatissimum* L.): Creation, Analysis and Prospects of use. Abstract of DSc. SPb. VIR, 2019; p 40 (in Russian).

93. Lyakh, V. A.; Mishchenko, L. Yu.; Polyakova, I. A. Genetic Collection of the Species *Linum usitatissimum* L.; Institute of oil crops of the Ukrainian Agronomic Academy of Sciences: Zaporozhye; 2003, p 60 (in Ukrainian).

94. Jaglo, M. N.; Lyakh, V. A. A Variety of Colorings of Seeds in Oil Flax and Features From Inheritance. *Bull. Belarus. State Agric. Acad.* **2015,** *3,* 70–73.

95. Lyakh, A.; Barron-Jimenez, R.; Dunayevskiy, I.; Go, R.; Kumar, C.; Patel, N. External Cavity Quantum Cascade Lasers with Ultra-rapid Acousto-optic Tuning. *Appl. Phys. Lett.* **2015,** *106,* 151101.

96. Korolev K. P.; Bogdan V. Z.; Bogdan T. M. Induced Mutagenesis as a Way to Create a New Source Material for Breeding Varieties of Intensive Type of various Cultures. *Agric. Sci.* **2016,** *4,* 11–16 (in Russian).

97. Singh, D. P.; Singh, P.; Kumar, A.; Sharma, H. C. Transpirational Cooling as a Screening Technique for Drought Tolerance in Oil Seed Brassicas. *Ann. Bot.* **1985,** *56*(6), 815–820.

98. Osmolovskaya, N. G.; Shumilina, Yu., S.; Grishina, T. V.; Didio, A. V.; Lukasheva, E. M.; Bilova, T. E.; Frolov, A. A. Modeling Drought in an Experiment and Evaluating its Effect on Plants. *J. Stress Physiol. Biochem.* **2017,** *14*(13), 5–15.

99. Trivedi, J. Effectiveness of Mobile Advertising on Gen Y's Attitude and Purchase Intentions. *Int. J. Mark. Bus. Commun.* **2015,** *4*(2), 17–30.

100. Bome, N. A. Intraspecific Diversity of Barley Cultivated (*Hordeum vulgare* L.) for Resistance to Chloride Salinity. In *Agrobiology*; Belotserkovsky National Agrarian University: Belaya Tserkov, 2014; Vol. 2, pp 16–22 (in Russian).

101. Belozerova, A. A.; Bome, N. A.; Bome, A. Ya.; Mostovshchikova, S. M. The Peculiarities of Seedlings Formation of *Triticum aestivum* L. In *Under Salinization Conditions. Biological Systems, Biodiversity and Stability of Plant Communities*; Weisfeld, L. I., Opalko, A. I., Bome, N. A., Bekuzarova, S. A., Eds.; Apple Academic Press Inc.: Oakville, Waretown, 2015; pp 361–384.

102. Bome, N. A.; Bome, A. Ya.; Kolokolova, N. N.; Trofimova, J. B. Problems of Caryopsis and Stability of Winter and Spring Forms of Cereals to Phytopathogenic Fungi of Genus *Fusarium* Link. In *Ecological Consequences of Increasing Crop Productivity. Plant Breeding and Biotic Diversity*; Weisfeld, L. I., Opalko, A. I.,

Bekuzarova, S. A., Bome, N. A., Eds.; Apple Academic Press: Canada, USA, 2015; pp 97–107.

103. Korolev, K. P.; Bome, N. A.; Kramar, K. V. Criteria for the Selection of Drought-Resistant Forms of Flax (*Linum usitatissimum* L.). "Plant Physiology-the Basis for Creation of Plants of the Future." Abstracts of Reports of the IX Congress of the Society of Plant Physiologists of Russia, September 18–24, 2019, Kazan. DOI Collection: 10.26907/978–5-00130-204-9-2019) (DOI Articles: 10.26907/978-5-00130-204-9-2019-229) (in Russian) https://congresskazan2019.ofr.su/speakers.

104. Bome, N. A.; Korolev, K. P.; Tetyannikov, N. V.; Bome, A. A.; Petrova, A. A. Modern Technologies for the Study and Conservation of Genetic Resources. The Combined Illustrated Catalog of Materials of International and all-Russian Exhibitions-Presentations of Scientific, Educational and Methodological publications and Educational Technologies; Russian Academy of Natural Sciences (International Association of Scientists, Teachers and Specialists), 2018; pp 14–15 (in Russian).

CHAPTER 5

Enzymatic Activity of Bacteria *Bacillus subtilis*: Assessment Using Phosphemidum

IRINA V. PAK[1*], OLEG V. TROFIMOV[1], LARISSA I. WEISFELD[2], and RIZVAN D. RUSTAMOV[1]

[1] Tyumen State University, 3 Pirogov St., 625043 Tyumen, Russia

[2] Emanuel Institute of Biochemical Physics, Russian Academy of Sciences, 4 Kosygin St., 119334 Moscow, Russia

[*] Corresponding author. E-mail: pakiv57@mail.ru

ABSTRACT

We performed evaluation of enzymatic activity of exogenic enzymes of *Bacillus subtilis* CABI, which are the base of numerous probiotic preparations. The mutagen phosphemidum was used in the experiments in five concentrations: 0.1%, 0.01%, 0.001%, 0.0001%, and 0.00001%. The following enzymatic activities of *B. subtilis* were determined after treatment with phosphemidum in the mentioned concentrations: amylolytic (EC 3.2.1), protease (EC 3.4.2), lipase (EC 3.1.1.3), xilanase (EC 3.2.1.8), CMC-ase (carboxymetylcellulase) (EC 3.2.1.203), and β-gluconase (EC 3.2.1.4). High fermentative effectiveness was shown at phosphemidum concentrations of 0.0001% and 0.00001%. Dynamics of changes in enzymatic activity with age of the culture was studied. It was shown that with increasing age of the cultures, the activity of the studied enzymes decreases.

5.1 INTRODUCTION

In recent years, probiotics based on bacteria *Bacillus subtilis* have been actively used in veterinary medicine, animal husbandry, and medicine. They are able to normalize the functioning of gastrointestinal tract of animals and humans, improve metabolic processes, increase productivity, and resistance to infectious diseases. *B. subtilis* attracts attention as an active agent that can effectively destroy organic pollution due to the fact that these bacteria produce and excrete a complex of active enzymes.

The positive experience of application of probiotics based on *B. subtilis* has been shown in the studies of various researchers: their ability to increase nonspecific resistance in fish, pigs, calves[1-5] to stimulate the human immune system[6,7] has been noted. Probiotics increase activity of digestive enzymes,[8,9] improve hematological and immunological parameters of calves,[5] strengthen the protective function of mucous membranes, and activate the effector of T- and B-lymphocytes.[10,11] Effectiveness of replacing antibiotics with probiotics in intestinal infections in pigs[12] and under the action of pathogens in poultry industry[13,14] was noted. The beneficial properties of *B. subtilis* are primarily associated with activity of exogenous proteolytic, amylolytic, and lipolytic enzymes. An increase in production of these enzymes in bacteria significantly increases effectiveness of probiotics.

To increase the genetic and phenotypic diversity of this object, we used phosphemid. The use of mutagens, the effect of which on *B. subtilis* has not been studied previously, opens up prospects for obtaining new properties in known producer strains.

The phosphemidum was first used as an antitumor drug in the All-Union Chemical and Pharmaceutical Research Institute (Moscow, USSR).[15] Its properties as a mutagen were shown on human and mouse fibroblasts,[16] and on the model object *Crepis capillaris* (L.) Wallr.[17,18] When using phosphemidum, new forms of winter and spring wheat were created.[19,20] In our previous studies on yeast *Candida maltose*, phosphemidum at concentrations of 0.1%, 0.01%, 0.001%, and 0.0001% induced an increase in the number and size of yeast colonies, a significant increase in biomass.[21]

The aim of this work was to identify new useful traits increasing the enzymatic activity of *B. subtilis*. The particular tasks were the following:

1. To study the effect of phosphemidum in different concentrations on activity of protease, amylase, lipase, xylanase, CMC-ase, and β-glucanase of *B. subtilis* CABI (Centre for Agriculture and Bioscience International).
2. To assess the dynamics in enzymatic activity with increasing age of the culture, obtained using phosphemidum.

5.2 MATERIAL AND METHODOLOGY

The study was conducted in 2017–2018 on the basis of the Center for Biotechnology and Genetic Diagnostics of Tyumen State University. A strain of *B. subtilis* CABI, which was used in this research, was established at the Probiotics Research Institute.

The cultivation of the strain was carried out in solid and liquid nutrient media. The liquid nutrient medium used in the study included the following components: 1% of peptone, 0.5% of yeast extract, 2% of glucose. pH of the medium was adjusted to 6.5 with a solution of NaOH. To obtain a solid nutrient medium, agar was added to the components of the liquid medium (up to 1.5%). Solid and liquid media were sterilized by autoclaving at temperature of +121°C and the pressure of two bars for 30 min.

The cultivation of bacteria on surface of solid nutrient media was carried out in Petri dishes with a diameter of 90 mm in a BD53 dry-air thermostat (Binder, German) at the temperature of 37°C for 12–24 h. Cultivation in liquid nutrient media was carried out in flasks in 100 ml of medium. The flasks with liquid medium were incubated in an Innova 43R thermostat-shaker (New Brunswick Scientific) at the temperature of 37°C and a stirring speed of 120 rpm for 24 h.

Determination of enzyme activity was carried out according to standard methods for amylase activity;[22] for determination of activity of other enzymes, namely the modified method of Ota, Yamada for the activity of lipase; determination of proteolytic activity according to Willstätter and Waldschmidt-Leutz; determination of reducing sugars according to Nelson-Somodi (β-gluconase, CMC-ase, xylanase activity) was carried out.[23]

Phosphemidum was synthesized in 2014 in the laboratory of physicochemical methods for analyzing the structure of matter at the Department of Chemistry of Lomonosov Moscow State University by Professor Eugene V. Babaev.

The experiments were carried out with five concentrations of mutagen, well proven in experiments with *Candida maltosa* CABI that aimed increase in the productivity of yeast cells[21]: 0.1%, 0.01%, 0.001%, 0.0001%, and 0.00001%. Phosphemidum (1% stock solution) was added to the nutrient medium until the desired concentration. Aliquots of the strain were sown in all prepared media containing different concentrations of mutagen (at the rate of about 10,000 cells per sample). Variants with mutagen-free medium were applied as control. Strains from different variants of the experiment were cultivated for 24 h at 37°C. To determine effectiveness of mutagenesis, standard tests for auxotrophy were applied.[25] To study the changes in enzymatic activity of bacteria with increasing age of the cultures, culture was transplanted every day. The activity of amylase, protease, and lipase was determined in 2, 10, 30, and 50 daily cultures. The activity of β-glucanase, CMC-ase, and xylanase was determined at the age of 2, 30, and 50 days. Statistical analysis of assay data was performed using standard methods in Microsoft Excel.

5.3 RESULTS AND DISCUSSION

The enrichment method with auxotrophic mutants, which is used for spore-forming bacteria, including bacteria of the genus *Bacillus*, is based on different sensitivity to elevated temperature of bacterial cells and their spores. When the temperature rises to 46°C in a liquid starvation medium, the plating on a full-fledged nutrient medium, auxotrophs germinate.

B. subtilis CABI responded to impact of phosphemidum by increasing number of auxotrophic cells in comparison with control using at two concentrations: 0.1% and 0.001%. In all other variants (0.01%, 0.0001%, and 0.00001%), fewer mutant surviving cells were observed.

Pure culture of the studied variants from Petri dishes was planted in a liquid nutrient medium and incubated at 37°C in a shaker-incubator.

At the end of incubation, 0.1 µl of obtained culture was planted on solid nutrient media for counting the number of living cells and measuring optical density of the culture at a wavelength of 600 nm (OD_{600}) using a spectrophotometer. As can be seen from Table 5.1, application of a 0.001%

solution of phosphemidum stimulated cell growth, which was displayed in an increase of the number of colonies and an increase of the number of colony-forming cells (CFU) in 1 ml of culture in comparison with the control. In all other mutant variants, these indicators were lower than in this variant and reliably indistinguishable from control, except for the variant with 0.00001%.

Results of determining activity of different enzymes in the original culture are presented in Table 5.1 and Figure 5.1. The effect of phosphemidum displayed an increase of the activity of lipase and protease. An increase in lipase activity 1.32 times in comparison with the control is observed in the variant with the lowest concentrations of phosphemidum: 0.0001% and 0.00001%.

TABLE 5.1 Effect of Phosphemidum on Growth of *Bacillus subtilis* CABI Colonies.

Phosphemidum concentration (%)	Number of colonies (pcs)	CFU per 1 ml (×10⁶)	Optical density of the culture
Control (without phosphemidum)	1600 ± 73	16.0 ± 0.7	1.531 ± 0.023
0.1	1440 ± 78	14.4 ± 0.8	$1.412 \pm 0.027^*$
0.01	1540 ± 68	15.4 ± 0.7	$1.138 \pm 0.018^{*\cdot}$
0.001	$1740 \pm 52^\cdot$	$17.4 \pm 0.5^\cdot$	$0.942 \pm 0.025^{*\cdot\#}$
0.0001	1450 ± 75	14.5 ± 0.7	$1.419 \pm 0.015^{*\cdot\#}$
0.00001	$1158 \pm 43^{*\cdot\#}$	$11.6 \pm 0.4^{*\cdot\#}$	$1.357 \pm 0.019^{*\cdot\#£}$

*Differences from control are statistically significant at $p < 0.05$; ·differences from the experiment variant 0.1% are statistically significant at $p < 0.05$; #differences from the experiment variant 0.01% are statistically significant at $p < 0.05$; £differences from experiment variant 0.0001% are statistically significant at $p < 0.05$.

An increase in enzymatic activity of proteases by 1.20–1.88 times in comparison with the control was noted when using 0.01%, 0.0001%, and 0.00001% solutions of phosphemidum. Phosphemidum in a concentration of 0.00001% increased the activity of amylases by 1.38 times compared with the control (Fig. 5.1, Table 5.2). When studying the effect of phosphemidum on activity of enzymes catalyzing the cleavage of nonstarch polysaccharides, a noticeable effect of increased activity was detected only for xylanase. The lowest concentration of phosphemidum, 0.00001%, had a significant stimulating effect, increasing the enzyme activity by 1.5 times compared with the control (Table 5.3 and Figure 5.2).

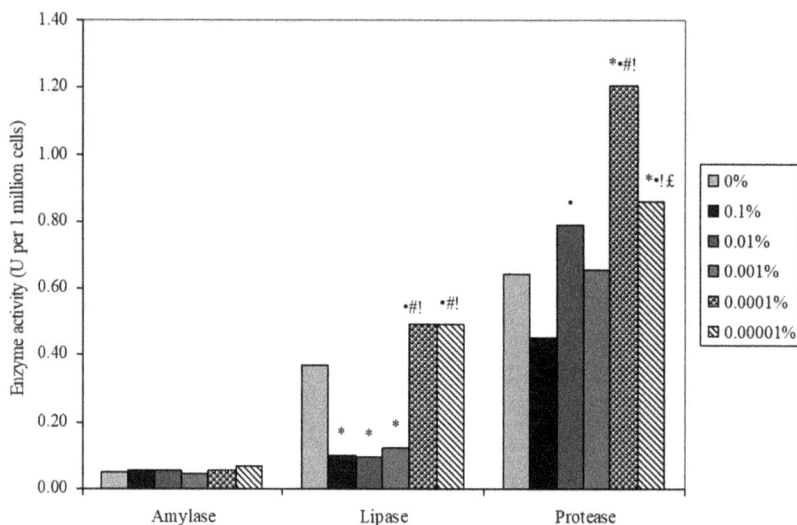

FIGURE 5.1   Enzyme activity (U per 1 million cells) of amylase, lipase, and protease in variants of the experiment using different concentrations of phosphemidum.

Note: *differences from control are statistically significant at $p < 0.05$; •differences from the experiment variant 0.1% are statistically significant at $p < 0.05$; #differences from the experiment variant 0.01% are statistically significant at $p < 0.05$; !differences from the experiment variant 0.001% are statistically significant at $p < 0.05$; £differences from the experiment variant 0.0001% are statistically significant at $p < 0.05$.

TABLE 5.2   Enzyme Activity Amylase, Lipase, and Protease (U Per 1 Million Cells) in Experiments with Different Phosphemidum Concentrations.

Phosphemidum (%)	Amylase	Lipase	Protease
Control (without phosphemidum)	0.050 ± 0.007	0.372 ± 0.052	0.640 ± 0.072
0.1	0.056 ± 0.004	$0.099 \pm 0.020^{*}$	0.451 ± 0.056
0.01	0.052 ± 0.005	$0.093 \pm 0.018^{*}$	$0.790 \pm 0.045^{*•}$
0.001	0.046 ± 0.006	$0.123 \pm 0.017^{*}$	0.654 ± 0.069
0.0001	0.055 ± 0.005	$0.492 \pm 0.056^{*•\#}$	$1.208 \pm 0.105^{*•\#!}$
0.00001	0.069 ± 0.006	$0.493 \pm 0.053^{*•\#}$	$0.857 \pm 0.049^{*•!£}$

Note: See the caption to Figure 5.1.

To clarify how long the changes in mutant variants are preserved with increasing age of the cultures, activity of the enzymes was evaluated in 2-, 10-, 30-, and 50-day cultures obtained in the variant with most effective concentrations of phosphemidum—0.0001% (Figs. 5.3 and 5.4). For this, the cultures obtained in experiments with phosphemidum were regularly transplanted every day.

TABLE 5.3 Enzyme Activity (U Per 1 Million Cells), β-Glucanase, CMC-ase, and Xylanase in Experiments with Different Phosphemidum Concentrations.

Phosphemidum concentration (%)	β-Glucanase	CMC-ase	Xylanase
Control (without phosphemidum)	0.032 ± 0.005	0.014 ± 0.003	0.122 ± 0.018
0.1	0.034 ± 0.006	0.010 ± 0.002	0.158 ± 0.015
0.01	0.031 ± 0.004	0.009 ± 0.003	0.134 ± 0.019
0.001	0.027 ± 0.006	0.012 ± 0.002	0.112 ± 0.014
0.0001	0.029 ± 0.007	0.011 ± 0.002	0.141 ± 0.011
0.00001	0.030 ± 0.005	0.015 ± 0.003	0.185 ± 0.016*

*Differences from control are statistically significant at $p < 0.05$.

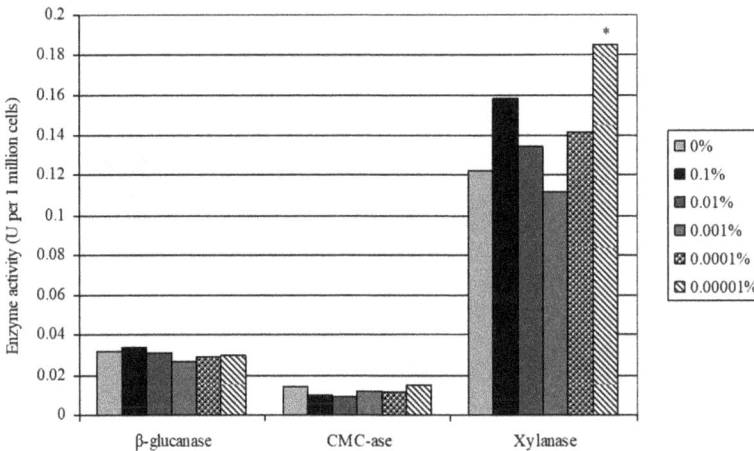

FIGURE 5.2 Enzyme activity (U per 1 million cells) of β-glucanase, CMC-ase, and xylanase in variants of the experiment using different concentrations of phosphemidum. *Note:* See note to Table 5.3.

Study of the dynamics of changes in activity of exogenous enzymes with increasing age of the cultures (Tables 5.4 and 5.5, Figures 5.3 and 5.4) revealed approximately the same pattern in the control and the variant with 0.0001% phosphemidum.

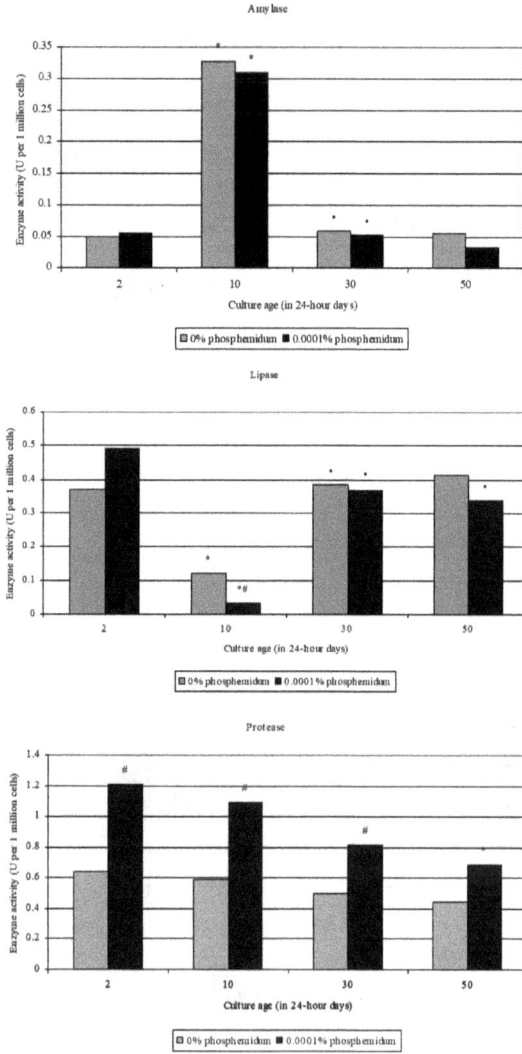

FIGURE 5.3 The change of enzyme activity (U per 1 million cells) of amylase, lipase, protease with increasing age of cultures after exposure to phosphemidum in concentration of 0.0001%.

Note: *differences between 2-, 10-, and 50-day cultures in control and in experiment variant with 0.0001% phosphemidum are statistically significant at $p < 0.01$; •differences between 10-, 30-, and 50-day cultures in control and in experiment variant with 0.0001% phosphemidum are statistically significant at $p < 0.01$; #differences between coetaneous cultures in control and in experiment variant with 0.0001% phosphemidum are statistically significant at $p < 0.01$.

In all studied enzymes, the activity decreases with increasing culture age. In protease, β-glucanase, CMC-ase, and xylanase, activity decreases smoothly, while amylase activity sharply increases in a 10-day culture, and then decreases just as sharply in a 30- and 50-day culture. Changes in lipase activity are also of a peculiar nature: they decrease sharply in a 10-day culture and increase in 30- and 50-day cultures (Fig. 5.3). Comparison of enzyme activity in mutant and control variants revealed advantage of mutant variant in three enzymes: activity of the protease, β-glucanase, and xylanase was higher in cultures of all ages (2-, 20-, 30-, and 50 days) in variant with 0.0001% phosphemidum solution in comparison with control (Tables 5.4 and 5.5, Figures 5.3 and 5.4).

TABLE 5.4 Changes in Enzyme Activity (U Per 1 Million Cells) of Amylase, Lipase, and Protease with the Increasing of the Cultures' Age.

Cultures' age (days)	Amylase	Lipase	Protease
Control			
2	0.050 ± 0.007	0.372 ± 0.052	0.640 ± 0.072
10	0.328 ± 0.042*	0.123 ± 0.022*	0.590 ± 0.061
30	0.059 ± 0.006•	0.386 ± 0.048•	0.494 ± 0.052
50	0.056 ± 0.008	0.413 ± 0.043	0.448 ± 0.057
0.0001% Phosphemidum			
2	0,055 ± 0.005	0,492 ± 0.056	1,208 ± 0.105#
10	0,310 ± 0.037*	0,034 ± 0.008*#	1,09 ± 0.097#
30	0,052 ± 0.007•	0,369 ± 0.047•	0,817 ± 0.098#
50	0,033 ± 0.012	0,341 ± 0.044•	0,688 ± 0.076*

*Differences between 2-, 10-, and 50-day cultures in control and in experiment variant with 0.0001% phosphemidum are statistically significant at $p < 0.01$; •differences between 10-, 30-, and 50-day cultures in control and in experiment variant with 0.0001% phosphemidum are statistically significant at $p < 0.01$; #differences between coetaneous cultures in control and in experiment variant with 0.0001% phosphemidum are statistically significant at $p < 0.01$.

An analysis of obtained data showed that phosphemidum in certain concentrations can cause changes leading to an increase in activity of enzymes. This is characteristic trait of alkylating compounds capable to cause useful mutation in single-gene mutations.[25] Our study clearly shows the stimulating effect of phosphemidum on activity of lipase, protease and xylanase at concentrations of 0.0001% and 0.00001% (see Figures 5.1 and 5.2), but this pattern is not regular, which requires additional studies to find out the reasons of such increase in activity. In experiments with *B. subtilis* CABI, the best options were those with the lowest phosphemidum concentrations (0.0001% and 0.00001%). Stimulating effect of mutagens in low and ultralow doses was noticed since the time of I.A. Rapoport.[26] This is due to the emergence of double genetic stimulation under the influence of mutagens: formation of a high heterozygosity after effect of the system and activation of enzymatic systems that have a positive effect on the manifestation of productive qualities. It is possible that the same mechanism is also displayed under increase of bacteria enzymatic activity when exposed to phosphemidum.

TABLE 5.5   Changes in Enzyme Activity (U Per 1 Million Cells) of β-Glucanase, CMC-ase, and Xylanase with Increasing of Cultures' Age.

Cultures' age (days)	β-Glucanase	CMC-ase	Xylanase
Control (without phosphemidum)			
2	0.032 ± 0.005	0.014 ± 0.003	0.122 ± 0.018
30	$0.007 \pm 0.001^*$	0.009 ± 0.003	0.082 ± 0.018
50	$0.012 \pm 0.003^*$	0.008 ± 0.001	$0.068 \pm 0.010^*$
0.0001% Phosphemidum			
2	$0.029 \pm 0.007^\#$	0.011 ± 0.002	0.141 ± 0.011
30	0.021 ± 0.004	0.008 ± 0.002	$0.094 \pm 0.016^*$
50	$0.008 \pm 0.002^{*\bullet}$	$0.005 \pm 0.001^*$	$0.082 \pm 0.015^*$

*Differences between 2-, 30-, and 50-day cultures in control and in experiment variant with 0.0001% phosphemidum are statistically significant at $p < 0.05$; •differences between 30- and 50-day cultures in experiment variant with 0.0001% phosphemidum are statistically significant at $p < 0.01$; #differences between 2-day cultures in control and in experiment variant with 0.0001% phosphemidum are statistically significant at $p < 0.01$.

5.4   CONCLUSIONS

Phosphemidum has been shown to affect the activity of vital enzymes: protease, amylase, lipase, β-glucanase, CMC-ase, and xylanase in *B. subtilis*. A high stimulating effect of low doses of phosphemidum (0.0001% and 0.00001%) was revealed. It was shown that activity of the enzymes both in control and mutant at 0.0001% phosphemidum solution variants decreases with increasing age of the cultures. Our study showed that, despite decline in some indicators on lipase, protease, CMC-ase, and xylanase with increasing age of the culture, the activity of protease, β-glucanase, and xylanase was higher in the variant with 0.0001% phosphemidum compared to control.

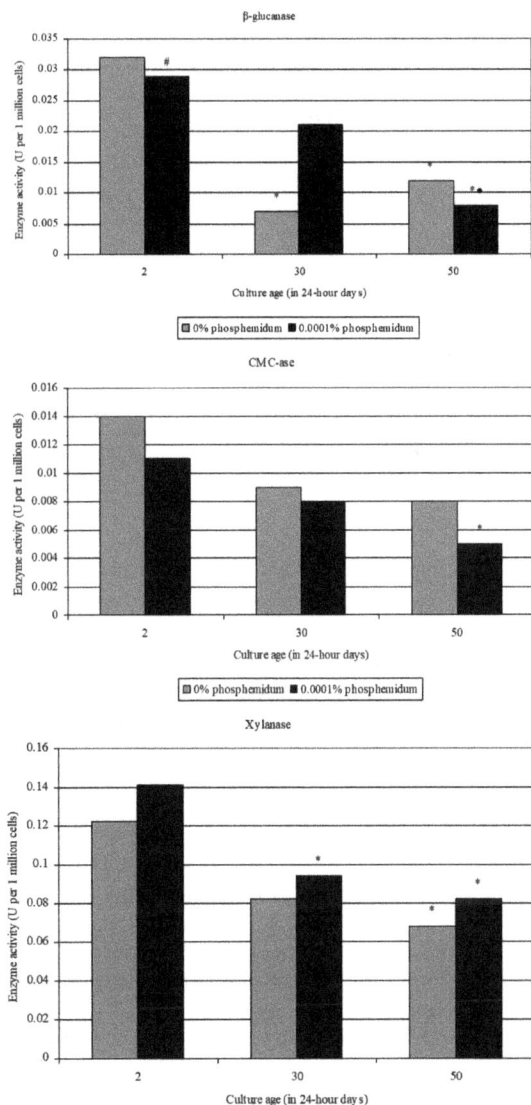

FIGURE 5.4 The change of enzyme activity (U per 1 million cells) of β-gluconase, CMC-ase and xylanase with increasing age of the cultures under effect of phosphemidum in concentration of 0.0001%.

Note: *Differences between 2-, 30-, and 50-day cultures in control and in experiment variant with 0.0001% phosphemidum are statistically significant at $p < 0.05$; •differences between 30- and 50-day cultures in experiment variant with 0.0001% phosphemidum are statistically significant at $p < 0.01$; #differences between 2-day cultures in control and in experiment variant with 0.0001% phosphemidum are statistically significant at $p < 0.01$.

KEYWORDS

- CMC-ase (carboxymetylcellulase)
- β-gluconase
- amylase
- protease
- lipase
- xylanase
- chemical mutagene

REFERENCES

1. Sorokulova, I. B. A Comparative Study of the Biological Properties of Biosporin and Other Commercial Preparations Based on Bacilli. *Microbiol. J.* **1997,** *59* (6), 43–49 (in Russian).
2. Wu, D. X.; Zhao, S. M.; Peng, N.; Xu, C. P.; Wang, J.; Liang, Y. X. Effect of Probiotic (*Bacillus subtilis* FY99-01) on the Bacterial Community Structure and Composition of Shrimp (*Litopenaeus vannamei*, Boone) Culture Water Assessed by Denaturing Gradient Gel Electrophoresis and High Throughput Sequencing. *Aquaculture Research* **2016,** *47* (3), 237–242.
3. Zhou, D.; Zhu, Y. H.; Zhang, W.; Wang, M.-L.; Fan, W. Y.; Song, D.; Yang, G. Y.; Jensen, B. B.; Wang, J. F. Oral Administration of a Select Mixture of Bacillus Probiotics Generates Tr1 Cells in Weaned F4ab/acR- Pigs Challenged with an F4+ ETEC/VTEC/EPEC Strain. *Vet. Res.* **2015,** *46* (1), 95 p.
4. Ramesh, D.; Vinothkanna, A.; Rai, A. K.; Vignesh, V. S. Isolation of Potential Probiotic *Bacillus* spp. and Assessment of Their Subcellular Components to Induce Immune Responses in *Labeo rohita* against *Aeromonas hydrophila*. *Fish Shellf. Immunol.* **2015,** *45* (2), 268–276.
5. Soto, L. P.; Astesana, D. M.; Zbrun, M. V.; Blajman, J. E.; Salvetti, N. R.; Berisvil, A. P.; Rosmini, M. R.; Signorini, M. L.; Frizzo, L. S. Probiotic Effect on Calves Infected with Salmonella Dublin: Haematological Parameters and Serum Biochemical Profile. *Benef. Microbes* **2016,** *7* (1), 23–33.
6. Lefevre, M.; Racedo, S. M.; Ripert, G.; Housez, B.; Cazaubiel, M.; Maudet, C.; Jüsten, P.; Marteau, P.; Urdaci, M. C. Probiotic Strain *Bacillus subtilis* CU1 Stimulates Immune System of Elderly during Common Infectious Disease Period: A Randomized, Double-Blind Placebo-Controlled Study. *Immun. Age.* **2015,** *12* (1), 24.
7. Aceti, A.; Gori, D.; Barone, G.; Callegari, M. L.; Di Mauro, A.; Fantini, M. P.; Maggio, L.; Meneghin, F.; Morelli, L.; Zuccotti, G.; Corvaglia, L. Probiotics for

Prevention of Necrotizing Enterocolitis in Preterm Infants: Systematic Review and Meta-analysis. *Ital. J. Pediatr.* **2015,** *41* (1), 199.

8. Bakulina, L. F.; Timofeev, I. V.; Perminova, N. G.; Polushkina, A. F.; Pechorkina, N. I. Probiotics Based on Spore-Forming Microorganisms of the Genus Bacillus and Their Use in Veterinary Medicine. *Biotechnology* **2001,** *2,* 48–56 (in Russian).

9. Hauville, M. R.; Zambonino-Infante, J. L.; Gordon, B. J.; Migaud, H.; Main, K. L. Effects of a Mix of *Bacillus* sp. as a Potential Probiotic for Florida Pompano, Common Snook and Red Drum Larvae Performances and Digestive Enzyme Activities. *Aquacult. Nutr.* **2016,** *22* (1), 51–60.

10. Cerezuela, R. L.; Guardiola, F. A.; Cuesta, A; Esteban, M. A. Enrichment of Gilt-head Seabream (*Sparus aurata* L.) Diet with Palm Fruit Extracts and Probiotics: Effects on Skin Mucosal Immunity. *Fish Shellf. Immunol.* **2016,** *49,* 100–109.

11. Frei, R.; Akdis, M.; O'Mahony, L. Prebiotics, Probiotics, Synbiotics, and the Immune System: Experimental Data and Clinical Evidence. *Curr. Opin. Gastroenterol.* **2015,** *31* (2), 153–158.

12. Tsukahara, T.; Tsuruta, T.; Nakanishi, N.; Hikita, C.; Mochizuki, M.; Nakayama, K. The Preventive Effect of *Bacillus subtilis* Strain DB9011 against Experimental Infection with Enterotoxcemic *Escherichia coli* in Weaning Piglets. *Anim. Sci. J.* **2013,** *84* (4), 316–321.

13. Fedorova, O. V.; Nazmieva, A. I.; Nuretdinova, E. I.; Valeeva, R. T. Probiotic Preparations Based on Microorganisms of the Genus *Bacillus. Bull. Technol. Univ.* **2016,** *19* (15), 170–173.

14. Li, Y.; Xu, Q.; Huang, Z.; Liu, X.; Yin, C.; Yan, H.; Yuan, J. Effect of *Bacillus subtilis* CGMCC 1.1086 on the Growth Performance and Intestinal Microbiota of Broilers. *J. Appl. Microbiol.* **2016,** *120* (1), 195–204.

15. Chernov, V. A.; Safonova, E. S.; Sazonov, N. V.; Kropacheva, F. F.; Klyuchareva, Z. D.; Erofeeva, T. A. (1973). *Medicine* **1973.** Invitation No. 301951, posted on 05.04.1973, Bulletin No. 17 (in Russian).

16. Weisfeld, L. I. Cytogenetic Effect of Phosphazine on Human and Mouse Cells in Culture. *Genetics* **1965,** *4,* 85–92.

17. Weisfeld, L. I. Damages of Chromosomes and Disruptions of Mitotic Activity in Seed-lings of *Crepis capillaris* by Alkylating Antineoplastic Preparation Phosphemidum. *J. Inf., Intell. Knowl.* **2012,** *4* (4), 295–307.

18. Weisfeld, L. I. Genetic Resources of Plants as a Factor of Estimation of the Stability of Agroecosistems. XIV. All-Russian Scientific and Practical Conference Dedicated to the Year of Ecology in Russia. *Tobolsk Sci.* **2017,** 95–99 (in Russian).

19. Bome, N. A.; Weissfeld, L. I.; Arsentiev, S. V. The Development of Winter Wheat under the Influence of the Chemical Mutagen Phosphemide. *Int. Res. J.* **2015,** *11* (42), 83–90. DOI:10.18454/IRJ.2015.42.064.

20. Bome, N. A.; Weissfeld, L. I.; Babaev, E. V.; Bome, A. A.; Kolokolova, N. N. Agrobiological Characteristics of Soft Spring Wheat (*Triticum aestivum* L.) during Seed Treatment with Chemical Mutagen Phosphemidum. *Agric. Biol.* **2017,** *52* (3), 570–579. DOI:10.15389/agrobiology.2017.3.570eng.

21. Pak, I. V.; Trofimov, O. V.; Weisfeld, L. I.; Rustamov, R. D.; Skvortsova, K. S. The Efficiency of Using Phosphemidum to Increase the Productivity of the Strain

Candida maltosa. Bull. Tyumen State Univ. Ecol. Nat. Manage. **2018,** *4* (3), 67–78. DOI:10.21684/2411-7927-2018-4-3-67-78 (in Russian).

22. National Standard of the Russian Federation. GOST R54330-2011. (2018). Enzyme Preparations for the Food Industry. Methods for the Determination of Amylolytic Activity (2018 Edition with Changes No. 1, 2 amended). Standardinform: Moscow, 2018. http://docs.cntd.ru/document/1200085581 (circulation date 01/28/2017) (in Russian).

23. Gracheva, I. M.; Grachev, Y. P.; Mosichev, M. S.; Borisenko, U. G.; Bogatkov, S. V.; Gernet, M. V. *Laboratory Workshop on the Technology of Enzyme Preparations*; Light and Food Industry: Moscow, 1982; 240 p. (in Russian).

24. Glaser, V. M.; Kameneva, S. V.; Mitronova, T. N. *Great Workshop on the Genetics of Microorganisms*; Moscow State University: Moscow, 1977; 120 p. (in Russian).

25. Lovelace, A. *Genetic Effects of Alkylating Compounds*; Science: Moscow, 1970; 254 p.

26. Rapoport, I. A. *Dual Genetic Stimulation Induced by Supermutagens*: *Mutation Selection*; Science: Moscow, 1977; pp 230–242 (in Russian).

CHAPTER 6

Cytogenetic Characteristics of Seed Seedlings of *Rhododendron ledebourii*, Introduced in the Botanical Garden of Voronezh State University

TATYANA V. VOSTRIKOVA[1*], JULIYA V. BURMENKO[2], and VLADISLAV N. KALAEV[1]

[1]*All-Russian Research Institute of Sugar Beet and Sugar, Federal Agency of Scientific Organizaions, 86, Ramonsky District, Voronezh Region, 396030, Russia*

[2]*Department of genetics and selection of fruit and berry crops, Federal State Budgetary Scientific Institution Federal Horticultural Research Center for Breeding, Agrotechnology and Nursery, 4 Zagor'evskaya St., Moscow, 115598, Russian Federation*

[*]*Corresponding author. E-mail: tanyavostric@rambler.ru*

ABSTRACT

The study of the seed progeny with different levels of the genetic stability of Rhododendron Ledebour (*Rhododendron ledebourii* Pojark.) was made. The micropreparations were examined according to following cytogenetic characteristics: mitotic activity, an indicator which is the mitotic index—a ratio of the number of dividing cells to the total number of cells counted, the percentage of cells in stages of mitosis, the level of mitosis pathologies, the level of cells with persistent nucleoli at metaphase and anaphase of mitosis, and the surface area of single nucleoli. The seedlings with high levels of mitosis pathologies were regarded as the mutable group, and with low pathologies were regarded with the low-mutable group. The seed progeny of the mutable group shows a decrease in mitotic activity, but an increase of the mitosis pathologies level, which indicates a high cytogenetic instability. The same parameters reveal the stability of maternal plants producing mutable and low-mutable seed progeny. The cytogenetic method may be used for identification of the seed progeny as the mode of separation the parental plants producing seed progeny with a high level of stability of the genetic material. These investigations may be appropriate for the assessment of the seed quality according to cytogenetic characteristics in parties with unknown origin.

6.1   INTRODUCTION

Currently, there are a number of studies showing that woody plants can produce seed progeny with different levels of cytogenetic disturbances.[1–9] A study of plant polymorphism is most often done at the genetic,[10–12] biochemical and cytogenetic[13–15] levels. However, the cytogenetic polymorphism by which we mean a variety of cytogenetic and nucleolar characteristics within a population of woody plants[16] is poorly understood. The cytogenetic polymorphism was previously recognized in terms of mitosis and nucleolar characteristics in seedlings of a weeping birch (*Betula pendula* Roth),[3] a pedunculate oak (*Quercus robur* L.)[4] and a Scots pine (*Pinus sylvestris* L.).[5] Seed progeny with a high level of cytogenetic disturbances are called mutable, progeny with opposite characteristics are called low mutable. It was revealed that the offspring with a high level of cytogenetic pathologies is characterized by worse growth rates (with comparing low mutable).[4] But these investigations haven't been applied for selection of the resistant to anthropogenic pressure seed progeny and high-quality seeds.

It should also be said that among the seed offspring with a low level of cytogenetic pathologies, a greater phenotypic diversity will probably be noted, which makes it possible to search for selection-valuable forms. Previously, the presence of forms with high and low levels of disturbances among seed progeny was shown in weeping birch and their cytogenetic characteristics were given.[3]

The timeframes of a single study often do not allow for the mass selection of selection-valuable trees based on anatomical and morphological features of several generations of the progeny. Experiments using herbaceous plants are faster. They allow to see results of research in several generations of offspring, to identify modified specimens, especially when using the cytogenetic method.[17,18] This is very valuable in artificial mutagenesis experiments.[18,19] Acceleration of obtaining the results of selection of mother plants producing seed progeny with different levels of stability of genetic material from *Rhododendron ledebourii* Pojark. can be achieved by analyzing cytogenetic indicators.[9,16]

Most often, the assessment of seed quality is carried out on the basis of study of phenotypic parameters. Selection in nurseries and among seeds can include sorting plants by size and by weight of seeds.[20] Sorting seeds and selecting them by absolute mass also does not allow to investigate the genetic characteristics of seed progeny and predict the growth abilities of seedlings obtained from them. Cytogenetic assessment of seed quality allows to carry out selecting and at the same time testing the quality of seed offspring at the early stages of the progeny development based on cellular and subcellular characteristics.

Using of cytogenetic method is effective, since it is fairly simple and fast, it makes possible to adequately determine stability of genetic material of parent plant and its seed offspring, and, at the same time, to assess environmental tension in the collection area[1,3–5,7,21,22] and seed quality. This property of the method is valuable because the environment affects genotype not only of plants but also of humans[23,24] and especially phenotype.[25,26]

The study goal was to reveal the influence of stressful environmental factors on the example of populations *R. ledebourii.*

6.2 MATERIALS AND METHODOLOGY

Approximately, 30–40-year-old phenotypically normal plants (without visual damage from parasites) of introduced population of *R. ledebourii* in B. M. Kozo-Polyansky Botanical Garden of Voronezh State University (Central Black Earth Region) were used.

The seeds germination was carried out at a temperature of +25°C. The roots of 0.5–1 cm length were fixed with the 3:1 mixture of ethanol and glacial acetic acid at 9 a.m., after that the material was stored in a refrigerator at +4°C. The consequently pressed micropreparations of the seedling's roots and Goyer's medium was prepared as previously described.[9,21,22]

A sample of 40 seedlings for cytological studies is considered to be representative.[3,4] We analyzed the cytogenetic characteristics of 40 seedlings. The micropreparations were studied by means of a LABOVAL-4 microscope (Carl Zeiss, Jena) with a total magnification of 40 × 10 × 1.5.

About 500–700 cells of each mixture (1 micropreparation corresponds to 1 root and 1 seedling) were analyzed. Total amount of about 25,000 cells was studied.

The micropreparations were examined according to following cytogenetic characteristics: mitotic activity, which indicator is mitotic index (MI)—a ratio of number of dividing cells to total number of cells counted (in %); percentage of cells in stages of mitosis; level of mitosis pathologies (as a ratio of number of cells with disorders to total number of dividing cells, in %); and level of cells with persistent nucleoli at metaphase and anaphase of mitosis (ratio of number of cells with persistent nucleoli to total number of cells at these stages, in %). Mitosis pathologies classification was carried out by I. A. Alov.[27]

To study the nucleolar characteristics in root meristem cells of the *R. ledebourii* seed progeny diameter of nucleoli was measured using an eyepiece micrometer (200 cells in each micropreparation were analyzed) and surface area of nucleolus (in μm^2) was calculated. Nucleolar activity was investigated by characteristics of surface area of the nucleoli.

We produced computerized statistical data processing using the "Stadia" 6.0 software (TechSmith Corporation, USA). The procedure of data grouping and processing is described by A. P. Kulaichev.[28] The samples' comparison on mitosis pathologies level and level of mitotic cells with persistent nucleoli was performed using *t*-criterion of Student and X-rank test of Van der Warden at $p < 0.05$, $p < 0.01$, $p < 0.001$, since this feature is not subject to the normal distribution.

Cluster analysis was performed using the normalized Euclidean distance metrics, a strategy of neighbor grouping. The matrix data for cluster analysis included the following cytogenetic indicators of the seed progeny: MI calculated considering the prophase cells (in %); MI calculated excluding the prophase cells (in %); percentage of the cells in prophase; percentage of the cells in metaphase; percentage of the cells in ana-telophase; level of mitosis pathologies (in %); level of the cells with persistent nucleoli at metaphase-telophase mitosis (in %); and the surface area of single nucleoli (μm^2). A discriminant analysis was performed using the Mahalanobis test.

6.3   RESULTS AND DISCUSSION

According to cluster analysis (based on the joint cytogenetic characteristics), the seedlings were divided into groups: mutable, low mutable, and intermediate, using the level of violations of mitotic division (mitosis pathologies level). The seedlings with high levels of mitosis pathologies were regarded as the mutable group, and with low pathologies—to the low-mutable group (Fig. 6.1). In addition, some of the intermediate groups were classified, as the characteristics of these samples were much close to mutable or low-mutable group. The mutable group (No. 4, presenting 20% of the *seedlings* under analysis) is characterized by the highest level of mitosis pathologies (4.0 ± 0.6%), cells with persistent nucleoli (13.4 ± 1.4%), the largest number of the prophase cells (55.3 ± 2.3%), and the lowest number of cells in the ana-telophase (31.4 ± 1.9%), low MI (6.7 ± 0.2%), particularly MI, calculated excluding the prophase cells (3.0 ± 0.2%), that indicates a delay of cells in prophase and large surface area of single nucleoli (79.4 ± 1.8 μm^2). In the spectrum of mitosis pathologies lagging chromosomes in anaphase and metakinesis prevailed (92.8%), the agglutination of chromatin (5.5%) was found out, and the bridges were rare.

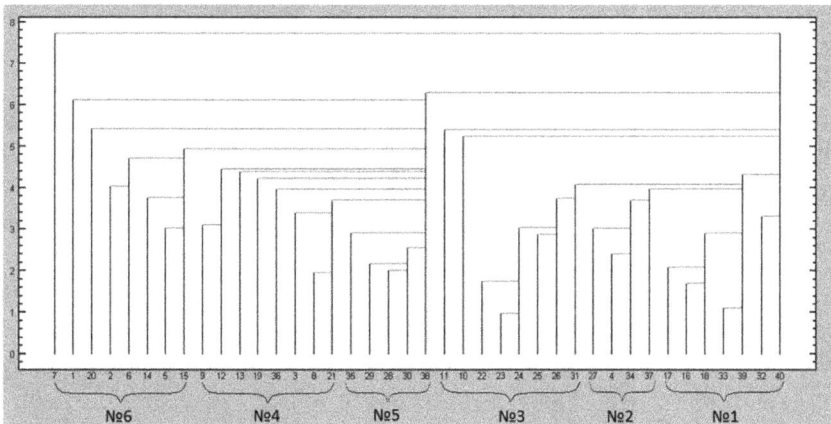

FIGURE 6.1 Cluster distances between seedlings of seed progeny of *Rhododendron ledebourii*, built on the basis of their cytogenetic characteristics.
Group of seedlings 1—low mutable, group 4—mutable, groups 2, 3, 5, 6—intermediate.

The low-mutable group (number 1 in Figure 6.1, presents 17.5% of total number of seedlings under analysis) is characterized: a high MI, calculated considering (9.1 ± 0.2%) and excluding the prophase cells (5.7 ± 0.1%), the highest number of cells in ana-telophase (45.9 ± 1.4%), the lowest number of cells in prophase (37.7 ± 1.3%), level of cells with persistent nucleoli (5.9 ± 0.3%) and abnormalities of mitosis (1.2 ± 0.2%), and smaller surface area of single nucleoli (61.7 ± 2.2 μm²) (differences from the mutable group are statistically significant) ($p < 0.01$). The spectrum of abnormalities of mitosis in cells of the low-mutable group was represented mainly by bridges (71.3%), which suggest a more active work of reparative system.[16,29] There were lagging chromosomes in anaphase and metakinesis (28.7%).

We have noted the decrease in MI (calculated considering and excluding the prophase) and the increase in the number of the prophase cells, which indicates chromosomal damage[27] in the mutable group of seedlings root meristem, and inhibition of growth processes in seed progeny of this group due to the damage of genetic material.

In cells of root apical meristem of *R. ledebourii* seedlings, an increase in the surface area of single nucleoli in the mutable group was revealed and that indicates increase in nucleolar activity.[30] The increase in nucleolar activity is considered to be an indicator of the growing intensity of metabolism in terms of stress.[31] The increase in nucleolar activity in the mutable group was noted in previous studies of seed progeny cytogenetic polymorphism of a weeping birch (*Betula pendula*),[3] a pedunculate oak (*Quercus robur*),[4] and a Scots pine (*P. sylvestris*).[5] It was revealed that the mutable groups present the increase in mitotic abnormalities and in surface area of single nucleoli in interphase.[16]

Normally, the nucleolus is absent in dividing cells. The phenomenon of the appearance of nucleolus-like structures connected to chromosomes in metaphase, anaphase, and early telophase of mitosis, when there is no nuclear membrane, chromosomes are still significantly reduced and nucleolus does not disappear during division, is described for many plants and animals.[3-5,7,9,16,32] These structures are called "persistent nucleoli" ("residual nucleoli") (Fig. 6.2).

The presence of persistent nucleoli in cell division was discovered by A. K. Butorina[33,34] and named a special case of amplification of nucleolar activity, and a mechanism of adaptation to the impact of adverse environmental conditions. Persistent nucleoli in cell division appear to be higher in seedlings of the mutable group that may indicate increased metabolic functions in the body.[9,16,34-36] According to the cytogenetic parameters, the closest to the mutable group appeared to be the intermediate group 5 (12.5% of the total seedlings under analysis). The intermediate group

6 (20% of total seedlings under analysis) differed from the mutable and the intermediate group 5, as it appeared to possess higher MI, calculated considering and excluding the prophase, and lower level of mitosis pathologies and cells with persistent nucleoli. Changes in these parameters indicate major cytogenetic stability of seedlings in this group in comparison to the mutable group.[16]

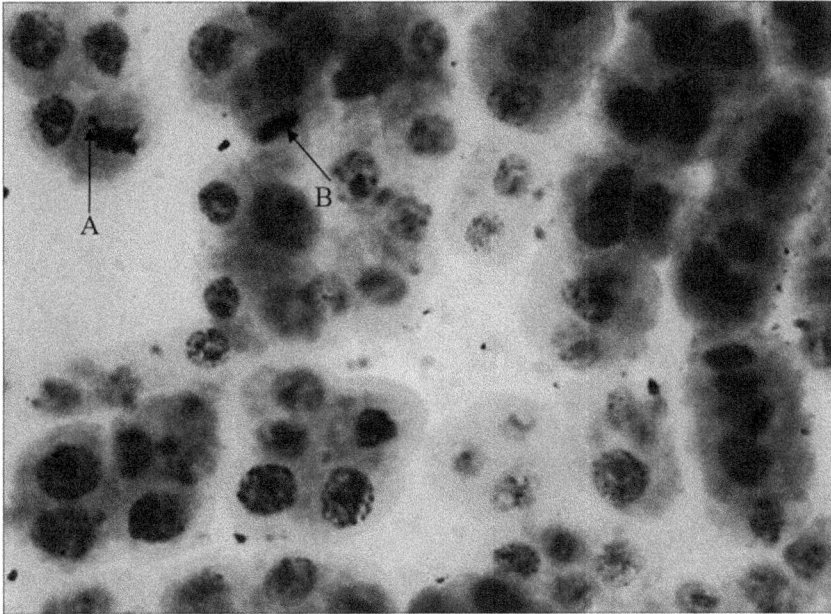

FIGURE 6.2 Persistent nucleolus on the stage of metaphase (a) and normal metaphase (b) of mitosis in root meristem cells of *Rhododendron ledebourii*.

The intermediate group 2 (10% of total seedlings under analysis) is the closest to the low-mutable group by its cytogenetic parameters, differing with lower MI (calculated considering and excluding prophase) and lower level of cells with persistent nucleoli. Unlike the low-mutable group, the intermediate group 3 (20% of the total seedlings under analysis) was characterized by lower MI (calculated considering and excluding prophase), lower number of cells in metaphase and ana-telophase, and higher levels of cells in prophase.

The increased occurrence of prophase cells can be regarded as the mitosis pathology associated with chromosomal aberrations, according to I. A. Alov.[27] This indicates a serious pathology of genetic material, inability of cells to move to the next stage of mitosis, and may happen due to a disturbance of protein synthesis of a mitotic spindle or chromosome damage. This phenomenon was noted previously in studies with other woody plants in conditions of anthropogenic pollution: a weeping birch, a pedunculate oak.[3–4,7,21,22] Higher levels of mitosis pathologies and cells with persistent nucleoli in the intermediate group 3 indicate lower cytogenetic stability of this group when compared to the low-mutable group and the intermediate group 2. The intermediate group 2 and group 3 differ from each other in number of prophase cells, level of cells with persistent nucleoli and level of mitosis pathologies, which were higher in the group 3. The increase in the level of cells with persistent nucleoli indicates an increase in nucleolar activity, which happens due to additional synthetic activity of ribosomal genes.[16,34]

Therefore, we can divide all the analyzed seedlings for two large groups (Table 6.1).

When comparing the results with similar works on cytogenetic study of polymorphism of native woody plants (*Betula pendula*,[3] *Quercus robur*,[4] and *P. sylvestris*[5]), there is a general trend in the degree of separation of seedlings by mutability, and ecologically clean territory presents the domination of mutable seedlings (*Betula pendula* and *P. sylvestris*). As for the mutable groups of woody plants seedlings, including *R. ledebourii*, the level of mitosis pathologies increases and their spectrum expands with prevalence of disturbances that indicates the fragmentation of chromosomes and low intensity of reparative processes.[29] The closest result in the comparative aspect is marked for *Quercus robur*,[4] in which mainly the seedlings of intermediate groups dominated at environmentally clean areas. This indicates heterogeneity and adaptation processes in populations from stress conditions, as well as the species specificity of cytogenetic characteristics.

Some authors[37–39] noted higher level of adaptability to environmental conditions of plants, the seed progeny of which have shown a lower level of chromosomal aberrations during the cytogenetic studies. A subsequent experiment on the growth parameters of seedlings of pedunculate oak, sampled at mature plants with different levels of mutability, prove their relationship.[4] It was found that seedlings with poorer growth capacity appear from acorns with smaller mass and diametrical size, moreover, they have low mitotic activity in root meristem, high levels of mitosis

TABLE 6.1 Cytogenetic Characteristics of Seed Progeny of *Rhododendron ledebourii* with Different Stability of Genetic Material.

Parameters	Mutable group	Low-mutable group
Level of mitosis pathologies	Medium or high (over 2.5%)	Low ($\leq$2.5%)
Mitotic activity	Mitotic index is $\leq$8%	Mitotic index is over 8%
Distribution of cells in stages of mitosis	% of cells in prophase—over 45%	% of cells in prophase—$\leq$45%
	% of cells in metaphase—over 25%	% of cells in metaphase—$\leq$25%
	% of cells in anaphase–telophase—$\leq$30%	% of cells in anaphase–telophase—over 30%
Nucleolus characteristics	High frequency of cells with persistent nucleoli (over 8%)	Low and medium frequency of cells with persistent nucleoli ($\leq$8%); low surface area of single nucleoli (55–75 μm^2)
	High surface area of single nucleoli (76–90 μm^2)	

pathologies and a large surface area (mutable seedlings).[16] Seedlings obtained from samples with increased mitotic activity, and low values of cytogenetic abnormalities (low mutable) showed the best growth activity.[4] We can assume that *R. ledebourii* seed progeny being low-mutable is more adapted to the conditions of introduction compared with other seedlings. The low-mutable copies can be used for landscaping the contaminated areas.[16]

According to mutability and estimation of cytogenetic characteristics, especially the level of mitosis pathologies, the seed quality can be assessed in parties with unknown origin. The seeds of low-mutable groups have high quality and conversely.

6.4 CONCLUSIONS

1. A cytogenetic analysis of seed progeny of *R. ledebourii* allowed us to divide the seedlings into two main groups and four intermediate groups.
2. Seedlings of the mutable groups are characterized by decrease in mitotic activity but increase of mitosis pathologies level that indicates high cytogenetic instability. Also, increase in number of cells with persistent nucleoli and in surface area of single nucleoli was revealed in this group.
3. Seed offspring of the group with low mutability has a higher mitotic activity, low rates of cytogenetic anomalies, a significantly smaller number of cells with constant nucleoli, and an increase surface in individual nucleoli.
4. Seedlings of the intermediate groups have both tendencies: two of them are similar to the mutable group and other—to the low-mutable group.
5. Analysis of lots of seedling groups indicates high level of genetic heterogeneity of seed progeny from maternal plants, growing in new environment. The high genetic heterogeneity provides the larger opportunity to adaptation.
6. Only genetically stable individuals are adapted to the conditions of introduction, for example, low mutable and closed to them.
7. Consequently, cytogenetic characteristics allow to identify compliance of heterogeneity of seed progeny.

KEYWORDS

- **woody plants**
- **cytogenetic polymorphism**
- **mitosis pathologies**
- **nucltoli**
- **mitotic index**
- **seed progeny**

REFERENCES

1. Oudalova, A. A.; Geras'kin, S. A. The Time Dynamics and Ecological Genetic Variation of Cytogenetic Effects in the Scots Pine Populations Experiencing Anthropogenic Impact. *Biol. Bull. Rev.* **2012,** *2* (3), 254–267. DOI:10.1134/S207908641203005X (in Russian).
2. Sedel'nikova, T. S.; Muratova, E. N.; Pimenov, A. V. Variability of Chromosome Numbers in Gymnosperms. *Biol. Bull. Rev.* **2011,** *1* (2), 100–109. DOI:10.1134/S2079086411020083.
3. Kalaev, V. N.; Karpova, S. S.; Artyukhov, V. G. Cytogenetic Characteristics of Weeping Birch (*Betula pendula* Roth) Seed Progeny in Different Ecological Conditions. *Bioremediat., Biodiversity Bioavailabil.* **2010,** *4* (1), 77–83 (in Russian).
4. Kalaev, V. N.; Popova, A. A. Cytogenetic Polymorphism of English Oak (*Quercus robur* L.) Seedlings from Areas with Different Levels of Anthropogenic Pollution. *Sil. Genet.* **2014,** *63* (6), 245–252 (in Russian).
5. Mashkina, O. S.; Kalaev, V. N.; Muray, L. P.; Lelikova, E. S. Cytogenetic Response of Seed Progeny of Scots Pine to Combined Anthropogenic Pollution in the Area of Novolipetsk Metallurgical Combine. *Ecol. Genet.* **2009,** *7* (3), 17–29 (in Russian).
6. Schwery, O.; Onstein, R. E.; Bouchenak-Khelladi, Y.; Xing, Y.; Carter, R. J.; Linder, H. P. As Old as the Mountains: The Radiations of the Ericaceae. *New Phytol.* **2015,** *207* (2), 355–367.
7. Kalaev, V. N.; Butorina, A. K.; Sheluchina, O. Y. Assessment of Anthropogenic Pollution in the Districts of Stary Oskol Based on Cytogenetic Indicators of Seed Seedlings of Weeping Birch. *Ecol. Genet.* **2006,** *4* (2), 9–23 (in Russian).
8. Yakymchuk, R. A. Cytogenetic After-Effects of Mutagen Soil Contamination with Emissions of Burshtynska Thermal Power Station. In *Ecological Consequences of Increasing Crop Productivity: Plant Breeding and Biotic Diversity*; Opalko, A. I., Weisfeld, L. I., Bekuzarova, S. A., Bome, N. A., Zaikov, G. E., Eds.; Apple Academic Press: Toronto, New Jersey, **2015;** pp 217–227.

9. Baranova, T. V.; Kalaev, V. N. Comparative Cytogenetic Analysis of Indigenous and Introduced Species of Woody Plants in Conditions of Anthropogenic Pollution. In *Heavy Metals and Other Pollutants in the Environment: Biological Aspects*; Zaikov, G. E., Weisfeld, L. I., Lisitsyn, E. M., Bekuzarova, S. A., Eds.; Apple Academic Press Inc.: Oakville, 2017; pp 241–254.

10. Motohashi, T.; Smirnov, S. V.; Kucev, M.; Shmakov, A. I.; Kondo, K. A. Study of Chromosome Numbers of *Petasites frigidus* (L.) Fries and *Petasites radiatus* (JF Gmel.) Toman (Asteraceae) of Altai, Russia. *Chromosome Botany* **2014**, *9* (3), 65–67. DOI:10.3199/iscb.9.65 (in Russian).

11. Kalendar, R. N.; Aizharkyn, K. S.; Khapilina, O. N.; Amenov, A. A.; Tagimanova, D. S. Plant Diversity and Transcriptional Variability Assessed by Retrotransposon-Based Molecular Markers. *Russ. J. Genet.: Appl. Res.* **2017**, *21* (1), 128–134 (in Russian).

12. Dar, T. H.; Raina, S. N.; Goel, S. Cytogenetic and Molecular Evidences Revealing Genomic Changes after Autopolyploidization: A Case Study of Synthetic Autotetraploid *Phlox drummondii* Hook. *Physiol. Mol. Biol. Plants* **2017**, *23* (3), 641–650.

13. Bukreeva, T. V.; Shavarda, A. L.; Morozov, M. A. *Aromadendranoids viridiflorol* and *leddiol* (Artifact) in *Ledum palustre* (Ericaceae) from the North-West Russia. *Plant Resour.* **2010**, *46* (4), 105–116 (in Russian).

14. Mizuta, D; Nakatsuka, A.; Kobayashi, N. Development of Multiplex PCR Markers to Distinguish Evergreen and Deciduous Azaleas. *Plant Breed.* **2008**, *127* (5), 533–535. DOI:10.1111/j.1439-0523.2008.01487.x.

15. Boubetra, K.; Amirouche, N.; Amirouche, R. Comparative Morphological and Cytogenetic Study of Five *Asparagus* (Asparagaceae) Species from *Algeria* Including the Endemic *A. altissimus* Munby. *Turk. J. Bot.* **2017**, *41* (6), 588–599.

16. Burmenko, J. V.; Baranova, T. V.; Kalaev, V. N.; Sorokopudov V. N. Cytogenetic Polymorphism of Seed Progeny of Introduced Plants on the Example of *Rhododendron ledebourii* Pojark. *Turczaninowia* **2018**, *21* (1), 164–173.

17. Weisfeld, L. I. About Cytogenetic Mechanism of Chemical Mutagenesis Station. In *Ecological Consequences of Increasing Crop Productivity: Plant Breeding and Biotic Diversity*; Opalko, A. I., Weisfeld, L. I., Bekuzarova, S. A., Bome, N. A., Zaikov, G. E., Eds.; Apple Academic Press: Toronto, New Jersey, 2015; pp 259–269.

18. Weisfeld, L. I. Importance of Discovery of I. A. Rapoport of Chemical Mutagenesis in the Study of Mechanism of Cytogenetic Effect of Mutagens. In *Ecological Consequences of Increasing Crop Productivity: Plant Breeding and Biotic Diversity*; Opalko, A. I., Weisfeld, L. I., Bekuzarova, S. A., Bome, N. A., Zaikov, G. E., Eds.; Apple Academic Press: Toronto, New Jersey, 2015; pp 271–274.

19. Weisfeld, L. I. Comparison of Chromosomal Rearrangements in Early Seedlings of *Crepis capillaris* L. after Treatment of Seeds by X-Rays and by the Chemical Mutagen. In *Chemical and Biochemical Technology: Materials, Processing, and Reliability*; Varfolomeev, S. D., Haghi, A. K., Zaikov, G. E., Eds.; Apple Academic Press: Toronto, New Jersey, 2015; pp 333–343.

20. Tsarev, A. P.; Pogiba, S. P.; Trenin, V. V.. *Forest Tree Breeding and Reproduction*; Logos: Moscow, 2003; 520 p. (in Russian).

21. Vostrikova, T. V. Instability of Cytogenetic Parameters and Genome Instability in *Betula pendula* Roth. *Russ. J. Ecol.* **2007**, *38* (2), 80–84 (in Russian).

22. Vostrikova, T. V.; Butorina, A. K. Cytogenetic Responses of Birch to Stress Factors. *Biol. Bull.* **2006,** *33* (2), 185–190.

23. Kuzemko, A. A. Influence of Anthropogenic Pressure on Environmental Characteristics of Meadow Habitats in the Forest and Forest-Steppe Zones. In *Temperate Crop Science and Breeding: Ecological and Genetic Studies*; Bekuzarova, S. A., Bome, N. A., Opalko, A. I., Weisfeld, L. I., Eds.; Apple Academic Press Inc.: Oakville, 2016; pp 385–404.

24. Chopikashvili, L. V.; Tsidaeva, T. I.; Skupnevsky, S. V.; Puкhaeva, E. G.; Bobyleva, L. A.; Rurua, F. K. Genetic Health of the Human Population as a Reflection of the Environment: Cytogenetic Analysis. *Temperate Crop Science and Breeding: Ecological and Genetic Studies*; Bekuzarova, S. A., Bome, N. A., Opalko, A. I., Weisfeld, L. I., Eds.; Apple Academic Press Inc.: Oakville, 2016; pp 287-302.

25. Opalko, A. I.; Opalko, O. A. Anthropo-Adaptability of Plants as a Basis Component of a New Wave of the "Green Revolution". In *Biologycal Systems, Biodiversity and Stability of Plant Communities*; Weisfeld, L. I., Opalko, A. I., Bome, N. A., Bekuzarova, S. A., Eds.; Apple Academic Press Inc.: Oakville; pp 3–18.

26. Bome, N. A.; Weisfeld, L. I.; Bekuzarova, S. A.; Bome, A. Y. Optimization of the Structurally Functional Changes in the Cultured Phytocoenoses in the Areas with Extreme Edaphic-Climatic Conditions. In *Biological Systems, Biodiversity and Stability of Plant Communities*; Weisfeld, L. I., Opalko, A. I., Bome, N. A., Bekuzarova, S. A., Eds.; Apple Academic Press Inc.: Oakville, 2015; pp 19–32.

27. Alov, I. A. *Cytophysiology and Pathology of Mitosis;* Medicine: Moscow, 1972; 263 p. (in Russian).

28. Kulaichev, A. P. *Methods and Means of Complex Data Analysis*; Forum Infra-M: Moscow, 2006; 521 p. (in Russian).

29. Simakov, E. A. About Post-irradiation Reparation of Cytogenetic Damages in Plantlets of Different Potatoes' Forms. *Radiobiology* **1983,** *23* (5), 703–706 (in Russian).

30. Arhipchuk, V. V. Using of Nucleolar Characteristics in Biotesting. *Cytol. Genet.* **1995,** *29* (3), 6–9.

31. Sobol, M. A. The Role of the Nucleolus in the Reactions of Plant Cells to the Action of Physical Environmental Factors. *Cytol. Genet.* **2001,** *35* (3), 72–84.

32. Sokolov, N. N.; Sidorov, B. N.; Durimanova, S. A. Genetic Control of DNA Replication in Chromosomes of Eukariotes. *Theor. Appl. Genet.* **1974,** *44* (5), 232–240. https://link.springer.com/article/10.1007%2FBF00274371?LI=true.

33. Butorina, A. K.; Isakov, Y. N. Puffing of Chromosomes in the Metaphase—Telophase of the Mitotic Cycle in *Quercus robur. Rep. Acad. Sci. USSR* **1989,** *308* (4), 987–988 (in Russian).

34. Butorina, A. K.; Kosichenko, N. E.; Isakov, Yu. N.; Pozhidaeva, I. M. The Effects of Irradiation from the Chernobyl Nuclear Power Plant Accident on the Cytogenetic Behaviour and Anatomy of Trees. In *Cytogenetic Studies of Forest Trees and Shrub Species*; Borzan, Z., Shlarbaum, S. E., Eds.; Croatian Forests: Zagreb, 1997; pp 211–226.

35. Nickols, W. W. Virus-Induced Chromosome Abnormalities. *Ann. Rev. Microbiol.* **1970,** *24,* 479–498.

36. Sheldon, S.; Speers, W. S.; Lehman J. Nucleolar Persistence in Embrional Carcinoma Cells. *Exp. Cell Res.* **1961,** *132* (1), 185–192.

37. Butorina, A. K.; Ermolaeva, O. V.; Cherkashina, O. N.; Mazurova, I. E.; Belousov, M V. Prospects for Using of Cytogenetic Analysis in Forestry for Example, Assessment of the Outlier Forests of Voronezh Region. *Biol. Bull. Rev.* **2008,** *128* (4), 400–408.
38. Skaptsov M. V.; Lomonosova M. N.; Kutsev M. G.; Smirnov S. V.; Shmakov A. I. The phenomenon of Endopolyploidy in Some Species of the Chenopodioideae (Amaranthaceae). *Bot. Lett.* **2017,** *164* (1), 47–53. DOI:10.1080/23818107.2016.1 276475.
39. Ramzan, F.; Younis, A.; Lim, K. B. Application of Genomic in Situ Hybridization in Horticultural Science. *Int. J. Genomics* **2017.** DOI:10.1155/2017/7561909.

CHAPTER 7

Anthropogenic Pollution Influence on the Antioxidant Activity in Leaves and on the Cytogenic Structures in the Seedlings of the Representatives of the *Rhododendron* Genus

TATYANA V. VOSTRIKOVA[1*], JULIYA V. BURMENKO[2*],
VLADISLAV N. KALAEV[3], and VLADIMIR N. SOROKOPUDOV[4]

[1]*All-Russian Research Institute of Sugar Beet and Sugar,
Federal Agency of Scientific Organizaions, 86, Ramonsky District,
Voronezh Region, 396030, Russia*

[2]*All-Russian Horticultural Institute for Breeding, Agrotechnology and
Nursery, 4 Zagor'evskaya St., Moscow, 115598 Russia*

[3]*Voronezh State University, Department of Genetics, Cytology and
Bioengineering, 1 University Sq., Voronezh 394018, Russia*

[4]*Russian State Agrarian University-Moscow Timiryazev Agricultural
Academy, 49 Timiryazevskaya st., Moscow 127550 Russia*

Corresponding author. E-mail: tanyavostric@rambler.ru

ABSTRACT

The species specificity of antioxidant activity (AOA) was found in members of the *Rhododendron* genus of the *Dauricum* series. The level of AOA varies within one conventional unit in different species (*Rhododendron mucronulatum* Turcz., *Rhododendron dauricum* L., *Rhododendron ledebourii* Pojark., *Rhododendron sichotense* Pojark.), as well as in the same species *Rhododendron ledebourii* growing on pollution-free and polluted area. The identified positive correlation of AOA level with the number of vacuolated cells in the apical root meristem, as well as with the level of mitosis pathologies, allow using these indicators for bioindication.

7.1 INTRODUCTION

Recently, the biochemical direction of research has been developing. The enzyme activity, component composition, antioxidant status of plants is studied in various conditions,[1–14] including influence of stress: oil, chemical, caused, for example, by heavy metals.[1–6] It is known that increasing in phenolic compounds' content is a physiological response to stress. Being influenced by heavy metal ions, changes occur at the cellular and subcellular levels, expressed in generation of reactive oxygen species and occurrence of oxidative stress,[15] various cell damages. On the other hand, it has been shown that an increase in the content of phenolic compounds and tocopherols, as well as activity of antioxidant enzymes, can significantly increase plant resistance to the toxic effect of heavy metal ions,[16] UV-B irradiation, and salinity.[17]

Phenolic compounds are special group of substances (secondary, stress metabolites), which are, on the one hand, reflecting the stress effect on the plant organism, on the other hand, many of them exhibit both allelopathic and protective, or antioxidant properties. Phenolic compounds are among the representatives of the secondary metabolites most widespread in the tissues of higher plants. They are involved in the basic life processes of plant cells: photosynthesis, respiration, the formation of cell walls, as well as protection from the effects of stress factors of biotic and abiotic origin.[18]

One-sided plant breeding for an increase in potential productivity usually results in a decrease in the resistance of cultivated plants to abiotic and biotic stress factors.[19–20] Species with higher potential productivity are more sensitive to environmental stressors; they are characterized by large amplitude of yield variability in adverse environmental conditions.[19,21–22] This trend in agriculture is quite understandable from the plant physiology's point of view. It is noted that productivity of fruit plant and its resistance to adverse factors are antipodal in nature, since the same metabolites are involved in their creation, but in different quantities.[23] Thus, productivity and resistance are formed on a base of the same photosynthesis products, but redistributed in different directions in accordance with the genetic regulation.[19]

Oxidative damage occurs when plants are exposed to various external factors or with a sharp change in the physiological state of the plant. Sufficient resistance to them is due to the existence in the plant cell of effective protective systems, which are based on antioxidants. The environmental aspect of the accumulation of antioxidants in plants begins to attract attention of researchers and needs further study. Antioxidant activity (AOA), reflecting the effect of all reducing agents of organic nature in the object, is associated with the effect of stress on the body.[24] The practical use of plants and products derived from them with AOA is becoming increasingly important in modern conditions. In connection with the search for sources of antioxidants for humans that protect the body from oxidative stress, and other biologically active substances that have an anticancerogenic effect; thus interest in this problem increases. A significant amount of such compounds is contained in different parts of conifers, plants of the families Ericaceae, Grossulariaceae, Amaranthaceae, and others. Low molecular weight phenolic compounds, for example, in *Amaranthus* leaves are represented by flavonoids, the dominant components of which are rutin, quercetin, and trifolene.[25] This genus plants' leaves contain a complex of water-soluble vitamins: ascorbic acid, niacin, vitamins of the B group. The green mass of Scarlet amaranth contains a significant amount of pectic substances. Flavonoids and pectins that make up amaranth leaves have AOA.[25] The antioxidant effect of flavonoid compounds due to their ability to bind free radicals and form compounds with metal ions (copper, iron), depriving them of their catalytic action in oxidation processes.[26] Pectins have a high biological activity: they affect cell growth, protect plants from drying out, increase their drought resistance, winter resistance, and protect them from pathogens effect.[27] The essential oils' component composition of some conifers[27] has been identified, in particular, representatives of the *Juniperus* genus,[7] as well as the leaves of the Erica representatives, that is *Ledum palustre*,[8] and previously unknown substances for these species have been extracted and identified.

The content of phenolic substances was noted in various plants, including valuable fruit of the *Ribes L.* genus (Grossulariaceae DC family),[28] as well as fancy species of the Erica family (Ericaceae Juss.) belonging to the *Rhododendron L.* genus.[9–14,29] The latter are rarely used in gardening, but some of them, for example, *Rh. ledebourii, Rh. mucronulatum, Rh. dauricum, Rh. sichotense, Rh. luteum,* are quite cold-resistant, drought-resistant, and at the same time exhibit biological activity. The composition of various plant parts of the *Rhododendron* genus contains essential oils (rhododendrol, rhododendren in amount of up to 0.1%), diterpenes, phenols, andromedotoxin, flavonoids, steroids, saponins, alkaloids, triterpenas, vitamins, phenolcarboxylic acids, benzoic acid, aldehydes, glycosides, glycosides.[30] High ability to synthesize secondary compounds is stated, including substances of a phenolic nature, quercetin, mircetin, hydroxybenzoic acid, hydroxycinnamic acid, chlorogenic acid, in some representatives of the genus: *Rh. luteum, Rh. japonicum, Rh. Smirnowii.*[30] The composition of the active compounds was established in species *Rh. dauricum, Rh. arboreum* Sm. *ssp. Nilagiricum.*[12–13] For example, in the north of the European part of the Russian Federation, leaves of naturally growing *Rhododendron adamsii* are used in herbal teas.[30] *Rhododendron luteum* exhibits allelopathic and cytostatic activity,[31] characterized by toxic substances content.[30] It was established that its allelopathic activity is due to the presence of substances of a phenolic nature (phenolcarboxylic acids, quercetin and rutin), amino acids, lectins, saponins. The latter have antibiotic and anticancer, and lectins exhibit cytostatic activity. Dzyuba noted greater cytostatic effect of the extracts of *Rh. luteum* compared with other species of the genus.[31]

The practical aspect of the research on AOA to protect against stress factors involves a combination of molecular genetics and biochemical analyzes. Sharoyko showed the biologically active action of a complex of extracts of golden rhododendron, mobilizing DNA repair and/or antioxidant protection systems, which led to the removal of the toxic action of lead.[32] Against the background of the action of stress factors on plant cells (radiation, temperature, nitrate, and nitrite anions). Fedorova established the modifying role of the golden rhododendron's extract in relation to antioxidant defense systems and the differential activity of the genome, leading to increase in their stability.[33]

Antioxidant characteristics of plants are offered to be used as test ones for monitoring. For example, when studying the accumulation of antioxidants (water-soluble antioxidants, carotenoids, anthocyanins) in plants under conditions of cadmium-contaminated urban environment, a decrease in antioxidant status was shown.[5] There is a positive correlation between the element content in plants and the intensity of traffic, as well as between the cadmium content and anthocyanin accumulation, a high negative correlation between the cadmium content in plants and their antioxidant status.[5]

Total content of phenols and flavonoids, as well as the antioxidant properties of crude extract and solvent fractions of *R. anthopogonoides* were determined using seven antioxidant analyzes. In addition, the protective effect of extracts was studied on hypoxia-induced damage in cells, and the correlation was analyzed between the content of phenols and flavonoids in extracts and their antioxidant properties.[34] Water and methanol extracts of *Rhododendron arboreum* leaves and *Rhododendron campanulatum* leaves were analyzed for phytochemical antioxidant and antiproliferative activity against cancer cell lines and repression of the growth factor in vascular endothelial cells, and high total phenol content, flavonoids, removal of free radicals, and decrease in energy activity were found in both plant extracts.[35] The ecological and geographical variability of the flavonoid composition of *Rhododendron parvifolium* was investigated depending on the edaphic conditions. It was shown that populations growing on rocky peaks differed from populations growing in swamps with a high content of flavonoids, including taxifolin glycosides, myricetin, quercetin, and free aglycons.[36]

However, information on the AOA of members of the *Rhododendron* genus of the *Dauricum* series under introduction conditions in the literature is fragmentary,[29,37,38] despite a detailed study of them in places of natural growth.[9-14] Complex works, which are now often carried out in various fields of science, have not been carried out according to the interrelation of cytogenetic and biochemical parameters of species of the *Rhododendron* genus in the Central Black Earth Region. The cytogenetic and molecular genetics features of the *Rhododendron* genus of the *Dauricum* series (subsections *Rhodorastrum* (Maxim.) Cullen sections of the Rhododendron section of the Rhododendron subgenus) were studied under conditions not typical of their natural growth.[39]

The purpose of this work was to identify the effect of anthropogenic pollution on the AOA in leaves and on cytogenetic indicators of seedlings in species of the *Rhododendron* genus of the *Dauricum* series.

7.2 MATERIALS AND METHODOLOGY

Species of the *Rhododendron* genus of the *Dauricum* series introduced in the Central Black Earth region were used as the objects to research, that are the pointed rhododendron (*Rh. mucronulatum* Turcz.), *Daurian* rhododendron (*Rh. dauricum* L.), *Ledebour* rhododendron (*Rh. ledebourii* Pojark). *Rh. sichotense* Pojark. in the botanical garden named after B.M. Kozo-Polyansky of the Voronezh State University. The sample consisted of five 30-year old plants of each species. Cytogenetic characteristics and AOA were investigated in *Rh. ledebourii* in non-polluted area (Botanical Garden of the Voronezh State University) and in the polluted area (Kominternovsky district of the Voronezh city) ranked in earlier studies.[40]

The material of the study was leaves of rhododendrons, which were collected at the end of September, when, according to literary data, second peak of AOA was noted in some woody plants.[29] Determination of AOA of rhododendrons was carried out according to the method for determining AOA of food products using Fe(III)/Fe(II) indicator system,[41] modified directly for species of the *Rhododendron* genus.[37,38] AOA was evaluated in arbitrary units compared with the standard; ascorbic acid was taken as a standard for our research.

For cytogenetic studies, seeds of a group collection of rhododendrons of each species were germinated in Petrie dishes at room temperature. When roots reached 0.5–1 cm length, they were fixed (at 9 o'clock in the morning) in aceto alcohol (a mixture of 96% ethanol and glacial acetic acid, 3:1), after which the material was stored in a refrigerator at a temperature of +4°C and prepared permanently pressed slide mounts according to the previously described method.[28]

The preparations (10–15 for each rhododendron species) were examined using LABOVAL-4 (Carl Zeiss, Jena) microscope with total magnification 40 × 1.5 × 10. About 500–700 cells were examined in each microsample (a slide mount corresponds to one root and one seedling). In total, about 38,500 cells of the studied species of the *Rhododendron* genus were examined.

The thickness of the pressed micropreparations can be different, which depends on several factors: degree of cell maceration, force of pressing, etc. To see cells at each level, it is necessary to use a micro screw to examine the pressed preparations (Figs. 7.1–7.3).

The following cytogenetic parameters were analyzed: mitotic activity (mitotic index (MI) is the ratio of number of dividing cells to total number of cells counted, %); percentage ratio of number of cells in mitosis stages; frequency of mitosis pathologies (ratio of number of cells with disorders to total number of dividing cells, %); proportion of vacuolation cells (ratio of number of vacuolation cells to total number of cells, %). The classification of pathological mitoses was performed according to Alov.[42] Computerized statistical data processing was performed using the Stadia 7.0 software package. The data grouping procedure and their processing are described by Kulaichev.[43]

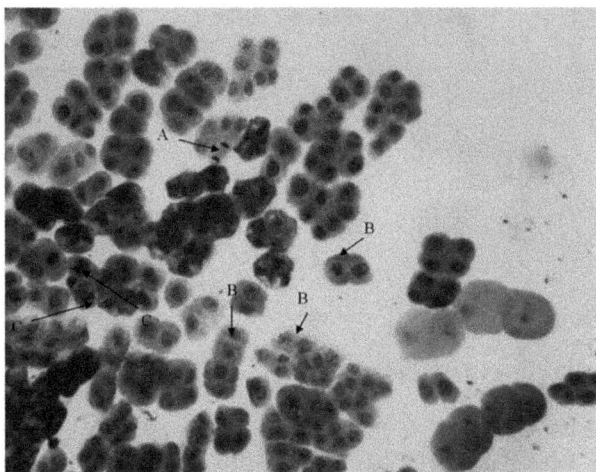

FIGURE 7.1 Stages of mitosis: anaphase (*A*), telophase (*B*), and pathological vacuolated cells (*C*) in the root meristem cells of *Rhododendron dauricum* L. with total magnification 40 × 1.5 × 10.

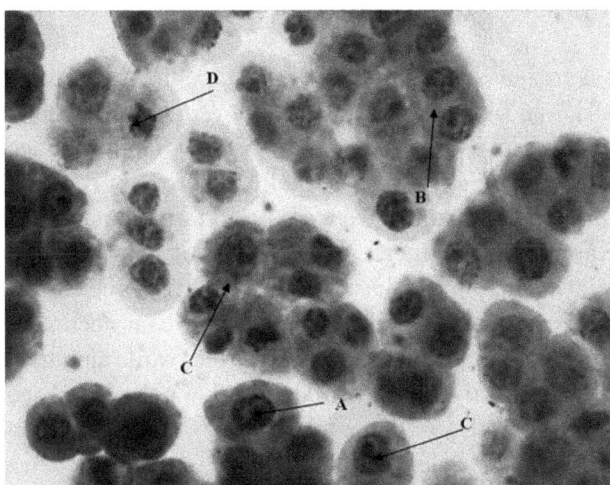

FIGURE 7.2 Prophase (*A*), telophase (*B*), pathologies vacuolated cells (*C*), and persistent nucleoli (*D*) in the root meristem cells of *Rhododendron mucronulatum* Turcz. with total magnification 40 × 1.5 × 10.

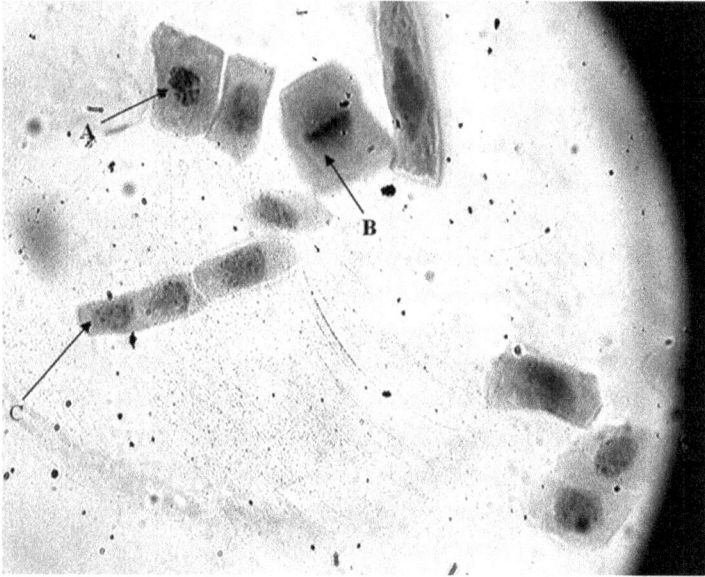

FIGURE 7.3   Stage of normal mitosis: prophase (*A*), metaphase (*B*) and telophase (*C*) in the root meristem cells of *Rhododendron sichotense* Pojark. with total magnification 40 × 1.5 × 10.

7.3   RESULTS AND DISCUSSION

Previously, a greater amount of phenolic compounds were found in leaves than in roots[31] and stems, except for *Rh. ledebourii*, who had almost the same content in all parts of plant.[29] The dynamics of accumulation of phenolic compounds in leaves of yellow European rhododendron and Asian Dahurian rhododendron were analyzed throughout growing season. There was a tendency to increase in content of phenolic compounds in leaves of yellow and Asian Dahurian rhododendron at the end of vegetation period.[9,10] Other authors also noted their significant accumulation in all tissues of stem, including core, to the rest period, while an important aspect of the study was not only the study of the level of polyphenol accumulation in plant tissues, but also the characteristics of their compartmentation.[29] Histochemical studies of leaves of annual shoots of rhododendrons taken at different stages of growing season (summer, winter) showed that phenolic compounds accumulated in epidermic, columnar mesophyll, and spongy parenchyma cells. In addition, polyphenols, predominantly of flavan nature, were found in the region of conducting beams, as well as in cells that form resin ducts and oil ducts.[29] Karpova and Karakulov[44] discovered a high degree of pairwise similarity in the composition of the phenolic compounds of the extracts and hydrolysates in pairs *Rh. sichotense—Rh. dauricum* and *Rh. adamsii—Rh. ledebourii*. Antioxidant activity and cytogenetic indices of *Rh. ledebourii* in the studied territories are presented in Table 7.1.

TABLE 7.1 Antioxidant Activity and Cytogenetic Indices ($\bar{x} \pm S\bar{x}$) of the *Rhododendron* Genus Representatives (*Rhododendron* L.).

Species	AOA	MI, %	MI exc. P, %	P, %	MP, %
Rh. dauricum	5.63 ± 0.02	7.6 ± 0.3	3.9 ± 0.2	48.9 ± 2.0	1.6 ± 0.4
Rh. ledebourii experiment	6.32 ± 0.03*	7.5 ± 0.4*	3.6 ± 0.5	52.2 ± 1.4**	5.5 ± 1.0*
Rh. ledebourii control	5.82 ± 0.03	6.1 ± 0.6	3.8 ± 0.4	37.9 ± 1.9	4.2 ± 1.1
Rh.mucronulatum	5.68 ± 0.02	7.7 ± 0.7	4.9 ± 0.7	37.5 ± 1.9	3.4 ± 0.3
Rh. sichotense	5.75 ± 0.03	5.6 ± 0.7	3.5 ± 0.4	37.5 ± 1.1	3.5 ± 0.5

AOA, antioxidant activity counted in conventional units (c.u.); 1 c.u., corresponds to the activity of 1 mg of ascorbic acid; MI, %, mitotic index, including prophase stage; MI exc. P, %, mitotic index, counted excluding prophase cells; MP, %, level of mitoses pathologies. *control differences are reliable ($p<0.05$). **control differences are reliable ($p<0.01$).

As it is follows from the data presented for *Rh. ledebourii*, an increase in mitotic index in cells of apical meristem of seedling roots was observed in the experiment compared to the control (differences are reliable ($p<0.05$)) due to a significant increase in number of cells at prophase stage (differences are reliable ($p<0.01$)), which indicates the delay of cells in prophase due to impaired mitotic apparatus and work of the checkpoint control system for integrity of genetic material.[45–47]

Other authors noted that technogenic dust pollution of growing area of parent plant of spear saltbush (*Atriplex patula* L.) led to a decrease in physiological parameters (germination, root length) of the offspring accompanied by an increase in the intensity of lipid peroxidation, lowing in molecular weight and activity of enzymatic antioxidants, also the rates of molecular genetics processes (DNA replication and repair, protein translation), against the background of a relatively small increase in mitotic Coy activity of seedling progeny seed cells compared to controls.[16] These changes are considered as adaptive reactions of spear saltbush

(*Atriplex patula* L.) to the effect of chronic anthropogenic stress factor of the environment in a certain interval of its intensity.[16]

The level of pathology of mitosis in seed progeny of *Rh. ledebourii* in the contaminated area was higher than in the environmentally clean (differences are reliable ($p<0.05$)) (Table 7.1). The increase in the proportion of cells at the prophase stage, according to Alov,[42] can be considered as a pathology of mitosis associated with chromosomal aberrations, which indicates serious damage to genetic material and the inability of cells to pass to the next stage of mitosis. The level of mitosis pathologies in cells of root meristem of seed progeny was determined as one of the criteria for cytogenetic monitoring, that is, bioindication index.[46–48] According to the classification of Alov,[42] all mitotic pathologies were divided in three groups: (1) pathologies related to chromosomal damage (delay of mitosis in prophase; disturbances of spiralization or despiralization of chromosomes; early disjunction of chromatids; chromosome fragmentation and pulverization; bridges; lagging chromosomes; formation of micronuclei; irregularities of chromosomal segregation; agglutinations of chromosomes); (2) pathologies related to injury of mitotic spindle (delay of mitosis in metaphase; C-mitosis; dispersion of chromosomes in metaphase; multipolar mitosis; three-group metaphase; asymmetric, monocentric mitosis); (3) disturbances of cytotomy (cytokinesis) (delay or absence of cytotomy; precocious cytotomy).[47]

We have noted an increase in number of vacuolation cells in seed progeny of *Rh. ledebourii* from the polluted area compared with the control (differences are reliable ($p<0.05$)). In the spectrum of mitosis pathologies lagging chromosomes in anaphase and metakinesis prevailed; agglutination of chromatin was found out in seed progeny of *Rh. ledebourii* from the polluted area, and bridges were rare (Table 7.1).

Target for impact of any extreme factor (at cellular level) are cell membranes and their components—proteins and lipids (at molecular level). At damage of membranes, there is an increase in release of some substances that previously could not overcome membrane barrier, for example, for phenols by increasing the permeability. On the other hand, damage to membrane leads to activation of enzymes involved in degradation of membrane components (lipids and proteins), and products increase total content of secreted substances. As a result of cell damage processes, lipids and proteins degrade.[49] Under influence of stress, volatile and nonvolatile compounds are formed in cell, which are referred to so-called stress metabolites, including terpenoids, phytohormones (gibberellinic and abscisic acids), phenols. At the same time, secondary metabolites are transformed into active compounds formed from a single precursor, phenylalanine. In physiologically active tissues of higher plants, phenolic compounds are found in vacuoles in the form of glycosides.[50] It has been established that plants have at least two vacuolar systems: one for storing spare substances, and the other for performing lytic functions.[51] Perhaps appearance of vacuolation cells in root meristem of seedlings, being an adaptive response of seed to stress, is caused by inclusion of a biochemical mechanism of antioxidant protection against damage.

Many authors believe that the presence of phenolic compounds in different parts of plants of the *Rhododendron* genus determines their biological activity.[10,41,52] The fact of localization of phenolic compounds in vacuoles[50] and presence in our studies of vacuolation cells in root meristem of seedlings (their higher number in seed progeny from an ecologically polluted area) leads to the conclusion that appearance of vacuolation cells (Figs. 7.2 C, 7.3 C) is a nonspecific cytogenetic reaction of woody plants we have noted in rhododendrons, as well as by other authors in Scots pine, for anthropogenic stress.[53] It has been shown that an increase in content of phenolic compounds and tocopherols, as well as activity of antioxidant enzymes, can significantly increase resistance of plants, determined by change in germination, root length to toxic effect of heavy metal ions.[16]

In our experiments, number of vacuolated cells in apical root meristem correlated with the level of mitosis pathologies ($r_s = 0.78$ ($p<0.05$)), therefore, this parameter, like the level of mitosis pathologies, may be a criterion for degree of contamination of the area. AOA correlated with number of vacuolation cells in apical root meristem ($r_s = 0.75$ ($p<0.05$)), as well as level of mitosis pathologies ($r_s = 0.83$ ($p<0.05$)), which allows using these indicators in bioindication.

Adaptation of plants to action of stressors depends both on functioning of antioxidant enzymes (catalase, peroxidase) and on accumulation in cells of low-molecular antioxidants (proline, ascorbic acid, phenols).[54,55] Under conditions of an urbanized environment, intensity of plant photosynthesis decreases, which affects content of ascorbic acid. Accumulation of ascorbic acid, as well as anthocyanin pigments in herbaceous plants of common oats, wild chervil, wild snakeroot, common dandelion, cocksfoot grass, common yarrow grown in soil pollution of railways and oil, was studied by Chupakhina and Maslennikov.[5] The authors proved that this increases effectiveness of antioxidant system of cell and there is an increase in plant resistance to effects of pollutants. The literature data are rather contradictory, which indicates species specificity of content and accumulation of phenolic compounds and antioxidants, as well as variability in reactions of organisms to stress factors.

7.4   CONCLUSIONS

1. The species specificity of AOA was found in representatives of the *Rhododendron* genus of the *Dauricum* series. The studied species of rhododendrons vary by content of antioxidant substances in leaves.
2. In different species, as well as the same species (*Rh. ledebourii*) growing in an ecologically clean and polluted area, the level of AOA varies within one conventional unit.
3. Direct correlations of AOA with the number of vacuolated cells in the apical root meristem, as well as with the level of mitosis pathologies were revealed, which allows using these indicators for bioindication.

KEYWORDS

- **essential oils**
- **phenols**
- **vacuolated cell**
- **micronuclei**

REFERENCES

1. Karimi, R.; Koulivand, M.; Ollat, N. Soluble Sugars, Phenolic Acids and Antioxidant Capacity of Grape Berries as Affected by Iron and Nitrogen. *Acta Physiol. Plant.* **2019,** *41*(7), 117.
2. Banaev, E. V.; Vysochina, G. I.; Kukushkina, T. A. Variability in the Content of Biologically Active Substances in the Leaves of *Nitraria sibirica* Pall. (Nitrariaceae). *Contemp. Problems Ecol.* **2014,** *7*(1), 90–96 (in Russian).

3. Kováčik, J.; Hedbavny, J. Ammonium Ions Affect Metal Toxicity in Chamomile Plants. *South Afr. J. Bot.* **2014**, *94*, 204–209.
4. Kalugina, O. V.; Mikhailova, T. A.; Shergina, O. V. Biochemical Adaptation of Scots Pine (*Pinus sylvestris* L.) to Technogenic Pollution. *Contemp. Problems Ecol.* **2018**, *11*(1), 79–88 (in Russian).
5. Chupakhina, G. N.; Maslennikov, P. V.; Maltseva, Ye. Yu.; Frolov, Ye. M.; Besserezhnova, M. I. Antioxidant Status of Plants Under Conditions of Cadmium Pollution in the Urban Environment. *Bull. Baltic Fed. Univ. I. Kant. Ser. Nat. Med. Sci.* **2011**, *7*, 16–23 (in Russian).
6. Wang, S.; Wang, L.; Zhou, Q.; Huang, X. Combined Effect and Mechanism of Acidity and Lead Ion on Soybean Biomass. *Biol. Trace Elem. Res.* **2013**, *156*(1–3), 298–307.
7. Zheljazkov, V. D.; Kacaniova, M.; Dincheva, I.; Radoukova, T.; Semerdjieva, I. B.; Astatkie, T.; Schlegel, V. Essential Oil Composition, Antioxidant and Antimicrobial Activity of the Galbuli of Six Juniper Species. *Ind. Crops Prod.* **2018**, *124*, 449–458.
8. Zhang, L.; Wang, H.; Wang, Y.; Xu, M.; Hu, X. Diurnal Effects on Chinese Wild *Ledum palustre* L. Essential Oil Yields and Composition. *J. Anal. Sci. Methods Instrum.* **2017**, *7*(02), 47.
9. Shrestha, A.; Rezk, A.; Said, I. H.; von Glasenapp, V.; Smith, R.; Ullrich, M. S.; H. Schepker.; Kuhnert, N. Comparison of the Polyphenolic Profile and Antibacterial Activity of the Leaves, Fruits and Flowers of *Rhododendron ambiguum* and *Rhododendron cinnabarinum*. *BMC Res. Notes* **2017**, *10*(1), 297.
10. Fandakli, S.; Yayli, N.; Kahriman, N.; Uzunalioğlu, E.; Çolak, N. U.; Yıldırım, S.; Yaşar, A. The Chemical Composition of the Essential Oil, SPME and Antimicrobial Activity of *Rhododendron caucasicum* Pall. *Rec. Nat. Prod.* **2019**, *13*(4), 316–323.
11. Shrestha, A.; Said, I. H.; Grimbs, A.; Thielen, N.; Lansing, L.; Schepker, H.; Kuhnert, N. Determination of Hydroxycinnamic Acids Present in *Rhododendron* Species. *Phytochemistry* **2017**, *144*, 216–225.
12. Cao, Y.; Chu, Q.; Ye, J. Chromatographic and Electrophoretic Methods for Pharmaceutically Active Compounds in *Rhododendron dauricum*. *J. Chromatography B.* **2004**, *812*(1–2), 231–240.
13. Kiruba, S.; Mahesh, M.; Nisha, S. R.; Paul, Z. M.; Jeeva, S. Phytochemical Analysis of the Flower Extracts of *Rhododendron arboreum* Sm. ssp. *nilagiricum* (Zenker) Tagg. *Asian Pac. J. Trop. Biomed.* **2011**, *1*(2), 284–286.
14. Wang, L.; Zhu, X.; Lou, X.; Zheng, F.; Feng, Y.; Liu, W.; Feng, F.; Xie, N. Systematic Characterization and Simultaneous Quantification of the Multiple Components of *Rhododendron dauricum* Based on High-Performance Liquid Chromatography with Quadrupole Time-of-Flight Tandem Mass Spectrometry. *J. Sep. Sci.* **2015**, *38*(18), 3161–3169.
15. Kaznina, N. M.; Titov, A. F.; Batova, Y. V.; Laidinen, G. F. The Resistance of Plants *Setaria veridis* (L.) Beauv. to the Influence of Cadmium. *Biol. Bull.* **2014**, *41*(5), 428–433 (in Russian).
16. Prokopiev, I. A.; Filippova, F. V.; Shein, A. A. Effect of Anthropogenic Pollution with Dust Containing Heavy Metals on Seed Progeny of Spear Saltbush. *Russ. J. Plant Physiol.* **2012**, *59*(2), 212–216 (in Russian).
17. Radyukina, N. L.; Toaima, V. I. M.; Zaripova, N. R. The Involvement of Low-Molecular Antioxidants in Cross-Adaptation of Medicine Plants to Successive

Action of UV-B Radiation and Salinity. *Russ. J. Plant Physiol.* **2012**, *59*(1), 71–78 (in Russian).

18. Zaprometov M. N. *Phenolic Compounds. Distribution, Metabolism and Function in Plants*; 'Nauka' (Science): Moscow, 1993; p 272 (in Russian).

19. Doroshenko, T. N.; Zakharchuk, N. V.; Ryazanova, L. G. *Adaptive Potential of Fruit Plants in Southern Russia*; Enlightenment—South: Krasnodar, 2010; p 140 (in Russian).

20. Rosielle, A. A.; Hamblin, J. Theoretical Aspects of Selection for Yield in Stress and Non-Stress Environment. *Crop Sci.* **1981**, *21*(6), 943–946.

21. Zhuchenko, A. A.; Ursul, A. D. *Strategy for Adaptive Intensification of Agricultural Production;* Ştiinţe: Chişinău, 1983; p 140 (in Moldova).

22. Diakov, A. B.; Trunova, M. V.; Vasilieva, T. A. Estimations of Yield Potentials and Drought-Resistance of Soybean Cultivars. *Oilseeds Sci. Tech. Bull. Russ. Res. Inst. Oilseeds* **2009**, *2*(141), 78–86 (in Russian).

23. Doroshenko, T. N. *Physiological and Environmental Aspects of Southern Fruit Growing;* Kuban SAU: Krasnodar, 2000; p 235 (in Russian).

24. Sorokina, I. V.; Krysin, A. P.; Khlyebnikova, T. B.; Kobrin, V. S.; Popova L. N. *The Role of Phenolic Antioxidants in Increasing the Stability of Organic Systems to Free Radical Oxidation*; SPSTL SB RAS: Novosibirsk, 1997; p 68 (in Russian).

25. Gins, M. S.; Gins, V. K.; Kononkov, P. F. Change in the Biochemical Composition of Amaranth Leaves During Selection for Increased Amaranthine Content. *Appl. Biochem. Microbiol.* **2002**, *38*(5), 474–479 (in Russian).

26. Faustova, N. M.; Kosman, V. M. Comprehensive Phytochemical Characteristics of *Amaranthus cruentus* (Amaranthaceae) Leaves. *Plant Res.* **2009**, *45*(4), 39–53 (in Russian).

27. Permyakova, G. V. Dinamics of Pectin Content in *Picea obovata* (Pinaceae) Bark. *Plant Res.* **2010**, *46*(4), 117–122 (in Russian).

28. Deineka, V. I.; Sorokopudov, V. N.; Deineka, L. A.; Shapashnik, E. I.; Burmenko J. V. Antioxidants of Belgorod State University Botanical Garden plants: *Ribes aureum* Fruits Anthocyanins. *Sci. Reports Belgorod State Univ. Series: Med. Pharm.* **2013**, *23*(18, 161), 225–228 (in Russian).

29. Kostina, V. M. Features of the Phenolic Metabolism of Plants of the Genus *Rhododendron* in vivo and in vitro: PhD Thesis in Biology; Moscow, 2009; p 22 (in Russian).

30. Aleksandrova, M. S. *Rhododendrons of the Natural Flora of the USSR;* "Nauka" (Science): Moscow, 1975; p 112 (in Russian).

31. Dzyuba, O. I. Physiological and Biochemical Features of *Rhododendron luteum* Sweet: Allelopathic Analysis: PhD Thesis; Kyiv, 2001; p 22 (in Ukrainian).

32. Sharoyko, V. V. Antioxidant and DNA Repair Systems in Protecting Cells from Exogenous and Endogenous Toxicants: Lead Cations, Phenols and Reactive Oxygen Species: PhD Thesis in Biology; Yakutsk, 2003, p 19 (in Russian).

33. Fedorova, A. I. The Role of Antioxidant and DNA Repair Systems in the Formation of the Response of Plant Cells Under the Influence of Stress Factors: Radiation, Temperature, Nitrate and Nitrite Anions: PhD Thesis in Biology; Yakutsk, 2004; p 18 (in Russian).

34. Jing, L.; Ma, H.; Fan, P.; Gao, R.; Jia, Z. Antioxidant Potential, Total Phenolic and Total Flavonoid Contents of *Rhododendron anthopogonoides* and its Protective

Effect on Hypoxia-Induced Injury in PC12 Cells. *BMC Complementary Altern. Med.* **2015,** *15*(1), 287.

35. Painuli, S.; Joshi, S.; Bhardwaj, A.; Meena, R. C.; Misra, K.; Rai, N.; Kumar, N. In vitro Antioxidant and Anticancer Activities of Leaf Extracts of *Rhododendron arboreum* and Rhododendron Campanulatum from Uttarakhand Region of India. *Pharmacogn. Mag.* **2018,** *14*(57), 294.

36. Karakulov, A. V.; Karpova, E. A.; Vasiliev, V. G. Ecological and Geographical Variation of Morphometric Parameters and Flavonoid Composition of *Rhododendron parvifolium. Turczaninowia* **2018,** *21*(2), 133–144 (in Russian).

37. Baranova, T. V Ways to Increase Resistance and Study the Antioxidant Activity of Representatives of the Genus *Rhododendron* L. *Bull. Krasnoyarsk State Agrar. Univ.* **2012,** *5*, 183–187 (in Russian).

38. Baranova, T. V; Sorokopudov, V. N.; Stupakov, A. G. The Antyoxidantive Activity of Same Introducents in Conditions of Black-Soil Region. *Sci. Rep. Belgorod State Univ. Series Nat. Sci.* **2012,** *21–1*(140), 78–81 (in Russian).

39. Baranova, T.; Kalendar, R.; Kalayev, N.; Sorokopudov, V.; Burmenko, J. Relationship Between Cytogenetic Characteristics and Molecular-Genetic Differences in Species of the Genus *Rhododendron* L. when Introduced. *Agric. Biol* **2018,** *53*(3), 511–520 (in Russian).

40. Kurolap, S. A.; Klepikov, O. V.; Dobrynina, I. V. Environmental Assessment of the Microclimate and Technogenic Pollution of the Air Basin of the City of Voronezh. *Prob. Reg. Ecol.* **2012,** *1*, 24–29 (in Russian).

41. Temerdashev, Z. A.; Khrapko, N. V.; Tsyupko, T. G.; Voronova, O. B.; Balaba, A. N. Determination of Anti-Oxidized Activity of Foodstuff with use of Fe(III)/Fe(II) Organic Reagent Indicating System. *Factory Laboratory. Material Diagnostics* **2006,** *72*(11), 15–19 (in Russian).

42. Alov, I. A. *Cytophysiology and Pathology of Mitosis;* Medicine: Moscow, 1972; p 263 (in Russian).

43. Kulaichev, A. P. *Methods and Means of Complex Data Analysis;* Forum Infra-M: Moscow, 2006; p 521 (in Russian).

44. Karpova, Ye. A.; Karakulov, A. V. Flavonoids of Some Rhododendron Species of Flora of Siberia and the Far East. *Chem. Plant Raw Mater.* **2013,** *2*, 119–126 (in Russian).

45. Lebedeva, L. I.; Fedorova, S. A.; Trunova, S. A.; Omelyanchuk L. V. Mitosis: Regulation and Organization of Cell Division. *Russ. J. Genet.* **2004,** *40*(12), 1589–1608 (in Russian).

46. Kalaev V. N.; Butorina A. K.; Sheluchina O. Yu. Assessment of Anthropogenic Pollution in the Districts of Stary Oskol Based on Cytogenetic Indicators of Seed Seedlings of Weeping Birch. *Ecol. Genet* **2006,** *4*(2), 9–23 (in Russian).

47. Baranova T. V.; Kalaev V. N. Comparative Cytogenetic Analysis of Indigenous and Introduced Species of Woody Plants in Conditions of Anthropogenic Pollution. In *Heavy Metals and Other Pollutants in the Environment: Biological Aspects*; Nourani, C. F., Zaikov, G. E., Weisfeld, L. I., Lisitsyn, E. M., Bekuzarova, S. A., Eds.; Apple Academic Press Inc.: Oakville, 2017; pp 241–254.

48. Butorina, A. K.; Kalaev, V. N.; Karpova, S. S. Cytogenetic Damage of Human Somatic Cells and Weeping Birch Cells in Voronezh Districts with Different Levels of Anthropogenic Pollution. *Russ. J. Ecol* **2002,** *6*, 438–441 (in Russian).

49. Malkhotra, S. S.; Khan, A. A. Biochemical and Physiological Effects of Priority Pollutants. In *Air Pollution and Plant Life*; Treshouyu, M., Eds.; Hydrometeoizdat: Leningrad, 1988; pp 144–189.

50. Fuksman, I. L. The Role of Substances of Secondary Metabolism in Indicating the State of Woody Plants Under Stress. *Plant Res.* **2003**, *39*(3), 153–161.

51. Alekhina, N. D.; Balnokin, Yu. V.; Gavrilenko, V. F.; Zhigalova, T. V.; Meychik, N. R.; Nosov, A. M.; Polesskaya, O. G.; Kharitonashvili, Ye. V.; Chub, V. V. In *Plant Physiology*; Yermakova, I. P., Ed.; Academy: Moscow, 2007; p 640 (in Russian).

52. Karpova, Ye. A.; Karakulov, A. V. Phenolic Compounds of Closely Related *Rhododendron* Species (Ericaceae). *Turczaninowia* **2011**, *14*(3), 145–149 (in Russian).

53. Belousov, M. V.; Shkurkina, N. N.; Mashkina, O. S.; Popov, V. N. Cytogenetic Reactions of Scots Pine to Lead Exposure as One of the Factors of Technogenic Stress. *Bull. Voronezh State Univ. Series Geogr. Geoecol.* **2011**, *1*, 129–131 (in Russian).

54. Cheeseman, J. M. Hydrogen Peroxide and Plant Stress: A Challenging Relationship. *Plant Stress* **2007**, *1*(1), 4–15.

55. Bukharina, I. L.; Zhuravleva, A. N.; Bolyshova, O. G. *Urban Plantations: Environmental Aspect*; Udmurt University Publishing House: Izhevsk, 2012; p 206 (in Russian).

CHAPTER 8

Cytogenetic Indices, Germination Ability, and Content of Total Protein in Seed Progeny of *Betula pendula* from Various Areas of Voronezh City with Different Levels of Anthropogenic Pressure

TATYANA V. VOSTRIKOVA[1*], OLGA A. ZEMLYANUKHINA[2], and VLADISLAV N. KALAEV[2]

[1]*All-Russian Research Institute of Sugar Beet and Sugar, Federal Agency of Scientific Organizations, 86, Ramonsky District, Voronezh Region, 396030, Russia*

[2]*Voronezh State University, Department of Genetics, Cytology and Bioengineering, 1 University Sq., Voronezh 394018, Russia*

Corresponding author. E-mail: tanyavostric@rambler.ru

ABSTRACT

The chapter presents results of study of cytogenetic indices, germination ability, and total protein content of seed progeny of *Betula pendula* in areas of Voronezh city (Central Black Earth Region of Russia) with different levels of anthropogenic pressure. The seed progeny of *Betula pendula* has an increased indicators of germination ability and total protein in areas with low pollution levels as compared to control group (seeds collected in ecologically clean territory) and to the same parameters for areas with high levels of anthropogenic pressure. The study established a high positive correlation between seed germination ability and amount of total soluble protein. The parameter "amount of total protein" serves as a seed germination marker determining their germination abilities. It is believed that stimulation of protein synthesis is happening in seed progeny of European white birch collected in the area with low pollution levels.

8.1 INTRODUCTION

Morphological,[1] geobotanical,[2] cytogenetic[3-5] studies, and modeling of plants' reactions to extreme factors in cultivated cenosis[6] are used to assess the environment situation. Modeling and extrapolation of results obtained on plants to humans have been used over recent years to forecast the environmental effect on health of population.[4,7-8]

European white birch (*Betula pendula* Roth) is a prevalent woody plant known for its medicinal properties and classified among main forest-forming species in European Russia. This species is a valuable material for the timber, wood chemical, pulp and paper, and other industries. It also has a high decorative value. Due to the content of biologically active substances, European white birch has a higher or lower phytoncidal activity in different periods of growing season[9-10] and is widely used in sustainable building.[11-13] The phytoncidal properties are especially important for planting in urban areas affected with technogenic pollution. Therefore, to fulfill this function, it is highly important to select plants resistant to anthropogenic pressure. European white birch is being actively studied in area of anatomy,[14-15] morphology,[12,13,16] biochemistry,[9-13] and cytogenetics.[3,4,7,17-19] As *Betula pendula* is a perennial plant, it may experience chronic impact from environmental mutagens, accumulate doses of mutagens, and be used in cytogenetic studies, being one of the most sensitive species for bioindication.[3-4,10,16,18,19]

The study of changing amount and composition of soluble protein in seeds is important as it is related to their quality and can be used to find the best storage conditions, and to study the agronomic features of woody plants. Earlier research showed an increase in highly soluble proteins in cytoplasm during the first hours of pollen tube growth.[20] The study of cytochemical properties of chromatin in germinating seeds in relation to their economically important traits has been conducted.[21] However, this study was primarily dedicated to agricultural crops: soy,[22] rape,[23] sunflower,[24] etc. Data about changing content of total protein in seeds of woody plants is scarce: *Acer negundo* L., *A. pseudoplatanus* L., *Fraxinus lanceolata* Borkh.[25] The correlation between speed of chromatin activation, germinating energy, and seed germination ability was established and special nature of these processes depending on the species and cultivars was found, which was demonstrated using various cultivars of peas, wheat, and corn.[21] Earlier research did not show a correlation between seed germination ability and amount of total soluble protein in seeds of woody plants.

Due to the deteriorating environmental situation in Voronezh city in 2010–2014 in particular[26] the effects of anthropogenic pressure on living organisms have been studied, including those on cellular and molecular level. The change in protection system indices (peroxidase activity, content of total protein, and aggregate number of sulfhydryl groups of protective peptides) in levels of European white birch and little-leaf linden under effects of automobile pollution was studied earlier in Nizhny Novgorod.[27] However, there was no comprehensive research of cytogenetic parameters, germination abilities, and amount of total protein in seed progeny, as well as an attempt to find a correlation between them. A comprehensive study of cytogenetic and biochemical indicators will allow a quick and adequate evaluation of seed quality, especially for plants whose seeds have a short storage life.

The purpose of this chapter is to study the amount of total soluble protein, germination abilities, and cytogenetic indicators for seed progeny of *Betula pendula* collected in ecologically clean territory and under anthropogenic pollution of different level.

8.2 MATERIALS AND METHODOLOGY

The research was conducted in two parts of Voronezh city (Central Black Earth Region) with different level of anthropogenic pollution: Central part, in Platonov street (area with a low pollution level) and Levoberezhny part on Leninsky avenue (area with a high pollution level). As a reference (ecologically clean) area we used the B. M. Kozo-Polyansky Botanical Garden of Voronezh State University, where environmental pollution is within permissible values.[28,29]

Seeds collected from 15 (5 in the ecologically clean and 5 in each of the polluted areas) trees of European white birch (*Betula pendula* Roth) were used as plant material for the research. Seeds were germinated and their germination abilities were assessed in accordance with the requirements of State standard.[30] The seeds were germinated at a constant temperature +25°C in three repeats 100 seeds each. The laboratory seed germination was determined as a ratio of number of germinated seeds to total number of seeds and was expressed in %, as recommended by the requirements of State standard.[30] In order to obtain protein preparations, a weighed portion of 1 seed (10 seeds from every tree were studied) was ground with glass sand in 0.1 M tris-NS1 buffer at pH 7.5, and centrifuged for 10 min at 20,000 g, 4°C, in a CM50 ELMI centrifuge (Latvia). The total soluble protein content was measured in supernatant following a standard method.[31] Bovine serum albumin (Sigma) was used as a standard.

When the roots reached a length of 0.5 to 1 cm, they were fixed at 9 a.m. in acetic alcohol: 3:1 mixture of 96% ethanol and glacial acetic acid, when mitotic activity of *Betula pendula* peaked.[32] After that, the material was kept refrigerated at a temperature of 4°C. Following the method described above, the roots of plantlets were used to make continuously squashed micropreparations using Hoyer's medium.[32]

The preparations were examined using a LABOVAL-4 (Carl Zeiss, Jena) microscope with a total magnification 40 × 1.5 × 10 и 100 × 1.5 × 10. The cytogenetic characteristics of 150 plantlets of *Betula pendula* (10 plantlets of seeds from each tree) collected in the ecologically clean and polluted areas were analyzed. About 500 cells were analyzed in each micropreparation (one preparation corresponds to one plantlet). In total, about 75,000 cells were analyzed.

The micropreparations were used to analyze the following cytogenetic parameters: mitotic activity by calculating the mitotic index (MI)—ratio of number of dividing cells to total number of calculated cells (%); percentage of cells at mitotic stages; level of mitotic pathologies: the ratio of number of aberrant cells with mitosis aberrations at stages of metaphase, anaphase, telophase to number of dividing cells, %; spectrum of mitotic pathologies: the percentage of every type of mitotic disturbances out of their total number, %; and level of cells with persistent nucleoli at stages of metaphase–anaphase of mitosis: ratio of number of cells with persistent nucleoli to total number of cells at the given stages, %. The classification of pathological mitosis was done according to Alov.[33] The results were processed statistically using the Stadia 7.0 software package. The procedure of data grouping and processing were described by Kulaichev.[34] A comparison of samples based on mitotic activity and number of cells in each mitosis stage was conducted using Student's t-test. A comparison of the samples was performed using the Van der Varden rank X-test, based on level of pathologies of mitosis, level of cells with persistent nucleoli, and amount of total soluble protein, as distribution of these indicators is nonparametric. The germination abilities of seeds were compared by Z-test for equality of frequencies. To estimate the influence of the "tree" on cytogenetic indices and amount of protein, a Kruskal–Wallis one-way parametric and nonparametric analysis of variance was performed. The power of influence was estimated according to Snedecor (in %). To calculate the correlation dependence, Spearman's rank correlation coefficient (r_s) was used.

8.3 RESULTS AND DISCUSSION

The content of total soluble protein in seeds of European white birch differed from tree to tree in the same area (in the same sample), that is, there were phenotypic differences in total soluble protein content (Table 8.1). This is in line with general idea that European white birch is a highly polymorphic species.[35] Furthermore, this assumption was proved by results of one-way ANOVA, which showed influence of the tree factor in all observed areas (Table 8.2). The biggest difference is between trees growing in clean and in the polluted areas, which might indicate that low stress can lead to synergistic effects that level the index, in clean and polluted environments there is an increase in phenotypic diversity in this characteristic.

We established that there is a considerable increase in the studied characteristic among trees growing in the area with low pollution as compared to the control area ($p<0.001$) and to trees in area with high anthropogenic pollution in Leninsky avenue (near "Voronezhsintezkauchuk") ($p<0.01$, $p<0.001$). The results obtained during our study are consistent with data presented by other researchers who pointed out that amount of total protein grows under stress.[37] For example, there is a strict correlation between content of soluble protein and polyphenols, whose character changes with increasing of technogenic pressure.[38] Moreover, stress-induced dehydrin proteins were detected in bud cells of *Betula pendula* in regions with contrastive climates.[39]

The largest amount of protein was found among birch trees from the area with low pollution (Table 8.1) which grow in conditions of low anthropogenic pressure. The following pollutants affect living organisms in Voronezh: SO_2, NO_2, and heavy metals of various concentrations depending on how polluted the area is. These areas were ranked (by the level of pollution) in earlier researches.[28,29,40]

TABLE 8.1 Content of Total Protein and Germination Abilities in Seeds of European White Birch (*Betula pendula*), Collected from Trees Growing in Different Ecological Environments.[36]

Tree number	Total protein in seeds, mg/mL	Total protein in plantlets, mg/mL	Germination capacity, %
	Botanical garden (control)		
1	0.168 ± 0.03	0.155 ± 0.03	28
2	0.182 ± 0.02*	0.160 ± 0.02	34
3	0.136 ± 0.01	0.148 ± 0.02	25
4	0.580 ± 0.01**	0.208 ± 0.03*	62
5	0.460 ± 0.06**	0.176 ± 0.03	55
Average	0.305±0.038	0.170 ± 0.02	40.8
	Platonov street (area with a low pollution level)		
1	0.730 ± 0.03*	0.164 ± 0.03	82
2	0.952 ± 0.05*	0.136 ± 0.02	94
3	0.606 ± 0.09	0.286 ± 0.04*	63
4	0.726 ± 0.07*	0.260 ± 0.03*	75
5	0.624 ± 0.05	0.225 ± 0.04**	68
Average	0.728 ± 0.035[b]	0.214 ± 0.03[a]	76.4[b]
	Leninsky avenue (area with a high pollution level)		
1	0.350 ± 0.02**	0.144 ± 0.04*	28
2	0.120 ± 0.01	0.160 ± 0.03*	12
3	0.380 ± 0.04**	0.135 ± 0.02	33
4	0.436 ± 0.03**	0.122 ± 0.03	44
5	0.558 ± 0.02**	0.118 ± 0.02	51
Average	0.369 ± 0.031[c]	0.136 ± 0.03[c]	33.6

Note: Differences between trees of the same sample are statistically significant at: *$p<0.05$; **$p<0.01$; Differences with the control are statistically significant at: [a]$p<0.01$; [b]$p<0.001$; [c]$p<0.001$.

TABLE 8.2 Influence of Factor "Tree" on Protein Content in Seeds and Plantlets of European White Birch from Areas of Voronezh City Differed in Level of Anthropogenic Pollution.

Area	Part of influence of factor "tree", %	
	In seeds	In plantlets
Botanical garden (control)	19.5**	18.3**
Platonov street (area with a low pollution level)	0.4*	0.6*
Leninsky avenue (area with a high pollution level)	19.4**	18.1**

Note: *influence is statistically significant ($p<0.01$); **influence is statistically significant ($p<0.001$).

Total protein content is considered to indicate ability of a biosystem to maintain normal balance between anabolic and catabolic processes.[27] It was shown that under stress, degradation of biomolecules predominate (predominance of catabolism over their biosynthesis, that is, anabolism) due to organism's increased energy demands to provide for phenotypic adaptation. Therefore, total protein content decreases as compared to control level.[27,41]

According to other authors, anthropogenic aromatic aerosols lead to an increase in soluble, enzyme-enriched protein fraction in leaves of resistant plants.[38] According to earlier research by other scientists, under regular impact by SO_2, NO_2, and heavy metals on woody plants, the correlation between albumins and globulins in seeds changes.[25] Two types of physiological and biochemical reaction responses were established. They are based either on conformational reorganizations in macromolecules without affecting the cell's gene regulation, or on adaptive change in biosynthesis. At the heart of these processes are protein and specific metabolisms, which can be affected by pollutants of different chemical natures.[38] In the other case, protein metabolism is shifted toward formation of high molecular proteins with low solubility, which is typical of senescent cells and tissues.[38] Environmental impact on maternal plants of European white birch and its seed progeny in the area with low pollution can be described as weak stress associated with moderate concentrations of SO_2 and NO_2 in the air and heavy metals in soil.[40] For example, according to Erofeyeva, [27] as a result of four seasons of monitoring (2010–2013), only in 2013 there was a monotonic increase in sulfhydryl groups in leaves and a nonmonotonic decrease in protein due to increased traffic. Other authors also observed a decrease in total protein[42] and an increase in aggregate number of sulfhydryl groups[43] due to various pollutants. In 2010–2012, different variants of paradoxical effects were revealed for these indices.[27] There can be two types of plant responses to environmental impact in terms of biochemical indicators: monotonic and nonmonotonic. Among nonmonotonic dose-effect relationships are hormesis[44–45] and paradoxical effects.[27,46–48] Hormesis is a two-phase dose-effect relationship at which small doses of influencing factor influence biological object in a stimulating (favorable) way and high doses of the factor affect the object in an inhibitive, that is, damaging way.[27,45,49]

Our study shows a strong stimulating effect on amount of total protein in the area with low pollution, which was increased 2.4 times as compared to the control group, and a weak stimulation in the area with high pollution (21% increase in the parameter as compared to the control group). Environmental impact on trees of European white birch and its seed progeny in areas with different ecological pressure and content of heavy metals[28,29] can be compared with the influence of microelements. For example, it was noticed that protein content in soy seeds affected by microelements increases by 1.5–5.1%. What is more, its maximum amount is accumulated under influence of boron, zinc, molybdenum, and a combined action of molybdenum and copper.[22]

It was shown that amount of total soluble protein in each studied tree is associated with seed germination ability: an increase in the first parameter leads to an increase in the second, and the other way round. It can be proved by high positive correlations between these indices ($r_s = 1$ ($p<0.01$)). The same way as for the index "amount of total soluble protein", birch trees in the area with low pollution had the highest seed germination ability, which was higher than the control value ($p<0,01$), which proved our assumption about stimulating effect of small pollutant doses on seed progeny. Positive correlation between the amount of total protein in seeds and their germination ability indicates that there is a relationship between intensity of growth processes (happening during seeds germination) and seed germination ability to protein supply. Therefore, the parameter "amount of total protein" (protein level) can be called a seed germination marker that determines its germination abilities. Evaluation of amount of total protein will allow to quickly and accurately determine seed germination abilities which are very important for plant industry, forestry, and other works that involve plant seeds with different storage lives.

Amount of total protein in plantlets characterizes metabolic processes that happen in seed progeny after seed germination. This parameter in plantlets of European white birch was lower (differences are significant ($p<0.001$)) than in seeds although trend to its changing in different environmental conditions remained (Table 8.1). Amount of total soluble protein in seeds collected in the area with high pollution was lower than for the control group and higher than in case of low pollution (Table 8.1). Presumably, synthetic processes in seed progeny collected in different environments do not differ in a qualitative way but differ in a quantitative way (in intensity) depending on how polluted the area is. To prove this assumption, cytogenetic characteristics of plantlets should be considered.

Our research showed that amount of total protein (biochemical parameter) is associated with cytogenetic characteristics. Cytogenetic parameters of seed progeny of *Betula pendula* under different ecological conditions are shown in Table 8.3.

The obtained data shows that *Betula pendula* has a decreased MI in the experiment as compared to the control group (differences are significant at $p<0.05$; $p<0.01$; $p<0.001$) and an increased level of mitotic pathologies and level of cells with persistent nucleoli (Fig. 8.1). Moreover, the percentage of mitotic pathologies in the area with high pollution was higher than in the area with low pollution (differences are significant ($p<0.01$)).

Therefore, the level of mitotic pathologies depends on level of pollution, which has been mentioned many times by us and other authors.[5,18,19,32] High correlations were established between the following indices: "MI calculated including cells at prophase stage" and "level of mitotic pathologies", $r_s = -0.85$ ($p<0.05$); "MI calculated including cells at prophase stage" and "level of cells with persistent nucleoli", $r_s = 0.95$ ($p<0.05$).

There is a change in duration of mitotic stages due to significant increase in number of cells at prophase stage (differences are significant ($p<0.001$)), which is more evident in the area with high pollution (Table 8.3). It means that cells are delayed at prophase due to mitotic apparatus disturbances and functioning of system of checkpoint control of genetic material integrity.[50] The delay of cells in prophase is associated with the following parameters: "MI calculated including cells at prophase stage" and "MI calculated excluding prophase cells" (delay of cells in prophase is accompanied by an increase in the former parameter and a decrease in the latter). There is a positive correlation between these parameters ($r_s = 0.9$ ($p<0.05$)). There is a simultaneous decrease in portion of cells in metaphase, anaphase, and

TABLE 8.3 Cytogenetic Parameters ($\bar{x} \pm S\bar{x}$) of Seed Progeny of *Betula pendula* in Different Areas of Voronezh.

Tree number	MI without P, %	MI, %	PM, %	PN, %	P, %	M, %	A, %	T, %
Botanical garden (control)								
1	3.1 ± 0.08	4.2 ± 0.1	2.5 ± 0.4	13.2 ± 1.1	27.3 ± 0.9	26.2 ± 1.2	10.9 ± 0.7	22.8 ± 0.9
2	3.0 ± 0.1	4.2 ± 0.2	4.2 ± 0.8	13.2 ± 1.2	27.9 ± 0.8	31.6 ± 1.3	12.2 ± 1.0	28.3 ± 0.8
3	2.7 ± 0.07	3.1 ± 0.08	2.3 ± 0.3	12.7 ± 1.0	29.7 ± 1.2	27.8 ± 1.2	10.2 ± 0.8	26.9 ± 0.8
4	2.8 ± 009	3.8 ± 0.09	4.6 ± 0.7	18.5 ± 1.3	27.9 ± 0.7	34.7 ± 1.4	12.1 ± 0.9	25.2 ± 0.9
5	2.8 ± 0.08	3.9 ± 0.1	4.4 ± 0.5	11.5 ± 1.1	27.7 ± 0.8	34.4 ± 1.3	12.0 ± 0.9	25.9 ± 0.7
Av	2.9 ± 0.07	4.0 ± 0.09	3.6 ± 0.5	13.8 ± 1.1	28.1 ± 0.4	31.0 ± 1.2	11.5 ± 0.4	25.8 ± 0.9
Platonov street (area with a low pollution level)								
1	1.8 ± 0.08	2.9 ± 0.1	16.2 ± 1.1	14.2 ± 1.2	36.0 ± 1.4	28.5 ± 1.2	8.9 ± 0.8	28.4 ± 1.3
2	1.8 ± 0.05	3.0 ± 0.07	16.8 ± 0.9	15.4 ± 1.3	39.4 ± 1.0	39.3 ± 1.0	8.5 ± 0.8	12.7 ± 1.1
3	2.4 ± 0.04	3.6 ± 0.06	10.0 ± 0.7	25.4 ± 1.0	35.5 ± 1.0	37.8 ± 1.1	9.2 ± 0.7	16.1 ± 1.2
4	2.1 ± 0.06	3.1 ± 0.08	12.8 ± 0.9	18.7 ± 1.1	32.2 ± 0.7	43.1 ± 1.1	6.3 ± 0.7	18.4 ± 1.2
5	2.4 ± 0.07	3.8 ± 0.1	11.7 ± 1.0	22.4 ± 1.1	35.5 ± 1.3	23.2 ± 1.0	13.7 ± 0.4	27.6 ± 1.3
Av	2.1 ± 0.1*	3.3 ± 0.2**	13.5 ± 1.3**	19.5 ± 1.2*	35.7 ± 1.1***	34.5 ± 1.3	9.3 ± 0.7	20.6 ± 1.2
Leninsky avenue (area with a high pollution level)								
1	1.8 ± 0.06	3.5 ± 0.08	19.5 ± 1.2	16.1 ± 0.8	48.4 ± 0.6	22.8 ± 1.1	5.4 ± 0.5	23.3 ± 1.1
2	2.1 ± 0.03	4.1 ± 0.04	16.4 ± 0.8	23.6 ± 0.8	49.3 ± 0.7	24.4 ± 1.1	10.9 ± 1.0	15.4 ± 1.1
3	1.7 ± 0.03	3.6 ± 0.06	21.0 ± 0.8	14.2 ± 0.6	52.6 ± 0.7	19.6 ± 1.0	5.4 ± 0.5	22.0 ± 1.0

TABLE 8.3 (*Continued*)

Tree number	MI without P, %	MI, %	PM, %	PN, %	P, %	M, %	A, %	T, %
4	1.8 ± 0.03	3.4 ± 0.07	26.6 ± 0.7	17.3 ± 0.9	48.6 ± 0.6	31.5 ± 1.1	6.9 ± 0.6	13.0 ± 0.9
5	1.5 ± 0.05	3.1 ± 0.07	26.6 ± 0.7	15.0 ± 0.4	51.0 ± 0.9	20.1 ± 1.0	9.2 ± 0.7	20.6 ± 1.0
Av	$1.8 \pm 0.1^{***}$	$3.5 \pm 0.2^{*}$	$22.3 \pm 1.1^{**}$[b]	$17.2 \pm 0.7^{*}$	$50.0 \pm 0.8^{***}$[c]	$23.7 \pm 1.1^{*}$[a]	$8.0 \pm 0.8^{**}$	$18.9 \pm 1.0^{**}$

Av, average value for the studied area; MI, %, mitotic index calculated including cells at prophase stage; MI without P, %, mitotic index calculated excluding prophase cells; MP, %, level of mitotic pathologies; PN, %, level of cells with persistent nucleoli; P, %, number of cells at prophase stage; M, %, number of cells at metaphase stage; A, %, number of cells at anaphase stage; T, %, number of cells at telophase stage.

Differences with the control are statistically significant at: $^{*}p<0.05$; $^{**}p<0.01$; $^{***}p<0.01$; differences with the low pollution area are statistically significant at: [a]$p<0.005$; [b]$p<0.01$; [c]$p<0.001$).

telophase in the area with high pollution, which is associated with the parameter "MI calculated excluding prophase cells" which, considering level of mitotic pathologies, indicates considerable changes in genetic material of maternal birch plants and their seed progeny (plantlets) as compared to those in the area with low pollution.

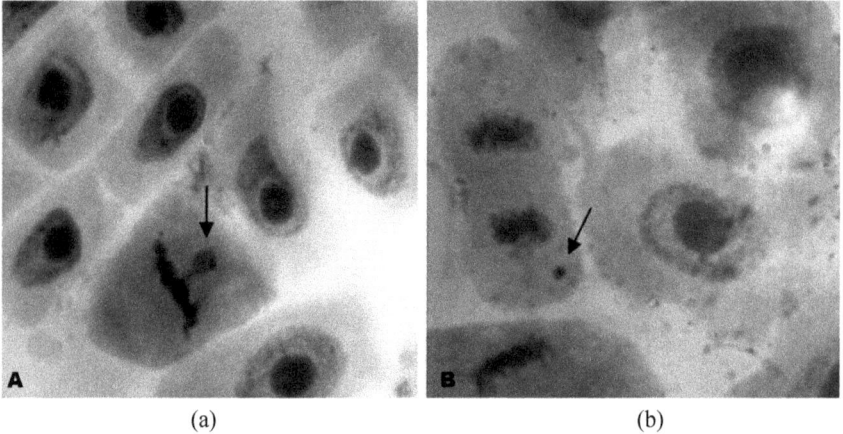

(a)   (b)

FIGURE 8.1   Persistent nucleoli at stages of metaphase (a), anaphase (b) of mitosis in root meristem cells of *Betula pendula*.

An increase in number of cells with persistent nucleoli in root meristem cells of plantlets of European white birch at metaphase, anaphase, and telophase stages of mitosis in the experiment areas as compared to the control group might be explained by influence of anthropogenic pollution on the studied objects. Persistent nucleoli look like round or drop-shaped structures connected with chromosomes at metaphase, anaphase, and early telophase of mitosis, when nuclear membrane is absent and chromosomes are significantly reduced. Phenomenon of appearance of persistent nucleoli at such mitosis phases as metaphase, anaphase and early telophase[51] is not typical for normal course of division, because, as a rule, nucleolus disappears at prophase and recovers only at late telophase.[52] According to Butorina et al.,[32,53] persistent nucleoli at metaphase, anaphase, and telophase of mitosis can be called an individual case of nucleolar activity enhancement, an adaptive mechanism for exposure to unfavorable environment. An increase in number of cells with persistent nucleoli indicates an enhanced nucleolar activity due to additional synthetic activity of ribosomal genes, similar to synthesis of "heat shock" proteins. Anthropogenic impact on maternal plants can be compared to influence of high temperature on living organisms which is stressful for them and leads to synthesis of stress proteins,[54] heat shock proteins, whose synthesis can be induced not only by heat shock but also by a number of other unfavorable impacts, including those of heavy metals.[37]

It was established that the tree factor influences cytogenetic indices of plantlets from seeds of European white birch growing in the areas of Voronezh with different levels of anthropogenic pollution (Table 8.4). Maximum effects can be seen in the area with a low pollution. The results can be explained by difference in individual reactions of seed progeny of some trees, which might mean that there are alternative cytogenetic mechanisms enabling stability of mitotic cycle. What is more, the difference is more evident in the case of a low level of pollution.

TABLE 8.4 Influence of the "Tree" Factor on Cytogenetic Indices of Plantlets from Seeds of European White Birch Growing in the Areas of Voronezh with Different Levels of Anthropogenic Pollution.

Indicators	Control	Area with low pollution	Area with high pollution
MI	–	***	*
MI without P	–	**	–
PM	–	–	–
PN	*	***	–
P	–	***	–
M	–	***	–
A	–	*	–
T	–	**	**

Note: Abbreviation see Table 8.3;—influence is statistically significant at: $*p<0.05$; $**p<0.01$; $***p<0.001$

Level of cells with persistent nucleoli at metaphase, anaphase, and telophase of mitosis, as well as amount of total protein for seed progeny of European white willow in the area with low pollution considerably exceeds control value. Earlier, other authors established that moderate doses of pollutants increase total protein content in plants activating synthesis of protective proteins (stress proteins, antioxidant system enzymes),[55] whereas large doses, on the contrary, decrease protein content which is associated with an increased production of reactive oxygen species which oxidize proteins.[27,56,57] We associate an increase in level of cells with persistent nucleoli at metaphase, anaphase, and telophase of mitosis with additional reserves enabling efficient protein synthesis during adaptation of the organism (at cellular level) to anthropogenic pollution.

This assumption is proved by high correlations between indicators "amount of total soluble protein" and "level of cells with persistent nucleoli at metaphase, anaphase, and telophase of mitosis," "amount of total soluble protein" and "MI" ($r_s=0.9$ ($p<0.05$)), as well as "amount of total soluble protein" and "level of mitotic pathologies" ($r_s=1$ ($p<0.01$)). Despite a higher level of mitotic pathologies in root meristem cells of plantlets of European white birch from areas with low pollution as compared to the control group (Table 8.3; $p<0.01$), the seeds have a high germination ability and have a better germination as compared to those in the control area ($p<0.01$) (Table 8.1).

Level of mitotic pathologies in root meristem cells of plantlets from areas with high pollution considerably exceeds the control parameter ($p<0.01$) and the same parameter for seed progeny for the area with low pollution (Table 8.3; $p<0.01$), although there is no difference in germination as compared to the control group (Table 8.1). However, it was noticed that MI during the experiment was lower than the control (Table 8.3), which means that germinating seeds, being living organisms, seek reserve growth capacities, that is, mechanisms responsible for adaptation to environmental factors are triggered. At cellular level, it results in an increased intensity of metabolic processes: an increase in amount of total protein, nucleolar activity (an increased number of cells with persistent nucleoli at metaphase, anaphase, and telophase of mitosis). Therefore, we can say that quality of seed material collected in the area with low pollution was lower than in the control area; however, it was higher than that in the area with high pollution. Due to a high level of mitotic pathologies in root meristem cells of plantlets of European white birch from polluted areas we can indicate genetic instability of maternal plants. Therefore, it is not recommended to collect pharmacognosy material and seeds for planting material from the analyzed trees. However, they can be used for selection purposes as an initial material for the production of new types (as mutable samples) in artificial mutagenesis.[58-61]

8.4   CONCLUSIONS

Abilities to germination and content of total protein in seeds of *Betula pendula*, collected from trees growing in different ecological environments, were investigated. Reliable increase content of total protein connected with germination ability was shown. Therefore, in seed progeny of European white birch collected in the area with low pollution levels might have stimulation of protein synthesis in plantlets in compare with control. We suggest that anthropogenic pollutants (heavy metals) in low doses play role as stimulating agents. These maternal trees and their progeny can be used as a source of biologically active substances and for collection of plant material for pharmacognostic purposes, as well as a stable species to be used for planting of urban areas.

A comprehensive study of cytogenetic and biochemical parameters allows one to evaluate quickly and adequately of seed quality of unknown origin which is especially important for plant industry and forestry as it allows selecting stable planting material and for collection of material for pharmacognostic purposes in areas with different ecological pressure. Cytogenetic indicators and estimation of amount of total protein make it possible to predict seed germination in European white birch.

KEYWORDS

- mitotic pathologies
- prophase
- nucleoli
- anthropogenic pollution

REFERENCES

1. Opalko A. I.; Opalko O. A. Anthropo-Adaptability of Plants as a Basis Component of a New Communities; Weisfeld, L. I.; Opalko, A. I.; Bome, N. A.; Bekuzarova, S. A., Eds.; Apple Academic Press Inc.: Oakville, 2015; pp 3–18.

2. Kuzemko, A. A. Influence of Anthropogenic Pressure on Environmental Characteristics of Meadow Habitats in the Forest and Forest-Steppe Zones. In *Temperate Crop Science and Breeding: Ecological and Genetic Studies;* Bekuzarova, S. A.; Bome, N. A.; Opalko, A. I., Weisfeld, L. I., Eds.; Apple Academic Press Inc.: Oakville; 2016, pp 385–404.

3. Kalaev, V. N.; Butorina, A. K.; Sheluchina, O. Yu. Assessment of Anthropogenic Pollution in the Districts of Staryy Oskol Based on Cytogenetic Indicators of Seed Seedlings of European White Birch. *Ecol. Genet.* **2006,** *4*(2), 9–23 (In Russian).

4. Baranova, T. V.; Kalaev, V. N. Comparative Cytogenetic Analysis of Indigenous and Introduced Species of Woody Plants in Conditions of Anthropogenic Pollution. *Heavy Metals and Other Pollutants in the Environment: Biological Aspects*; Zaikov, G. E.; Weisfeld, L. I.; Lisitsyn, E. M.; Bekuzarova, S. A., Eds.; Apple Academic Press Inc.: Oakville, 2017; pp 241–254.

5. Yakymchuk, R. A. Cytogenetic After-Effects of Mutagen Soil Contamination with Emissions of Burshtynska Thermal Power Station. In *Ecological Consequences of Increasing Crop Productivity: Plant Breeding and Biotic Diversity*; Opalko, A. I.; Weisfeld, L. I.; Bekuzarova, S. A.; Bome, N. A.; Zaikov, G. E., Eds.; Apple Academic Press Inc.: Toronto, New Jersey, 2015; pp 217–227.

6. Bome, N. A.; Weisfeld, L. I.; Bekuzarova, S. A.; Bome, A. Y. Optimization of the Structurally Functional Changes in the Cultured Phytocoenoses in the Areas with Extreme Edaphic-Climatic Conditions *Biological Systems, Biodiversity and Stability of Plant Communities;* L. I. Weisfeld; A. I. Opalko; N. A. Bome; S. A. Bekuzarova, Eds.; Apple Academic Press Inc.: Oakville, 2015; pp 19–32.

7. Chopikashvili L. I.; Tsidaeva, T. I.; Skupnevsky,, S. V.; Puкhaeva, E. G.; Bobyleva, L. A.; Rurua, F. K. Genetic Health of the Human Population as a Reflection of the Environment: Cytogenetic Analysis. In *Temperate Crop Science and Breeding: Ecological and Genetic Studies;* Bekuzarova, S. A.; Bome, N. A.; Opalko, A. I., Weisfeld, L. I., Eds.; Apple Academic Press Inc.: Oakville, 2016; pp 287–302.

8. Butorina, A. K.; Kalaev, V. N.; Karpova, S. S. Cytogenetic Damage of Human Somatic Cells and European white Birch Cells in Voronezh Districts with Different Levels of Anthropogenic Pollution. *Russ. J. Ecol.* **2002,** *33*(6), 413–416.

9. Vetchinnikova L. V. *Birch: Questions of Variability (Morphophysiological and Biochemical Aspects)*; Science: Moscow, 2004, p 183 (in Russian).

10. Bukharina, I. L. Features of the Dynamics of the Content of Ascorbic Acid and Tannins in the Shoots of Woody Plants Under the Conditions of the City of Izhevskю. *Plant Res.* **2011,** *47*(2), 109–117.

11. Neverova, O. A. Ecological Assessment of the State of Woody Plants and Environmental Pollution of an Industrial City (by the Example of Kemerovo): Abstract of Doctor of Biology. M., 2004; p 37 (in Russian).

12. Bukharina, I. L.; Zhuravleva, A. N.; Bolyshova, O. G. *Urban Plantations: The Environmental Aspect: A Monograph*; Udmurt University Publishing House: Izhevsk, 2012; p 206 (in Russian).

13. Erofeeva, Ye. A.; Naumova, M. M. Seasonal Dynamics of the Morphophysiological Indicators of the Sheet *Betula pendula* (Betulacea) During Road Pollution. *Plant Res.* **2012,** *48*(1), 59–70 (in Russian).

14. Makhnev, A. K.; Degtyarev, E. S.; Migalina, S. Intraspecific Variability of *Betula pendula* Roth in Triterpenes in Leaves. *Sib. J. Ecol.* **2012**, *19*(2), 237–244 (in Russian).
15. Migalina, S. V. Changes in the Morphology and Structure of the Leaf *Betula pendula* Roth and *Betula pubescens* Ehrh. When Adapting to Climate: PhD Thesis in Biology, Ekaterinburg, 2011; p 21 (in Russian).
16. Erofeeva, E. A. Fluctuating Asymmetry of the Sheet Betula pendula (Betulacea) in the Conditions of Road Pollution (Nizhny Novgorod). *Plant Res.* **2015**, *51*(3), 366–383.
17. Butorina, A. K.; Kormilitsyna, E. V. Cytogenetic and Anatomical Features of Birch Hanging from a Thirty-Kilometer Zone of the Novovoronezh NPP. *Ontogenez* **2001**, *32*(6), 428–433 (in Russian).
18. Vostrikova, T. V. Instability of Cytogenetic Parameters and Genome Instability in *Betula pendula* Roth. *Russ. J. Ecol.* **2007**, *38*(2), 80–84. DOI: 10.1134/S1067413607020026 (in Russian).
19. Kalaev, V. N.; Karpova, S. S.; Artyukhov, V. G. Cytogenetic Characteristics of European white Birch (*Betula pendula* Roth) Seed Progeny in Different Ecological Conditions. *Bioremediation Biodivers. Bioavailab.* **2010**, *4*(1), 77–83 (in Russian).
20. Ilchenko, K. V. The Change in the Quantity and Quality of the Protein Composition in the Subcellular Compartments of Growing Tobacco Pollen Tubes. 2 Congress Vses. Islands of Physiologists, Minsk, September 24–29, 1990: Proc. Report Part 2. M., 1990, 1992; p 84 (in Russian).
21. Troyan, V. M.; Savorona, T. A.; Ilchenko, L. N.; Kalinin, F. L. Investigation of the Cytochemical Properties of Chromatin of Cells of Germinating Plant Seeds in Connection with Economically Important Traits. 2 Congress All. Islands of Physiologists, Minsk, September 24–29, 1990: Proc. report Part 2. M., 1990, 1992; p 212 (in Russian).
22. Korsunova, M. N.; Leplyavchenko, L. P.; Onishchenko, L. M. Determination of Protein and Fat Content in Soybean Seeds on the Background of Micronutrients. *News High. Educ. Inst. Food Technol.* **2000**, *2–3*, 11–12 (in Russian).
23. Murtazaliyeva, M. K.; Abakargadzhieva, P. R. The Study of the Fractional Composition of Proteins of Rapeseed Izvestia of Dagestan State Pedagogical University. *Nat. Exact Sci.* **2011**, *3*, 29–30 (in Russian).
24. Minakova, A. D.; Shcherbakov, V. G.; Lobanov, V. G. Biochemical Changes in Proteins During Storage of Sunflower Seeds. *News High. Educ. Inst. Food Technol.* **1996**, *1–2*, 16–18 (in Russian).
25. Yusypiva, T. I.; Koval, Yu. P. The Influence of Technogenesis on the Total Protein Content and the Ratio of Protein Fractions in the Seeds of Woody Plants. *Vestnik Dnipropetrovsk Univ. Biol. Med.* **2011**, *2*(3), 123–127 (in Ukrainian).
26. Report *"On the State of Sanitary and Epidemiological Welfare of the Population in the Voronezh Region in 2015"*. Voronezh: Office of the Federal Service for Supervision of Consumer Rights Protection and Human Welfare in the Voronezh Region, 2016; p 209 http://36.rospotrebnadzor.ru/download/apxiv/gd2015.pdf (appeal date 10.03.2018) (in Russian).

27. Erofeeva, E. A. Hormesis and Paradoxical Effects in Plants Under Conditions of Motor Pollution and Under the Action of Pollutants in the Experiment: Dis. for the Degree of Doctor of Biology. Nizhny Novgorod, 2016; p 184 (in Russian).
28. *Medical and Environmental Atlas of the Voronezh Region*; Kurolap, S. A., Mamchik, N. P., Klepikov, O. V., Eds.; Publishing House "Istoki": Voronezh, 2010; p 167 (in Russian).
29. Kurolap, S. A.; Klepikov, O. V.; Dobrynina, I. V. Environmental Assessment of the Microclimate of Technogenic Pollution of the Air Basin of the City of Voronezh. *Problems Reg. Ecol.* **2012**, *1*, 24–29 (in Russian).
30. GOST 13056.6-97. *Seeds of Trees and Shrubs. Methods for Determining the Germination*; Publishing House of Standards, 1997; p 28 (in Russian).
31. Bradford, V. V. A Rapid and Sensitive Method for the Quantities of Protein Utilizing the Principle of Protein-Dye Binding. *Anal. Biochem.* **1976**, *72*(4), 417–422.
32. Vostrikova, T. V.; Butorina, A. K. Cytogenetic Responses of Birch to Stress Factors. *Biol. Bull.* **2006**, *33*(2), 85–190. DOI: 10.1134/S1062359006020142 (in Russian).
33. Alov, I. A. *Cytophysiology and Pathology of Mitosis*; Medicine: Moscow, 1972; p 232 (in Russian).
34. Kulaichev, A. P. Methods and Tools for Integrated Data Analysis. M.: FORUM: INFA-M, 2006; p 512 (in Russian).
35. Popov, V. K. *Birch Forests of the Central Forest-Steppe of Russia*; Voronezh Publishing House State University: Voronezh, 2003; p 424 (in Russian).
36. Vostrikova, T. V.; Zemlyanukhina O. A.; Kalaev, V. N. Use of Physiological-Biochemical, Cellular and Sub-Cellular Characteristics of *Betula pendula* as Markers of Seed Germination and Monitoring Territory Pollution. *Periódico Tchê Química* **2019**, *16*(32), 1034–1044.
37. Chirkova, T. V. *Physiological Basis of Plant Resistance: A Manual for Students of Biological Faculties of Universities*; SPb: SPSU, 2002; p 244 (in Russian).
38. Sidorovich, E. A.; Getko, N. V. Protein and Specific Metabolism-the Basis of the Formation of Tolerance of Plants in Man-Made Environment. 2th Congress All Islands of physiologists, Proc. Report, Part 2, Minsk, 1992; p 192 (in Russian).
39. Tatarinova, T. D.; Bubyakina, V. V.; Vetchinnikova, L. V.; Perk, A. A.; Ponomarev, A. G.; Vasilyeva, I. V. Stressful Dehydrin Proteins in the Birch Buds in Climatically Contrasting Regions. *Cell Tissue Biol.* **2017**, *11*(6), 483–488. DOI: 10.1134/S1990519X17060098.
40. Fedorova, A. I.; Shunelko, E. V. Pollution of the Surface Horizons of the Soil of the City of Voronezh with Heavy Metals. *Prossiding Voronezh State Univers. Series Geogr. Geoecol.* **2003**, *1*, 74–82 (in Russian).
41. Garkavi, L. Kh.; Kvakina, E. B.; Ukolova, M. A. *Adaptive Reactions and Body Resistance*. Rostov-on-Don: Because of the Rostov University, 1990; p 375 (in Russian).
42. Singh, S.; Bhatia, A.; Tomer, R.; Kumar, V.; Singh, B.; Singh, S. D. Synergistic Action of Tropospheric Ozone and Carbon Dioxide on Yield and Nutritional Quality of Indian Mustard (*Brassica juncea* (L.) Czern.). *Environ. Monit. Assess.* **2013**, *185*(8), 6517–6529.
43. Ding, X.; Jiang, J.; Wang, Y. Y.; Wang, W. Q.; Ru, B. G. Bioconcentration of Cadmium in Water Hyacinth (*Eichhornia-crassipes*) in Relation to Thiol-Group Content. *Environ. Pollut.* **1994**, *84*(1), 93–96.

44. Calabrese, E. J. Hormesis: Why it is Important to Toxicology and Toxicologists. *Environ. Toxicol. Chemi.* **2008,** *27*(7), 1451–1474.
45. Cedergreen, N.; Streibig, J. C.; Kudsk, P.; Mathiassen, K.; Duke, S. O. The Occurrence of Hormesis in Plants and Algae. *Dose Response* **2007,** *5*, 150–162.
46. Schatz, A. More on Paradoxical Effects. *Fluoride* **1999,** *32*(1), 43–44.
47. Batyan, A. N.; Frumin, G. T.; Bazylev, V. N. Basics of General and Environmental Toxicology; SPb: SpecLit., 2009; p 352.
48. Smith, S. W.; Hauben, M.; Aronson, J. K. Paradoxical and Bidirectional Drug Effects. *Drug Saf* **2012,** *35*(3), 173–189. DOI: 10.2165/11597710-000000000-00000.
49. Eidus, L. Kh. On the mechanism of the Nonspecific Reaction of Cells to the Action of Damaging Agents and the Nature of Hormesis. *Biophysics* **2005,** *50*(4), 693–703.
50. Lebedeva, I. N.; Fedorova, S. A.; Trunova, S. A.; Omelyanchuk, L.V. Mitosis. Regulation and Organization of the Division of the Cell Nucleus. *Genetics* **2004,** *40*(12), 1589–1608.
51. Sokolov, N. N.; Sidorov, B. N.; Durimanova, S. A. Genetic Control of DNA, Replication in Chromosomes of Eukaryotes. *Theor. Appl. Genet.* **1974,** *44*, 232–240.
52. Butorina, A. K.; Isakov, Yu. N. Puffing of Chromosomes in the Metaphase – Telophase of the Mitotic Cycle in *Quercus robur. Reports Acad. Sci. USSR* **1989,** *9*(4), 987–988 (in Russian).
53. Butorina, A. K.; Kosichenko, N. E.; Isakov, Yu. N.; Pozhidaeva, I. M. The Effects of Irradiation from the Chernobyl Nuclear Power Plant Accident on the Cytogenetic Behaviour and Anatomy of Trees. In *Cytogenetic Studies of Forest Trees and Shrub Species*; Zagreb: Croatia, 1997; pp 211–226.
54. Kulaeva, O. N.; Mikulovich, T. P.; Khokhlova, V. A. Stress Proteins of Plants. *Modern Biochemistry Issues/Academy of Sciences of the USSR. Institute of Biochemistry*, 1991; p 229 (in Russian).
55. Srivastava, S.; Mishra, S.; Dwivedi, S.; Baghel, V. S.; Verma, S.; Tandon, P. K.; Rai, U. N.; Tripathi R. D. Nickel Phytoremediation Potential of Broad Bean *Vicia Faba* L. and its Biochemical Responses. *Bull. Environ. Contam. Toxicol.* **2005,** *74*, 715–724.
56. Arvind, P., Prasad, M. N. V. Cadmium–Zinc Interactions in a Hydroponic System Using *Ceratophyllum demersum* L.: Adaptive Ecophysiology, Biochemistry and Molecular Toxicology. *Braz. J. Plant Physiol.* **2005,** *17*, 3–20.
57. El-Khatib, A. A.; Hegazy, A. K.; Abo-El-Kassem, A. M. Bioaccumulation Potential and Physiological Responses of Aquatic Macrophytes to Pb Pollution. *Int. J. Phytoremediat.* **2014,** *16*(1), 29–45. DOI: 10.1080/15226514.2012.751355
58. Weisfeld, L. I. About Cytogenetic Mechanism of Chemical Mutagenesis. *Ecological Consequences of Increasing Crop Productivity: Plant Breeding and Biotic Diversity*; Opalko, A. I.; Weisfeld, L. I.; Bekuzarova, S. A.; Bome, N. A.; Zaikov, G. E., Eds.; Apple Academic Press Inc.,: Toronto, New Jersey, 2015, pp 259–269.
59. Weisfeld, L. I. Importance of Discovery of I. A. Rapoport of Chemical Mutagenesis in the Study of Mechanism of Cytogenetic Effect of Mutagens. *Ecological Consequences of Increasing Crop Productivity: Plant Breeding and Biotic Diversity*; Opalko, A. I.; Weisfeld, L. I.; Bekuzarova, S. A.; Bome, N. A.; Zaikov, G. E., Eds.; Apple Academic Press Inc.: Toronto, New Jersey, 2015; pp 271–274.
60. Weisfeld, L. I. Comparison of Chromosomal Rearrangements in Early Seedlings of *Crepis capillaris* L. After Treatment of Seeds by X-Rays and by the Chemical

Mutagen. In *Chemical and Biochemical Technology: Materials, Processing, and Reliability*; Varfolomeev, S. D.; Haghi, A. K.; Zaikov, G. E., Eds.; Apple Academic Press Inc.: Toronto, New Jersey, 2015; pp 333–343.

61. Weisfeld, L. I. Plant Genetic Resources as a Factor in Assessing the Stability of Agroecosystems. *Radiation and Chemical Monitoring, Bioindication of the State of the Environment*: Mat. Scientific Practical Conf. (from the International Part), Devoted to the Year of Ecology of Russia (Tobolsk, November 16–17), Tobolsk, 2017; 95–99 (in Russian).

PART III

Diagnosis of Environmental Status and the Adaptive Potential of Plants

CHAPTER 9

Photoindication of Plant Nitrogen Nutrition

RAFAIL A. AFANAS'EV, GENRIETTA E. MERZLAYA, and
MICHAIL O. SMIRNOV*

*D. N. Pryanishnikov All-Russian Scientific Research Institute of
Agrochemistry, 31A Pryanishnikov St., Moscow, 127550 Russia*

Corresponding author. E-mail: User53530@yandex.ru

ABSTRACT

The advantages of photometric diagnostics of plant nitrogen nutrition
in comparison with chemical methods are considered. The experimental
substantiation of photometric diagnostics while using ground and remote
methods of diagnostic works in sowings of cereal and other agricultural
crops is given.

9.1 INTRODUCTION

Express diagnostics of plant nitrogen nutrition is an important part of
modern agricultural technologies of agricultural crops' cultivation,
especially of winter and spring crops, vegetables, potatoes. It makes it
possible to timely identify the need of crops in nitrogen, since it is almost
impossible to determine it by other methods. Not only the productivity of
the phytocenosis depends on the timely diagnostics, but also the ecology
of the environment, since excess of nitrogen, when applying fertilizer to
the nutrition "by eye," is a powerful destabilizing factor of metabolism in
the soil–plant systems, causing the underdevelopment or lodging of plants,

nitrate-nitrite toxicosis, contamination of soils, products, and groundwater with nitrates and nitrites, carcinogenic nitrosamines. In nitrogen-overfilled plants, that is often observed in practice, they are more damaged by fungal diseases and various pests. The lack of nitrogen nutrition has a negative impact on crop yields, as well as on the quality of products, especially its availability with protein, essential amino acids, many vitamins, other biologically active substances (carotenoids, flavonoids, enzymes), causes an excess of indigestible fiber (in feed).

Of great importance is the economic component of the rational use of nitrogen fertilizers, since both mineral fertilizers and nitrogen-containing organic fertilizers at the cost of their acquisition, transportation and application currently occupy up to 30–40% of the farm production cost. The amount of production associated with the yield of culture; its cost is economic categories. In addition, the quality of agricultural products is usually also an important economic criterion for its sale. Thus, accurate diagnostic determination and application of nitrogen fertilizer doses for cultivated crops to optimize their nitrogen status from an economic point of view is a priority technological method.

9.1.1   CHEMICAL METHOD FOR DIAGNOSING NITROGEN NUTRITION AVAILABILITY OF CROPS

In previous years, to determine the security of crop nitrogen nutrition, mainly winter and spring cereals, so-called soil and plant diagnostics was used.[1] It consisted in selection of soil samples on the field by layer to a 1 m depth in the autumn or early spring period and chemical analysis of their content of nitrogen mineral forms—nitrates and ammonium or only nitrates. In the shooting phase of cereal crops, then earing—flowering plant samples were selected in a field, and concentration of nitrate nitrogen in the segments of plant stems was determined by semiquantitative method on the intensity of their coloring with a solution of diphenylamine in strong sulfuric acid. In the phase of earing—the beginning of grain filling selection of plant samples (flag leaves from cereals) in the field and their chemical analysis for total nitrogen content are required again. As a result of the stage-by-stage soil and plant diagnostics, doses of nitrogen fertilizing were established, if required, to increase the yield and quality of agricultural products. Despite the relatively high accuracy of the described diagnostic techniques, due to the results of long-term researches of many research groups, they are not widespread, neither in our country nor in other countries due to the high complexity of implementation, significant expenditures of manual labor, and unsafe for the health of laboratory personnel associated with the use of highly concentrated acids and alkalis, and also because these methods require special training of personnel, laboratory equipment, cost, time, which is limited by the timing of both the actual diagnostics and fertilizing plants with fertilizers. Because of these difficulties, the practical possibilities of rapid diagnostics of plant nitrogen nutrition are extremely limited, and now its scale in the industrial conditions is actually reduced to zero.

9.1.2 PHOTOMETRIC METHOD FOR ESTIMATION OF NITROGEN NUTRITION

The way out of this situation was found by switching from chemical methods of diagnostics of plant nitrogen nutrition to physical, namely to photometry of both individual plants and crops as a whole.[2]

Photometric equipment based on determination of intensity of plant green color can be used to identify the supply of plants with nitrogen nutrition, since this indicator is closely related to the level of nitrogen availability of plants. In this case, the most informative is light reflected from the vegetating plants within red and near infrared region of spectrum (Fig. 9.1).

Photometry should also take into account the stages of plant ontogenesis, since the content of chlorophyll in the leaves, that is, the intensity of their green color, depends on the age of the plants.[4]

The highest accuracy of nitrogen nutrition diagnostics is determined with so-called normalized differentiated vegetation index (NDVI), which is calculated by points as the ratio of the difference between the photometry indexes in the near infrared (NIR) and red (RED) regions of the radiation spectrum to the sum of these parameters: "nir—red/nir + red." Depending on the subject, the value can range from −1 to +1. As a rule, the reflection from uncovered soil varies from −0.05 to + 0.05 points, from plants—from 0.4 to 0.9 points, from sowing with incomplete projective soil coverage—from 0.3 to 0.8. According to the diagnostic survey of crops in certain phases of vegetation, and according to the results of testing (calibration) of certain photometric devices, that is, by determining the nitrogen nutrition availability of plants with help of indicators (points) of a particular type of photometer, doses of nitrogen fertilizing can be assigned.

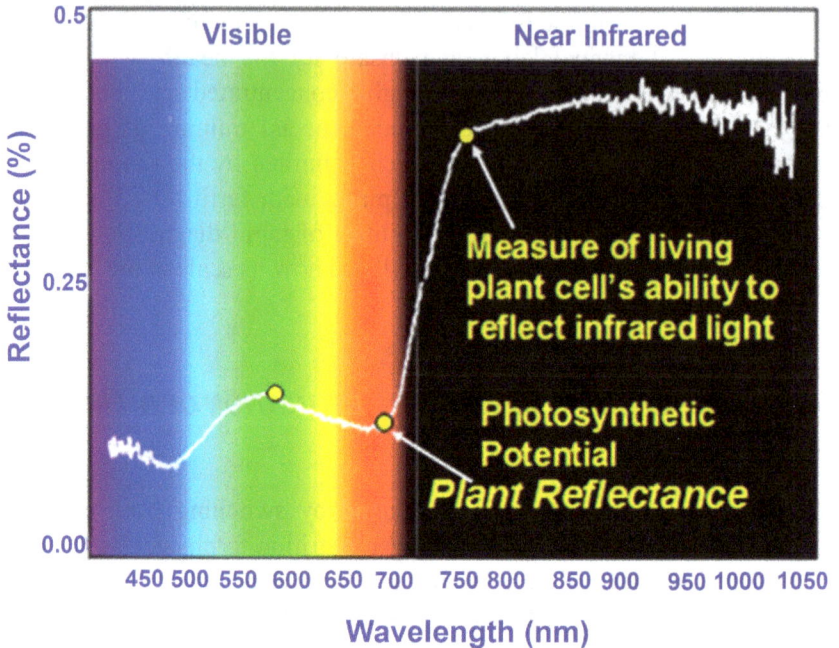

FIGURE 9.1   The nature of the reflection of sunlight by plants.[3]

The advantage of this technique is that by rapid diagnostics, nitrogen nutrition of plants is measured using a portable device to determine the need of plants for nitrogen fertilization in industrial conditions. The need for nitrogen nutrition of plants is established depending on the ratio of chlorophyll fluorescence of the plant leaf to its light permeability, which is determined by the photometric characteristic of the leaf on at least 40 plants recorded by the device. Monitoring of nitrogen supply is carried out by clamping the leaf of the diagnosed plant in the appropriate unit of the device and its transmission by optical radiation in the range $\lambda = 400 \pm 50$ nm with registration of optical characteristics of light transmission through the plate of one developed plant leaf. Monitoring is carried out on 40 randomly selected plants, in different randomly selected areas of the same crop and variety, which allows us to give a conclusion about the need for urgent application of nitrogen fertilizer. With the normalized transmittance values less than 1.00 ± 0.01 registered by the device, a conclusion is given about the need for urgent application of nitrogen fertilizer in doses up to

60–90 kg/ha of the active substance with an indicator of 1.00 ± 0.01 and more, nitrogen fertilizer is not required. The necessary result is achieved by the fact that the proposed method and device provide for leaf diagnostics of nitrogen nutrition of agricultural crops, since green leaves are integral bodies of information about mineral, including nitrogen nutrition of plants. The method is based on the determination of the chlorophyll concentration in the plant leaves by the intensity of its fluorescence and transparency (light permeability) of the leaf blades, which also depends on the level of chlorophyll content.[5]

Indicator organs of individual plants can be diagnosed with portable photometers, and crops—by land portable, mobile, or remotely—from satellites and aircraft. However, despite the significant advantages over traditional chemical diagnostics, both ground and space indication of nitrogen supply of crops have certain limitations: ground—on the scale of coverage, aerospace—on the time parameters.

9.1.3   APPLICATION OF UNMANNED AERIAL VEHICLES

In this regard, according to the authors, the most promising is the use of unmanned aerial vehicles (UAVs) for the rapid diagnostics of plant nitrogen nutrition.

The widespread use of UAVs in modern agriculture is due to a number of their advantages. UAVs are able to collect planting information which is sufficient for accurate application of pesticides and herbicides where chemicals are needed. This promises farmers the opportunity to save on the use of agrochemicals and also saves the environment. UAVs allow you to create a cartographic basis with the exact coordinates of all objects, which will further conduct a visual analysis of objects with a resolution of up to several centimeters per pixel. On this basis, it will be possible to apply vector layers: fields, infrastructure, and roads. Such a framework allows the calculation of accurate area, distances, resource needs, etc., and also to define the objective area of arable lands, hayfields, pastures, lea lands, fallows, fall-plowed lands, crops, thin sowings, and additional sowings.

Aerial photography with UAV more detailed than a space picture. Spatial resolution of images is possible even in centimeters, due to changes in altitude above the ground. In addition, UAVs allow shooting even in cloudy conditions, which is not available to satellites and makes it difficult to use aviation, and it is possible to adjust the flight of UAV in real time, if the customer needs it.

This clearly shows the advantage of aerial photography over space photography, for which such an atmospheric phenomenon is a serious obstacle. UAV performance reaches 30 km^2 in 1 h with area survey and up to 35 km/h for linear objects.

When using UAVs agricultural workers provide significant savings in research costs and gain in time compared to all other types of surveillance: ground survey, satellite photos, and the use of manned aircraft. Another advantage—efficiency: the footage taken in the morning by lunch was at the disposal of agronomists and farm managers.

The data obtained allow us to judge the state of crops, the moisture content in the soil, the amount of fertilizers that still need to be applied, and even predict the yield. A key role is played with the NDVI, calculated on the basis of spectral survey. The main purpose of our research was to establish quantitative dependences of vegetation indices (NDVI) on the availability of nitrogen nutrition for its regulation by carrying out appropriate nitrogen fertilizing of vegetative sowings of winter wheat.

9.2　MATERIALS AND METHODOLOGY

In 2016 and 2017, during the growing season, observations were made on the growth and development of winter wheat plants (cv. Moscow) in field experiments with increasing doses of nitrogen fertilizers laid at the Central experiment station of the D. N. Pryanishnikov All-Russian Scientific Research Institute of Agrochemistry. The tillering coefficient at the stage of plants entering the tube was 1.5; the number of ear-bearing stems according to the variants of the field experiment was 450–520 pieces per 1 m^2. In variants of the field experience put in threefold replication, the supply of plants with nitrogen nutrition was determined by chemical method (stem diagnostics) and two physical methods using a portable photometer "Yara" and multispectral equipment of an UAV of "Agrodrongroup" limited liability company (LLC) (Fig. 9.2). Mathematical processing was conducted according to Ref [6].

According to the research results, orthophotos of field experiments were obtained (Fig. 9.3).

FIGURE 9.2 Survey of field experience with UAV manufactured by "Agrodrongroup" (2017). (Picture of the authors)

(A)	(B)

FIGURE 9.3 The orthophotos of field experiments conducted at the Central experiment station of D. N. Pryanishnikov All-Russian Scientific Research Institute of Agrochemistry 2016 (A) and 2017 (B) on the diagnostics of nitrogen nutrition of winter wheat using UAVs LLC "Agrodrongroup." (Picture of the authors).

Since the main objective of the research was to study the possibility of using UAVs for rapid diagnostics of crop nitrogen nutrition, the experiment used a periodic flight of the experimental sowing of winter wheat as a test culture by an UAV fixing the reflection from sowing in the green (550 nm), red (660 nm), near (735 nm), and far infrared (790 nm) regions of the electromagnetic spectrum. Along with remote diagnostics to verify the adequacy of ground-based methods and for calibration of the UAV equipment photometric diagnostics with the use of N-tester "Yara" and stem diagnostics with the application of diphenylamine[7–9] was conducted in parallel. In addition, the relationship of photometric indicators with the biological activity of the soil was studied by quantifying the intensity of carbon dioxide emissions, depending on the level of nitrogen fertilizer application for winter wheat. This chapter describes the scientific experience and practical proposals for the calibration and use of photometric devices installed on the UAV.

9.3   RESULTS AND DISCUSSION

Large-scale survey of winter wheat crops in field experiment with unmanned aircraft of LLC "Agrodrongroup" revealed a close relationship of the magnitude of the NDVI (a measure of the amount of photosynthetic active biomass, which is calculated by points as ratio of the difference between the photometry indexes in the NIR and RED regions of the emission spectrum to the sum of these indicators: "nir − red/nir + red"), calculated according to the results of shooting with the UAV, with the points of ground-based photometry (Fig. 9.4).

Doses of nitrogen fertilizers: N0—0 kg/ha; N30—30 kg/ha; N60—60 kg/ha; N90—90 kg/ha N120—120 kg/ha.

The coefficient of pair linear correlation (r) between the vegetation index (NDVI) and nitrogen doses was 0.84, between the vegetation index and photometry scores was 0.97. According to the classification accepted in domestic agronomy, it is considered that at r <0.3 correlation dependence between indicators is weak, at r = 0.3–0.7—average, at r >0,7-strong.[6] Despite minor differences in yields in a number of field experience options, the high dependence of yield and other diagnostic indicators of agrocenosis on the value of vegetation indices (NDVI), indicates a high sensitivity of the vegetation index to changes in the state of sowing of winter wheat. Thus, remote, using UAVs, the definition of NDVI to determine the availability of nitrogen nutrition for plants is statistically significant. The connection of vegetation indices with the data of stem diagnostics and winter wheat yield was high (Table 9.1).

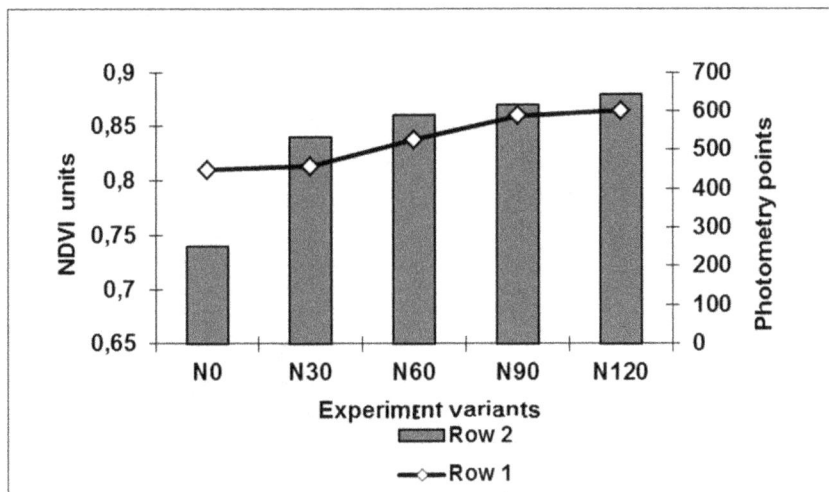

FIGURE 9.4 Dependence of NDVI (normalized differentiated vegetation index) and "Jara" photometry points on doses of nitrogen fertilizers applied to winter wheat. Legend: Row 1—points of the photometry, row 2—units of NDVI obtained by the apparatus of the unmanned aerial vehicle (UAV).

TABLE 9.1 Linkage of Diagnostics Indicators of Nitrogen Nutrition of Winter Wheat, Performed with the use of UAVs, with Other Indicators of Nitrogen Supply in the Shooting Phase.

NDVI indications, points	Dose of nitrogen, kg/ha	Photometer "Yara" indications, points	Data of stem diagnostics, indexes	Emission of CO_2, g/m² per day	Winter wheat yield, t/ha
0.65	0	356	0.00	7.7	2.83
0.81	30	511	0.87	8.2	4.41
0.84	60	541	1.40	10.1	4.99
0.85	90	580	2.50	9.7	5.00
0.86	120	620	2.70	12.3	5.10
Correlation coefficient (r)	0.84	0.97	0.86	0.73	0.99

Note: indices of stem diagnostics: (1) strong need for nitrogen; (2) weak need for nitrogen; (3) nitrogen nutrition is sufficient.

It is noteworthy that the vegetation index calculated from the UAV remote survey results was associated even with the biological activity of the soil, determined by the emission of carbon dioxide from the soil at a pair linear correlation coefficient r = 0.73 (Fig. 9.5).

This relationship is due to the fact that nitrogen fertilizers applied into the soil, enhanced the processes of mineralization of soil organic matter, consisting, as is known, more than half of the carbon, which during oxidation and turned into carbon dioxide.

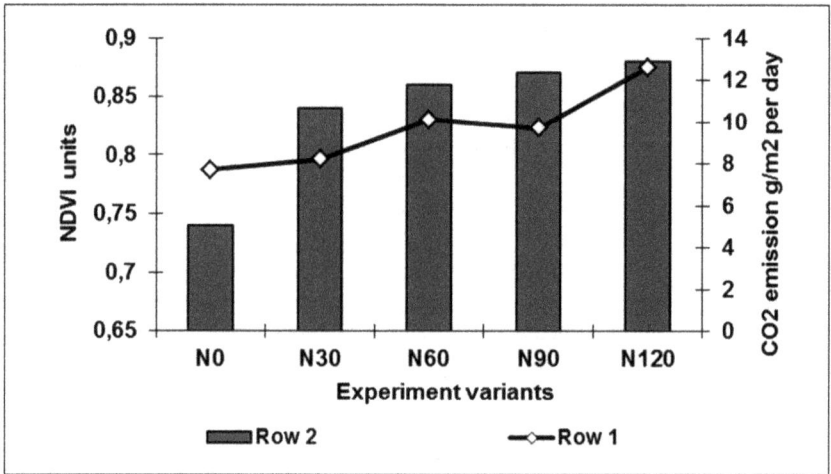

FIGURE 9.5   The dependence of the value of the normalized differentiated vegetation index (NDVI), obtained with the help of UAV "Agrodrongroup," on the biological activity of the soil, diagnosed by the intensity of carbon dioxide emission.

Note: (1) Emission of CO_2, 2—units NDVI (2017). Doses of nitrogen fertilizers: N0—0 kg/ha; N30—30 kg/ha; N60—60 kg/ha; N90—90 kg/ha N120—120 kg/ha.

These data clearly indicate the effectiveness of remote diagnostics of grain crop nitrogen state using UAVs equipped with appropriate photometric equipment. Calibration of the photometric device, that is, creation of scales which should show levels of plant supply with nitrogen nutrition—the most labor-intensive, important, and responsible operation in all technological chain of practical diagnostics of plant nitrogen nutrition. The installation of such scales should be carried out with the participation of competent specialists, mainly agronomic (agrochemical) profile, as it should be carried out by putting on crops field experiments with increasing doses of nitrogen fertilizers. Field test schemes should consist of variants representing doses of ammonium nitrate or other nitrogen fertilizers in the active substance, that is, in nitrogen (N), from zero dose (N0) to a presumably optimal or exceeding optimum, for example, up to N 150–180 kg/ha. Since the calibration of the instruments does not require large areas of sowing, the field (micro field) experiments are laid with plots for each variant with an area of several square meters, for example 16 m^2 (4 m × 4 m). If field experiments have any more tasks, such as environmental or economic, then each fertilizer variant of such scientific experiment is putted in the triple–four replications.

If the task is limited only to calibration of the photometric device, then putting of so-called industrial (scientific and production) field experiment is made in one-double repetition for simplification of technological operations. In particular, the most acceptable is the following scheme of the field experiment with six variants of nitrogen fertilizers for calibration of the photometer on winter wheat sowing in the conditions of the non-Chernozem zone with the planned yield of 50–60 t/ha:

1—N0 (control); 2—N30; 3—N60; 4—N90; 5—N120; 6—N150.

We used this scheme by us in the described experiment.

The area of plots of scientific field or industrial experiments can be much larger than the above example. The minimum area should provide shooting of each plot in such a way that it can take the whole frame when processing the image and through the NDVI value characterize the level of fertilizing that is characteristic for this variant of the experience. For large size plots of square or rectangular shape is used the appropriate program of shooting and processing of the images, determined by the height and trajectory of the UAV. Technology of deployment of field research and industrial experiments is described in the respective teaching guides.

Survey of field experiments with UAV is made in the main (critical) phases of plant growth and development: for cereals in the phase of spring tillering, shooting, earing-flowering, the beginning of grain filling. To do this, select the typical weather for this period, preferably dry and windless or not very windy, daytime from 11 to 15 h. Based on the results of field experiments, a scale of dependence of NDVI value on increasing doses of nitrogen fertilizers applied into the soil in the form of a table is constructed (Table 9.2). If in the field experiment the survey was carried out for several variants of the same name, then the values of NDVI averaged for each variant are entered in the table.

TABLE 9.2 The Scale Scores of Readings of the Photometer UAV According to Remote Survey Results of Field Experience in a Non-Chernozem Zone with a Planned Yield of 50–60 kg/ha.

Dose of nitrogen, kg/ha	Photometer "Yara" indications, points
0	0.4
30	0.5
60	0.6
90	0.7
120	0.8
150	0.9

Created scale scoring can be used to determine the doses of nitrogen application in production winter wheat crops in soil-climatic conditions adequate to the conditions of the calibration field experiments. To do this, a table is compiled in which the recommended doses of nitrogen fertilizers in the active substance (N) are put down in the order opposite to the increasing doses of field experience (example Table 9.3), that is, the minimum doses of experience correspond to the maximum doses of nitrogen fertilizers recommended for application and vice versa.

TABLE 9.3 Recommended Doses of Nitrogen for Winter Wheat Nutrition at the Planned Grain Yield of 5–6 t/ha.

NDVI indications, points				
< 0.40	0.41–0.50	0.51–0.60	0.61–0.70	>0.7
Dose of nitrogen, kg/ha				
150 (90 + 60)	120 (90 + 30) 90		60	0–30

Note: Increased doses of nitrogen fertilizers (150 and 120 kg/ha N) apply in two stages with a 2-weeks interval.

This takes into account the existing scientific and industrial experience of fertilizing winter wheat. In particular, it is considered advisable not to apply nitrogen fertilizers to the fields with winter wheat more than 90 kg/ha of nitrogen a.s. at a time. If necessary, increased doses of nitrogen fertilizers are applied in two stages with two week interval, so as not to cause a violation of biochemical processes in plants. To further increase the yield of winter wheat and the content of protein in the grain, the total dose of nitrogen can be increased to 180 kg/ha with its application in three to four stages: in autumn and/or spring tillering, shooting, earing-flowering grain filling. It should also be borne in mind that in the initial periods of vegetation—tillering, the beginning of the shooting—due to incomplete projective cover of the soil with plant biomass, the parameters of UAV photometers may not exceed 0.5 points, although the supply of plants may be high, which does not require fertilization. In such cases, remote diagnostics of plant nitrogen nutrition is desirable to complete with the ground diagnostics, using portable N-testers such as "Yara" or "Spad."

An example is the results of remote sensing of winter wheat sowing in the spring 2017 with the quadrocopter of LLC "Agrodrongroup" in the field of the central experiment station of the Pryanishnikov All-Russian Scientific Research Institute of Agrochemistry (Fig. 9 6).

According to the results of the remote field survey, the average value of the nitrogen supply score of plants was established automatically by a special program—0.67 according to the calculation formula: "nir − red/nir + red" (the ratio of the difference between the photometry index in the near infrared (nir) and red (red) regions of the radiation spectrum to the sum of these indicators). The scale of the figure also shows the area of sowing with different nitrogen supply of plants, which can be used for differentiated (within the area of the field) nitrogen application on precision farming technology.

According to Table 9.3, we find that for a score of 0.67, the appropriate dose of nitrogen for fertilizing the sowing of winter wheat in order to obtain a grain yield of 5–6 t/ha is 60 kg/ha. This level of productivity (~5 t/ha) was achieved in the field experiment with increasing doses of nitrogen in this field.

In the absence of special survey equipment, remote sensing of crops can be performed using UAVs equipped with conventional equipment used for color amateur surveys. You can use the technology, developed at D. N. Pryanishnikov All-Russian Scientific Research Institute of Agrochemistry and LLC "Agrodrongroup." It consists in the fact that photographing vegetative crops is carried out with UAVs equipped with conventional digital photo or video cameras, followed by the transfer of color photo images through color printers of personal computers on paper conventional typewritten format (A4). From the obtained photographs, portable photometers such as "Yara" or "Spad" can determine the availability of nitrogen in order to regulate nitrogen nutrition of plants. To do this, the paper medium is cut into strips about 2 cm wide, which are examined by a portable photometer along their entire length, as shown in Figure 9.7.

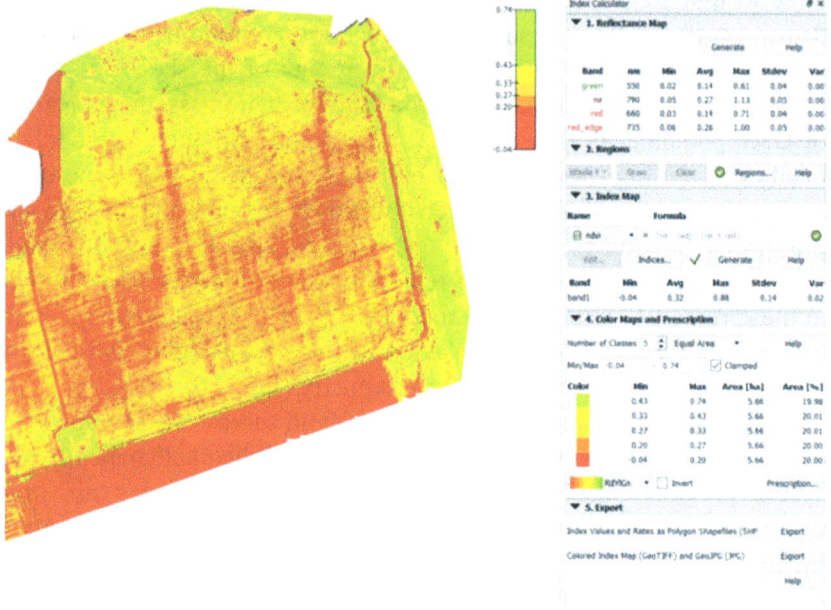

FIGURE 9.6 Results of remote sensing of the field with winter wheat at the central experiment station of the Pryanishnikov All-Russian Scientific Research Institute of Agrochemistry with the quadrocopter LLC "Agrodrongroup" (2017).

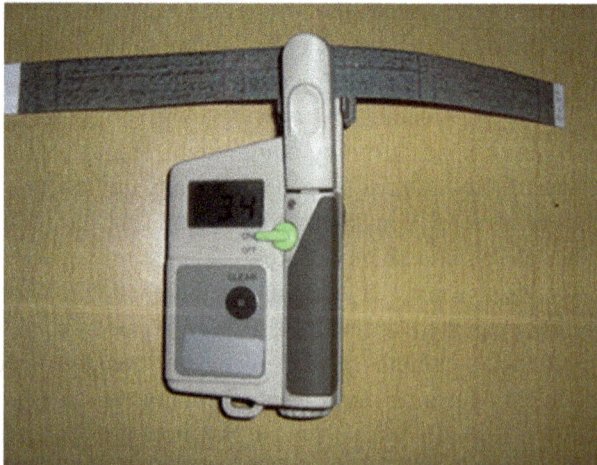

FIGURE 9.7 Using of portable photometer "Yara" to determine the supply of winter wheat sowing with nitrogen nutrition. Strips of paper 2 cm wide are visible at the top of the picture.

The intensity of their green color on the examined fragments indicates the nitrogen status of plants (this technology has received the corresponding registration certificate[9]). Comparison of the direct determination of the plant nitrogen status in the field of vegetating plants with determining it from this field photographs allows to determine with no less accuracy the level of plant nitrogen nutrition (Table 9.4).

TABLE 9.4 Indicators of Diagnostics of Winter Wheat Nitrogen Nutrition, Obtained by Various Methods in the Phase of Plant Shooting.

Dose of applied nitrogen, kg/ha	Diagnosis by photometer "Yara"		NDVI (received via UAV)	Stem diagnostics	CO_2 emissions, g/m^2 for days	Yield, t/ha
	By photo	By plants				
			Points			
0	2	366	0.65	0	7.7	2.83
30	22	511	0.81	0.87	8.2	4.41
60	26	541	0.84	1.4	10.1	4.99
90	30	580	0.85	2.5	9.7	5.6
120	37	620	0.86	2.7	12.3	5.1
Coefficients of pair linear correlation of indicators on a photograph with other indicators (r)	0.99		0.97	0.93	0.85	0.97

These diagnostic indicators and their close linkage, expressed by high values of the coefficients of paired linear correlation, indicate the possibility of photo camera processing obtained by remote shooting of actively vegetating, that is, green plants, by portable N-testers such as "Yara" or "Spad".

9.4 CONCLUSIONS

Recommendations on the use of UAVs for the diagnostics of agricultural crop nitrogen nutrition prepared based on field studies are aimed at further improving the technologies of their cultivation in modern conditions. The practical significance of these recommendations is that the development of unmanned aircraft (drones) for the agricultural interests can significantly reduce the cost of agricultural production by replacing the previous labor-intensive operations for the diagnostics and optimization of plant mineral nutrition with modern high-performance methods.

KEYWORDS

- **wheat**
- **nitrogen**
- **aerial photography**
- **photometry**
- **phytocenoses**

REFERENCES

1. Tserling, V. V. *Diagnostics of Crop Nutrition*; Agropromizdat ("Agro-industrial Publishing House" in Russian): Moscow, 1990; 235 (in Russian).
2. Sychev, V. G.; Afanas'ev, R. A.; Ermolov, I. L.; Kladko, S. G.; Voronchikhin V. V. Diagnostics of Nitrogen Nutrition of Plants Using Unmanned Aerial Vehicles. *Fertility* **2017,** *5*(98), 2–4 (in Russian).
3. Afanas'ev, R. A.; Belenkov, A. I.; Berezovsky, E. V.; Shuklina, O. A. Photometric Diagnostics of Nitrogenous Nutrition of Plants as Factor of Precision Agriculture Robotization. *Fields Russ.* **2016,** *6*(139), 68–71 (in Russian).
4. Korolyov, K.; Bome, N.; Weisfeld, L. A Comparative Study of Morphophysiological Characteristics of Flax in Controlled and Natural Environmental Conditions. *Zemdirbyste Agric.* **2019,** *106*(1), 29–36.
5. Belousova, K. V.; Litvinskiy, V. A.; Shchuklina, O. A. Influence of Different Forms of Nitrogen Fertilizers on the Pigment Content Under Different Lighting Conditions of Spring Triticale and Spring Barley Seedlings. "Proceedings of the 46th International Scientific Conference of Young Scientists and Specialistsю Efficiency of Application of Chemicals in Modern Technologies of Cultivation of Agricultural Crops"; Pryanishnikov All-Russian Scientific Research Institute of Agrochemistry Publishing House: Moscow, 2012; 14–17 (in Russian).
6. Dospekhov, B. A. *Technique of Experimental Case*; Ear: Moscow, 1979; 416 (in Russian).
7. Afanas'ev, R. A. Accounting Within-Field Variability of Soil Fertility to Optimize Differentiated Fertilizer Application. In *Temperate Crop Science and Breeding: Ecological and Genetic Studies*, 1st ed.; Bekuzarova, S. A., Bome, N. A., Opalko, A. I., Weisfeld, L. I., Eds.; Apple Academic Press: Oakville, Waretown, 2015; 476–489.

8. Methods of Field Experiments on Optimization of nitrogen nutrition of Grain Crops, Sugar Beet and Potatoes on the Basis of Operational Soil and Plant Diagnostics. Pryanishnikov All-Russian Scientific Research Institute of Agrochemistry Publishing House: Moscow; 1985; 92 (in Russian).

9. Sychev, V. G.; Afanas'ev, R. A.; Voronchikhin V. V. Invention № 2661458 (2007) Method of Photometric Diagnostics of Nitrogen Nutrition of Plants with the use of Unmanned Aerial Vehicles (UAVS). Request № 201713126; Registered of the Federal Service for Intellectual Property, Patents and Trademarks 16.07.2018. Date of Publication: 16.07.2018. Bull. 20. Int. CI G01C 11/02 (2006.01) (in Russian).

CHAPTER 10

Revegetation and Launching Self-Restoration Process of the Disturbed Landscape Along the Transport Corridor in Western Siberia

NINA A. BOME[1*], NIKOLAI G. IVANOV[1], MARINA V. SEMENOVA[1], LEE A. NEWMAN[2], and ALEXANDER Y. BOME[1,3]

[1]*Department of Botany, Biotechnology and Landscape architecture, Institute of Biology, University of Tyumen, 6 Volodarsky St., 625003 Tyumen, Russia*

[2]*Environmental and Science Biology, State University of New York College of Environmental Science and Forestry, 1 Forestry Drive, 13210 Syracuse, NY, USA*

[3]*Exeter Produce and Storage ltd, 149A Thames Rd. W, N0M 1S3, Exeter, ON, Canada*

Corresponding author. E-mail: bomena@mail.ru

ABSTRACT

The chapter presents the results of vegetation monitoring of the road interchange slopes Tyumen—Krivodanovo since the commissioning (1996) to the present time. To develop of a plant formation was included grass mixture with four species plants (*Bromopsis inermis* L., *Festuca pratensis* L., *Brassica napus*, *Amaranthus cruentus* L.) and wild plants. The monitoring was on 30 discount areas (plots) regularly spaced in the low, middle, and top layers for the south and the north slopes. The increasing of herbaceous plant species and the emergence of trees and shrubs were found out. There were found major differences in a ratio of species, biomass, phenological stages of plants in deferent layers of the both slopes. More favorable conditions for plants and the formation of seeds were on the southern slope of the site.

10.1 INTRODUCTION

Disturbances produce patterns and lead to ecological consequences that need significant research. Wu and Loucks consider disturbances as a natural agent of changes. In this term, any transport corridor which integrate with ecological system and any natural society becomes a driver of temporal heterogeneity. Transport corridors may be considered as disturbances that have ecological impact and need study[1-4] give the definition of disturbances as "any relatively discrete event that disrupts the structure of an ecosystem, community, or population, and changes resource availability of the physical environment." According to their definition, we propose that the construction of transport corridor is an agent of a landscape disturbance as a discrete event which changes the ecological system. In our research, we consider transport corridors as agents of disturbances.

The development and presence of transportation corridors results in three primary consequences: (1) reduced landscape permeability, (2) habitat loss, and (3) increased habitat fragmentation. These fundamental changes to landscape structure, which can occur during the construction and post-construction phases of corridor development, can all have profound cascading ecological implications.[5] On the other hand, ecological management of these sites may maintain roadside native-plant communities in areas of intensive agriculture, reduce the invasion of exotic (non-native) species, attract or repel animals, enhance road drainage, and reduce soil erosion. Ponds, wetlands, ditches, berms, varied roadside widths, different sun and shade combinations, different slope angles and exposures, and shrub patches can offer a variety of roadside species richness.[6]

This study of a transport corridor was delineated by the two slopes (southern and northern facing), and their effect on natural community dynamics and impacts on space structure, the age structure, the relative abundances of species with time.

The territory of the Tyumen region, occupying 1.43 million sq. km, is characterized by a high contrast and complexity of specific soil and climatic features. Plant species composition, their growth and development is determined by a number of soil and climatic factors, including but not limited to: the amount and timing of precipitation, effective temperature ranges, duration of the growing season, and level of soil fertility.

In recent decades, extreme conditions of plant habitat are intensified under the influence of anthropogenic factors. Some anthropogenic impacts lead to destruction of herbaceous plants as well trees and shrubs. Khoteyev[7] pointed out than the area of disturbed lands in the Tyumen region for the beginning of the 21th century is 1.73 thousand sq. km. In the southern portion of the Tyumen region, numerous roads and overpasses are actively under construction, and this has a significant impact on competitive and adaptive properties of plants; their ability to survive and reproduce. Therefore, a study of a plant response to impacts of unfavorable factors has a high ecological value.

The purpose of this research is the study the impact of road construction on native plants communities immediately adjacent to the construction, and the impact of the introduction of the nurse plants on the reestablishment of native plant communities. The study further examines the impact of other factors (slope orientation, sunlight, drainage) to the success of the reestablishment protocols. A monitoring of a natural revegetation of the disturbed area and the experimental work on a selection of plant species for enhancement of this revegetation is a paramount objective of this complex research. This study is based on reducing the negative impact of the disturbance to the environment and also revegetation with an aesthetic approach.

Research stages:

- Developmental plan for launching self-restoration process on a disturbed landscape.
- Presentation of results of 20 years observation of the plant society.
- Introduction of a plant mixture created favorable conditions for site restoration.

10.2 MATERIAL AND METOLOGY

10.2.1 SITE DESCRIPTION

The study site is placed in Tyumen, Russia, and its geographical coordinates are 57.108164° N, 65.80540° E. The study site is the slopes along the road between Tyumen and Krivodanovo, adjacent to a rail line. The slopes are roughly 700 m in length, with a height that varies from 1.2 m to 12 m (Fig. 10.1). This area was classified and plant hardiness zone five according to the USDA hardiness zones classification.

FIGURE 10.1   Location of the study site on Tyumen region, Russia.
Note: Road interchange of the slope of the Tyumen—Krivodanovo highway—Krivodanovo highway. In the photo above, is the density of vegetation in the year 2005.

10.2.2   PLANTS SELECTED

In 1996, annual (*Brassica napus, Amaranthus cruentus* L.) and perennial plants (*Bromopsis inermis* L., *Festuca pratensis* L.) were found on the slopes during the initial site survey and selected for the further work. It was decided, however, that rather than introducing the plants individually, to create mixtures of the plants including both annual and perennial plants, annuals only, or perennials only (Table 10.1). There was the expectation that that the inclusion of fast-growing annual plants would better prevent erosion of the soil during initial phases of restoration due to the rapid formation of a dense root layer, and perennial plants, as dominants on the slopes, will be the predominant plant populations after overwintering. Consequently, some of combination was introduced as a mixture of annual and perennial plants. The most diverse plant combination was represented by *Bromopsis inermis* (40%), *Festuca pratensis* (15%), *Brassica napus* (30%), *Amaranthus cruentus* (15%). In contrast, plant mixtures were homogeneous groups and presented by annual or perennial plants only. And finally, plant mixtures were heterogeneous groups of annuals and perennials, and represented by two plant genera.

TABLE 10.1 Grass Mixtures for the Revegetation of the Slopes (*Bromopsis inermis* L., *Festuca pratensis* L., *Brassica napus*, *Amaranthus cruentus* L.)

Mixtures	Combination	Ratio, %
Mixture A	B. inermis + F. pratensis + B. napus + A. cruentus	40/15/30/15
Mixture B	B. inermis + F. pratensis	60/40
Mixture C	B. napus + A. cruentus	50/50
Mixture D	B. inermis + B. napus	60/40
Mixture E	F. pratensis + A. cruentus	60/40

10.2.3 SAIT PREPARATION

Because of the potential for the soil along the slopes to be inhospitable to plant establishment, preliminary soil analysis was conducted. Fresh soil was added to both slopes across the top and middle regions to improve soil quality. The lower region of the slopes did not have any soil amendments added. Prior to examination of the site for plant establishment success, the soils were analyzed for texture by dry sieving method of Screen analysis in the laboratory according to Kachinsky.[8]

10.2.4 SEEDLING

The study site was originally divided into two major study areas based on the direction the slopes faced; north or south. This was important as the amount of light reaching the plants would be very different on the two slopes. Each slope was further divided by the height of the planting area on the slope; in general, divided into the lower, the middle, and top sections.

Each of five combinations of the plants listed above were seeded in plots along transects. A plots size was 1 m² and each plot was spaced 30 m from any other plots. Each mixture of plants had three replicates on each slope; one each at the low, middle, or top sections of the slope. This resulted in 30 plots across the two slopes according to the design by Uglov.[9] The planting scheme of the site is showed in Figure 10.2.

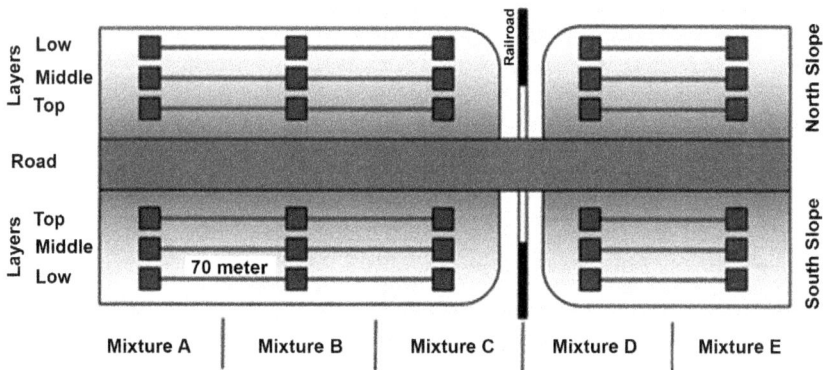

FIGURE 10.2   Planting scheme of the study site.

Note: Dark green squares denote planting areas on the slopes. The north facing and south facing slopes were divided into top, middle, and low levels and different seed/plant mixtures were installed on each slope and location.

10.2.5   *EVALUATIN OF PLANT SUCCESS*

The research work was done in three stages. Between 1996 and 1999, preliminary research involved characterization of grass covering existing on the slopes before seeding (natural plant cover), description of species composition, features of growth and development of plants of each of three layers for the two slopes. After site characterization, plots were cleared, and plants or seeds from the groups listed in table were installed to enhance revegetation and provide for stabilization of the slopes. Plots were surveyed the following year to determine which seed/plant mixture resulted in the best survival rates for the plants after an initial winter season. Subsequent site evaluations were done during the years 2004–2006, and final site evaluation was completed in 2016–2017. These site evaluations consisted of determining maximal survival of seeded plants both inside and outside the original plots, and also overall site evaluation for naturally establishing plants.

10.2.6 STATISTICS

Statistical analysis was performed using table processor Microsoft Excel and STATISTICA 6.0 software (StatSoft).

10.3 RESULTATS AND DISCUSSION

The revegetation analysis of the slopes was determined by examining a number of factors. Among these were soil analysis to determine both soil texture (grain size) and soil water holding capacity following the addition of the soil amendments. The texture of the soils along the top and middle layers was classified as a sandy loam sand. Soils along the lower layer were classified as a clay loam. This difference reflects the fact that the upper layers of the slopes received soil amendments, while the lower layer did not. This also shows that there was minimal, if any, migration of the amended soil down the slopes over the time of the study.

A preliminary objective of the project was the selection of optimal plant genera to be used in site restoration; their ability to decrease erosion processes and maximizes revegetation of the site were primary considerations.

As stated, there were five artificially created mixed populations of plants (five mixtures), which were installed on the slopes. These mixtures contained both plants originally seen on the site, as well as additional plant genera believed to enhance erosion control. During the growth season of the second year a dense root layer had been formed by the annual plants. This layer provided the maximum slope stabilization and reinforcement in the most vulnerable places.

During the 1997 evaluation of plant establishment, Mixture A, containing four different plant genera, clearly had the best plant establishment rates. During the plant survey, we found on average 987 plants per m^2 for plots seeded with mixture A. This efficiency of establishment for mixture A was 4.0–68.1% greater than for the other plant mixtures.

Plant health and potential for prolonged survival on the site was determined by examining the plant for hardiness based on the color of leaves and condition of growth apex as described by Kuperman and Ponomarev.[10]

While all plants in mixture A were present in the other mixtures, the combination of the four genera together seems to confer better survival of the plants; Kuperman and Ponomarev rank plant survival on a scale of 1–5, with 5 being the best survival rates. Plants in mixture A were ranked as a 5.

The construction of the road affected natural plant populations. The 2004 survey also found trees and bushes starting to colonize the site: Silver birch (*Betula pendula* Roth.), European aspen (*Populus tremula* L.), Scots pine (*Pinus sylvestris* L.), and Wooly twig willow (*Salix dasyclados* Wimm). The plant survey in 2016 showed those trees and shrubs growing on both the northern and southern slopes, although the location of the plants followed and interesting pattern. The location of the trees and shrubs on the slopes is given in Table 10.2. What can be seen is that the willow, which has a higher water requirement, was better established on the northern slope, which received less sunlight, and thus would have lower evapotranspiration rates, while the pine, which prefers drier soils, was better established on the sunnier and thus drier southern slope.

TABLE 10.2   Trees and Shrubs Found on the Slopes.

Northern slope		Southern slope	
Variety	Number	Variety	Number
Betula pendula Roth.	2	*Betula pendula* Roth.	1
Salix × dasyclados Wimm.	32	*Salix × dasyclados* Wimm.	–
Populus tremula L.	20	*Populus tremula* L.	5
Pinus sylvestris L.	–	*Pinus sylvestris* L.	2

The plant surveys in the unplanted regions for years 1996 and 2004 showed that the diversity of herbaceous plant species increased from 15 to 18 genera. The herbaceous vegetation not only had changes in the number of genera found on the site, but also the species ratio. In 1996, *Poa pratensis* L., *Trifolium pratense* L., *Achillea millefolium* L. were most frequently found on the slopes, but in 2004 the two dominant species were *Trifolium pratense* L. and *Achillea millefolium* L., but numbers of *Poa pratensis* L. had decreased in comparison. Additionally, *Phalaris arundinacea* L. which was one of the new genera on the slopes was now also one of the dominant genera.

Ecological conditions of the slopes had an impact on the occurrence of species. For example, in 1996, *Elytrigia repens* (L.) Desv. ex Nevski was common in the southern slope of this site (112 plants per 1 sq. m), but this plant was very rare on the northern slope. Another example is *Phalaris arundinacea* L., which was more adapted to the northern slope. In 2004, *Bromus inermis* Leyss. (52 plants per 1 sq. m) and *Carex acuta* L. (65 plants per 1 sq. m) were dominant on the northern slope, but those species were not seen on the southern slope.

In years 1997 and 1998, the variability of quantitative traits of the dominant species of *Trifolium pratense* L. on the northern and southern slopes was studied. The response of plants in the changing environmental conditions was determined based on properties of the stem: length, weight, and leaf mass. The condition of the lower slopes, which suffered the least amount of disturbance, also had the least detrimental impact on plant growth. There are differences in the phenotypic traits observed on different slopes. The vegetative mass of 50 plant shoots samples was in 60.1% more on the southern slope, but the weight of leaves from 50 shoots was more on the northern slope (increase of 33.5%) (Table 10.3).

TABLE 10.3 Vegetative Mass of 50 Shoots *Trifolium pratense* L.

Trifolium pratense L.	Southern slope	Northern slope
Vegetative mass	279.5 g	168.0 g
Leaves mass	46.5 g	139.0 g

Table 10.4 illustrated that plants of the southern slope were taller and had less height variations (CV = 13.0–27.4%) than plants of the northern slope (CV = 20.3–42.6%).

TABLE 10.4 The Effect of the Exposure of Slopes and the Degree of Landscape Disturbances on the Maximum Height of *Trifolium pratense* L. (Blossom Set).

Slopes	Height variations for slope levels, cm				
	Lower level		Middle level		Top level
	1997	1998	1997	1998	1997
The Southern	37.8 ± 1.8	40.5 ± 2.4*	33.8 ± 3.0*	38.1 ± 4.3	41.4 ± 1.8#
The Northern	34.2 ± 2.2	30.7 ± 3.5	55.2 ± 2.2	35.3 ± 2.5	41.8 ± 3.1#

Note: (1) Differences are significant: *—by location of slopes; #—by location of levels. (2) The data on plant height for top level 1998 was not presented, because there were insufficient plants for statistically significant analysis.

The middle and top levels of the slopes were subjected to maximum anthropogenic impacts. The imported soil to form these parts of the slopes was a sandy loam texture. In this context, it was important to study the response of *Trifolium pratense* L. due to environmental factors based on colonization of plants on the slopes.

In 1997, the average height of the *Trifolium pratense* L. population decreased with height up the southern slope and increased with height up the northern slope. The weight of 50 shoots was 204 g for plants on the northern slope and 111 g for plants on the southern slope. The differences in mass and height (Table 10.3 and Table 10.4) were associated with the high day time temperature for the southern slope and increasing soil temperature to the depth of 15–20 cm of 1.5–3.70°C as compared to the northern slope. The increasing temperatures aided more rapid water evaporation from the soil on the southern slope, which improved growth condition for *Trifolium pratense* L.

The morphometric analysis of the plants of the top level also revealed differences between northern and southern slopes on the vegetative mass of plants (115 and 126 g, RESP.). Indicators of temperature changes, soil moisture, and light at the top level of the slopes were similar irrespective on exposure of the slopes. However, the weight of leaves in the phytomass proportion was higher for the northern slope (33.5%), than the weight for the southern slope (Table 10.3).

According to published data, reproductive organs of plants showed less variability than other physical traits when plants are exposed to varying environmental factors. However, a change of floral traits usually indicates a presence of an adaptive process.[11] Increasing relative weight of biomass can indicate differences in adaptability of flowering plants.[12] Essential at a morphological level is the number of flowers in an inflorescence, which is positively associated with the number of seeds formed.[13]

Trifolium pratense L. plants formed greater amounts of inflorescences at all levels of the southern slopes than plants on the northern slope (Table 10.5). Statistically significant differences in the number of flowers in an inflorescence were not found for the lower level of slopes. In the middle of the slopes, the floral number is much higher on the northern slope. The inflorescences of plants growing on the top level of the southern slope were larger by morphometric parameters (length and width), and had the advantage over the northern slope in the number of flowers. For the middle level of the slope this plant characteristic was much higher on the northern slope. Inflorescences of plants on top level of the southern slope were larger by the same morphometric parameters, and had the advantage in the number of flowers over the northern slope.

TABLE 10.5 The Parameters (in numbers) of Inflorescences of *Trifolium pratense* Typus L., 1997–1998.

Slopes	Lower level		Middle level		Top level	
	Inflorescences	Flowers	Inflorescences	Flowers	Inflorescences	Flowers
The Southern	83	64.6 ± 5.3	62	43.4 ± 4.0#	57	75.8 ± 6.2
The Northern	74	60.2 ± 2.6	54	80.4 ± 2.7*#	49	57.7 ± 4.3*

Note: Differences are significant: *—by direction of the slopes; #— by location on slope.

Analysis of the results showed that the reaction of *Trifolium pratense* L. to changing environmental conditions depends on the degree of disturbance of natural landscape and showed a variability of quantitative traits that determine growth and development of plants, including reproductive traits. The most favorable conditions for *Trifolium pratense* L. were on the lower level of the southern slope. Variations of traits were studied in extreme environmental conditions of the middle and top level of the slopes can be an indicator of adaptive properties of *Trifolium pratense* L. plants.

The character of variability of traits of vegetative and reproductive organs of *Trifolium pratense* L. plants are presented in the Table 10.6. There was an increase in some traits of *Trifolium pratense* L. such as a weight of shoots (but decreasing leaf mass), height of plants, width of inflorescences, a number of inflorescences, and a proportion percentage of *Trifolium pratense* L. in phytocenosis of the middle level of the northern slope.

Six of nine studied traits were characterized as increasing. However, the phytomass, number of inflorescences and number of flowers for a same inflorescence decreased for the top levels of the both slopes.

The plants on the top level of the southern slope showed a lag in the development of vegetative organs and an increase the morphometric parameters of inflorescences compared with plants of the lower level.

The colonization of *Trifolium pratense* L. on disturbed landscapes depends on the phenotypic expression of quantitative traits. For example, in 1996 *Trifolium pratense* L. was found on all three levels of the both slopes. The proportion of *Trifolium pratense* L. was 61.0 and 49.5% in the middle and the top levels of the southern slope and was 46.8 north and 37.2% for the same levels of the northern slope. In accordance of the records, *Trifolium pratense* L. was the dominant species.

TABLE 10.6 Changing the Quantitative Characteristics of Vegetative and Generative Organs of Plants *Trifolium pratense* L.

Characteristic of organs	Northern slope				Southern slope			
	Middle level		Top level		Middle level		Top level	
	$X_{avg.}$	CV	$X_{avg.}$	CV	$X_{avg.}$	CV	$X_{avg.}$	CV
Weight of 50 shoots, g	+	nc	-	nc	-	nc	-	nc
Including leaves, g	-	nc	-	nc	-	nc	-	nc
Stems, g	+	nc	-	nc	-	nc	+	nc
Plant height, cm	+/+	-/-	+	+	-/+	+/-	+	=
Inflorescences length, cm	-	+	+	-	-	-	+	+
Inflorescences width, cm	+	+	+	-	No diffe-rences	+	+	-
Number of flowers in inflorescences	+	-	-	+	-	+	+	-
Number of inflorescences	-	nc	-	nc	-	nc	-	nc
% of *Trifolium pratense* L. in plant formation	+	nc	-	nc	+	nc	-	nc

Note: /-above the diagonal is 1997, under the diagonal is 1998; + increasing; -decreasing; = no significant differences; nc—CV (coefficient of variation, %) was not calculated.

Summer of 1998 assessment of phytocenosis also noted the relatively high value of colonization of clover plants in the lower and middle levels of the slopes (38.5 and 42.1% for the southern slope, 46.1 and 63.5% for the northern slope). Conditions on the top level of the slope were unfavorable for over wintering plants; that may be associated with a small snow level and in some places a lack of snow due to wind removal. As a result, in 1998, only isolated plants are preserved of the top level of the both slops. Thus, *Trifolium pratense* has a high ability to survive and can adapt to changing environmental conditions, but within a certain limit, which was confirmed by our data.

The analysis of the species plant composition for the lower, the middle and the top levels in 2004–2005 periods is presented in the Table 10.7.

TABLE 10.7   The Species Compositions of Plants During 2004–2005 Period.

Plant species	The Southern slope (pcs)			The Northern slope (pcs)			Total
	Lower level	Middle level	Top level	Lower level	Middle level	Top level	
Achillea millefolium L.	15	29	15	16	68	28	171
Artemisia vulgaris L.	7	6	20	0	2	2	37
Bromus inermis Leyss.	0	0	0	0	75	135	210
Carex acuta L.	0	0	0	107	8	0	115
Chamenerium angustifolium L.	7	6	0	12	10	2	37
Convallaria majalis L.	0	0	0	4	0	0	4
Digraphis arundinac L.	85	65	56	43	0	90	339
Eguisetum hiemale L.	0	0	0	35	31	0	66
Equisetum arvense L.	0	12	12	30	3	0	57
Medicago sativa L.	4	6	11	0	0	0	21
Poa annua L.	45	5	32	3	0	1	86
Rubus saxatilis L.	0	0	0	3	0	0	3
Sonchum arvensis L.	0	0	0	1	6	0	7
Tanacetum vulgare L.	1	0	0	0	0	0	1
Taraxacum officenale Wigg.	0	0	0	12	6	6	24
Trifolium pratense L.	34	60	5	52	61	0	212
Tussilago farfara L.	0	0	0	5	5	11	21
Vicia cracca L.	36	21	5	0	0	11	73
Total (by levels)	234	210	156	323	275	286	1484
Total (by slopes)	–	600	–	–	884	–	
Genera (by levels)	9	9	8	13	11	9	–
Genera (by slopes)	–	26	–	–	33	–	–

The analysis of the species composition of plants during the 2004–2005 period showed that the lower levels of the slopes were characterized with the highest diversity of plants. Plants surveys showed 17 species of plants across the site. The species composition was different on the northern and the southern slopes. There were 9 described species for the southern slope, and 13 species for the northern slope. *Phalaris arundinacea* L. and acute sedge had maximum number of plants; *Phalaris arundinacea* L. was 85 pcs/sq. m on the southern slope and *Carex acuta* L. was 107 pcs/sq. m on the northern slope. There was no *Phalaris arundinacea* L. on the northern slope, and no *Carex acuta* L. on the southern slope.

For the middle levels there were 9 spp. on the southern slope and 11spp. on the northern slope. At the same time, *Trifolium pretense* L. were regularly distributed on the both slopes (60 and 61 pcs. plants per 1 sq. m). For the southern slope, the dominant plant was again species of *Phalaris arundinacea* L. (65 pcs. per sq. m), while the dominant plants on the northern slope were species of *Achillea millefolium* L. (68 pcs. per sq. m.) and *Bromus inermis* Leyss (75 pcs. per sq. m).

Plants surveyed on the top level of the slopes showed that the number of different species decreased. There were eight described species for the southern slope, and nine species for the northern slope. The dominant were *Phalaris arundinacea* L. (56 and 90 plants per sq. m., for both the southern and northern slopes, respectively) and *Bromus inermis* Leyss (135 pcs. per sq. m.) on the northern slope.

A study of two species (*Trifolium pretense* L. and *Epilobium angustifolium* L.) shows a variability of plant height in different years of the study (Table 10.8). The plants of both species had the maximum height in the flowering phase in 1996, which may be associated with a lower density of plants in the first year of the road interchange, compared with the later period (2004–2005).

TABLE 10.8 The Height of Plants *Trifolium pretense* L. and *Chamerion angustifolium* L. in Different Years of the Research (Flowering Phase)

Slopes	1996	2004	2005
	Trifolium pratense L.		
The Southern	37.2 ± 2.81	19.7 ± 1.30*	26.8 ± 1.23*
The Northern	40.5 ± 1.52	32.6 ± 3.24*	29.2 ± 2.53
	Chamerion angustifolium L.		
The Southern	87.9 ± 5.30	66.7 ± 10.70*	38.2 ± 6.00*
The Northern	103.0 ± 7.32*	83.0 ± 10.3*	33.0 ± 6.60

Note: Differences are significant: *—by years of research; *—by location of slopes.

The height of plants of the southern slope of the site was less as compared to the northern slope on most plots that indicated the differences of soil water availability.

This shows how weather condition effects on the height of plants. For example, in the hot and arid 2004, plants grew higher on the northern slope, but in the more humid and slightly cooler growing year 2005, higher plant growth was found on the southern slope, except for the genera of *Trifolium pratense* L. and *Epilobium angustifolium* L.

The height of the plants was dependent on weather; in the hot and arid year (2004) the plants grew on the northern slope, and in the more humid and warm year (2005), the plants of the Southern slope were higher, except *Trifolium pratense* L. and *Chamaenerion angustifolium* (L.) Scop.

The data analysis of 2005 on total biomass for transects shows the differences on plant species, slopes and its levels (Table 10.9). Productivity of plants on the Southern slope was higher (2010 g/sq. m.) as compared to plants on the northern slope (1429 g/sq. m). Maximum vegetative mass formed on the middle and top levels of the Southern slope, and the lower level of the northern slope.

TABLE 10.9   The Plant Productivity Analysis of the Slopes in 2005.

Characteristic of productivity	The Northern slope		
	Lower level	Middle level	Top level
Total productivity		1429 g per sq. m	
Plant productivity (g per sq. m.)	*Trifolium pretense* (147)	*Bromus inermis* (332)	*Bromus inermis* (262)
	Carex acuta (309)		*Carex acuta* (191
	Tanacetum vulgare (248)		
	The Southern slope		
Characteristic	Lower level	Middle level	Top level
Total productivity		2010 g per sq. m	
Plant productivity (g per sq. m)	*Carex acuta* (303)	*Carex acuta* (350)	*Carex acuta* (695)
		Medicago sativa (606)	*Medicago sativa* (297)

Note: In parentheses—the number of plants in the area.

The proportion of inflorescences in biomass structure on the southern slope (*Carex acuta* L., *Trifolium pretense* L., *Bromus inermis* L.) increased with slope height. On the northern slope the proportion of inflorescences increased with slope height for *Carex acuta* L. and decreased for *Bromus inermis* Leyss (Table 10.10).

TABLE 10.10   Percent Biomass of Plant Structures Growing on Both Slopes During, %, 2005.

Species	The Northern slope								
	Lower level			Middle level			Top level		
	IF	LVs	STM	IF	LVs	STM	IF	LVs	STM
Carex acuta	11.9	41.5	46.6	12.9	39.1	48.0	16.7	37.0	46.3
Trifolium pretense	15.3	46.2	38.5	19.4	43.8	36.8	10.6	42.9	46.5
Achillea millefolium	8.8	50.8	40.4	–	–	–	–	–	–
Bromus inermis	–	–	–	12.7	28.0	59.3	14.9	26.6	58.5
	The Southern slope								
Carex acuta	7.8	55.8	36.4	15.6	36.7	44.7	24.0	23.2	52.8
Trifolium pretense	4.0	37.0	59.0	0.6	52.1	47.3	–	56.5	43.5
Achillea millefolium	–	100.0	–	1.5	89.5	9.0	–	100.0	–
Bromus inermis	–	–	–	14.0	26.7	59.3	12.8	30.0	57.2

Note: IF, inflorescence; LVs, leaves; STM, stems.

The most favorable conditions for the formation of seeds were realized on the southern slope, as evidenced by kernel size index. The weight of 1000 seeds of *Achillea millefolium* L. and *Bromus inermis* L. was 0.19 g and 10.1 g on the southern slope and 0.15 g and 5.3 g on the northern slope, respectively.

Plant composition along with phytomass and percent size of plant organs for each species (including leaves, stems, and inflorescences) changed depend on exposure of the slopes and its height. The weight of 1000 seeds on the southern slope was higher as compare to seed of the same species on the northern Slope.

The analysis of the data confirms the hypothesis that the studied slopes of the road interchange Tyumen—Krivodanovo are a modified geocomplex with disturbances which is in the stage of self-restoration. Among positive trends were colonization by new species plants (including natural colonization of slopes with trees and shrubs), self-development of phytocoenosis depends on external factor effects (humidity and temperature), side of slopes (southern and northern) and levels (lower, middle, and top).

There is a lot of research that concerns revegetation of disturbed landscapes. For example, Mirkin and Mironova[13] studied the diversity of vascular plant species toward plant formations, diversity of plant formations, and total value of species in the disturbed area of North-Western and Southern of Yakutia; this depends on the stage during the progressive succession. The researchers concluded that self-restoration process proceeds more rapidly in lower levels in warm climate conditions. Similar results were obtained in our work where there was considerable slope gradients at the road interchange. In our work, plants *Trifolium pratense* L., *Phalaris arundinacea* L., *Bromus inermis* Leyss., *Chamerion angustifolium* L., and *Medicago sativa* L. were more common in our study area due to their presence in border plant formations. The results confirm research work with restoration of plant formations of a mining industry disturbance landscape in Kola Peninsula.[14] Restoration activities may be designed to replicate a pre-disturbance ecosystem or to create a new ecosystem where it had not previously occurred. Restoration ecology is the scientific study of repairing disturbed ecosystems through human intervention.[15]

The basin is sometimes perceived as a region of homogeneous ecosystems. Our results demonstrate the widespread differentiation of the population associated with the habitat, and local adaptation. Plants obtained from local sources are likely to adapt with rates and values that are directly related to the success of the restoration, and our results indicate that certain key features and environmental variables should be prioritized in future plant assessments in the study region. Using a broad literature review of common garden studies published between 1941 and 2017, we documented the commonness of these three signatures in plants native to North America's Great Basin, an area of extensive restoration and revegetation efforts, and asked which traits and environmental variables were involved.[16]

10.4 CONCLUSIONS

Natural overgrowth of the slopes of the road interchange Tyumen—Krivodanovo provide positive results during 20 years of observation. Remediation and restoration must be performed during the initial period of establishment of vegetation, especially on steep slopes, to prevent erosion and accelerate self-restoration processes. Including annual plants in grass mixtures along with perennial grasses (*Brassica napus* L. and

Amaranthus L.) provides relatively rapid formation of plant mass and the root systems during the sowing year that consequently provide more favorable conditions for overwintering perennial grasses.

ACKNOWLEDGMENTS

This research was partially supported by Gennady Filippovich Kutsev, former provost of Tyumen State University. We thank Andrei Tolstikov, vice provost of Tyumen State University who helped to start the collaboration with Dr. Newman and Dr. Lanza from SUNY ESF (SUNY College of Environmental Science and Forestry).

We thank Dr. Newman from SUNY ESF for her wisdom and assistance with finishing this chapter, and Dr. Lanza from SUNY ESF for comments that greatly improved the manuscript. We are also grateful to everyone from SUNY ESF and TSU who helped and supported us with their comments and discussions.

Any errors are our own and should not tarnish the reputations of these esteemed persons. This research did not receive any specific grant from funding agencies in the public, commercial, or not-for-profit sectors.

KEYWORDS

- **disturbed landscape**
- **transport corridor**
- **revegetation**
- **plant survey**
- **slopes**

REFERENCES

1. Pickett, S. T. A.; White, P. S. *The Ecology of Natural Disturbances and Patch Dynamics*; Orlando: Academic Press, 1985; p 472.
2. Wu, J.; Loucks, O. L. From Balance of Nature to Hierarchical Patch Dynamics: A Paradigm Shift in Ecology. *J Quart. Rev. Biol.* **1995,** *70*(4), 439–466.

3. Turner, M. G.; Gardner, R. H.; O'Neill, R. V. *Landscape Ecology in Theory and Practice: Pattern and Process*; New York: Springer-Verlag, 2001; p 332.
4. Perry, J. N.; Liebhold A. M.; Rosenberg, M. S.; Dungan, J.; Miriti, M.; Jakomulska, A.; Citron-Pousty, S. Illustrations and Guidelines for Selecting Statistical Methods for Quantifying Spatial Pattern in Ecological Data. *Ecography* **2002,** *25*(5), 578–600.
5. Bennett, V. J.; Smith, W. P.; Betts, M. G. Toward Understanding the Ecological Impact of Transportation Corridors (General Technical Report). U.S. Department of Agriculture, Forest Service, Pacific Northwest Research Station, Portland, OR, 2011; p 40.
6. Forman, R. T. T.; Lauren, E. A. Roads and Their Major Ecological Effects. *Ann. Rev. Ecol. Syst.* **1998,** *29*, 207–231
7. Hoteev, V. V. The Formation of Vegetation in Oil-Contaminated Territories of Various Post-Climatic Zones of the Tyumen Region (PhD. thesis). Tyumen State University, Tyumen, 2003 (in Russian). https://www.dissercat.com/content/formirovanie-rastitel-nosti-na-neftezagryaznennykh-territoriyakh-razlichnykh-pochvenno-klimat
8. Kachinsky, N. A. *Soil Physics*; Higher School: Moscow, 1965; p 83 (in Russian).
9. Uglov, V. A. *Landscape-Ecological Variability of the Concentration of Phenes in Plant Populations. Phenetics of Populations*; Science: Moscow, 1982; pp 119–125 (in Russian).
10. Kuperman, F. M.; Ponomarev, V. I. *Diagnostics of Winter Hardiness of Winter Crops*; Moscow, 1971; p 134 (in Russian).
11. Zhuchenko, A. A. *The Adaptive Potential of Cultivated Plants: Ecologist, Foundations*; Academy of Sciences of the Moldavian SSR. Institute of Ecology and Genetics: Chisinau, Stinza, 1988; p 767 (in Russian).
12. Magomedmerzaev, M. M.; Huseynova; Z. A. *On Adaptive Strategies of Induced Species of Cultivated Alfalfa. Production Resources of Mountain Crop Production.* Makhachkala, 1996; pp 120–132 (in Russian).
13. Mirkin, B. M.; Mironova, S. I. On Biodiversity in Some Technogenic Successive Systems in Yakutia. *Ecology* **1999,** *4*, 226–270 (in Russian).
14. Kapelkina, L. P. On the Natural Development and Regulation of the Outer Lands of the North. *Adv. Mod. Nat. Sci.* **2012,** *11*, 98–102.
15. Vaughn, K. J.; Porensky, L. M.; Wilkerson, M. L.; Balachowski, J., Peffer, E., Riginos, C.; Young, T. P. Restoration Ecology. *Nat. Educ. Knowl.* **2010,** *3*(10), 66.
16. Baughman, O. W.; Agneray, A. C.; Matthew, L. F.; Kilkenny, F. F.; Espeland, E. K.; Fiegener, R.; Horning, M. E.; Johnson, R. C.; Kaye, T. N.; Ott, J. E.; Leger, E. A. Strong Patterns of Intraspecific Variation and Local Adaptation in Great Basin Plants Revealed Through a Review of 75 Years of Experiments. *Ecol. Evol.* **2019,** *9*(11), 6259–6275.

CHAPTER 11

Pigment Content in Plant Leaves as a Bio-Indicator of Adaptability to Growing Conditions

EUGENE M. LISITSYN*

N. V. Rudnitsky Federal Agricultural Research Center of the North-East, 166a Lenin Street, Kirov, 610007 Russian

Corresponding author. E-mail: edaphic@mail.ru

ABSTRACT

In the chapter, problem is considered of application of data on the content of pigments in plant leaves for (1) diagnostics of resistance to stressful abiotic factors; (2) differentiation of closely related taxonomic groups of plants on the example of two subspecies of oats: covered oats (*Avena sativa* subsp. *sativa* L.) and naked oats (*Avena sativa* subsp. *nudisativa* L.). It is shown that under influence of a stress of aluminum ions, conditions of harvesting of light energy change in high degree whereas processes of its transformation into organic substances in the reaction centers (RC) are rather protected from a stressor. At decrease in pH value of soil solution the content of chlorophyll *b* most decreased (by 53.22%). The total content of chlorophyll *a* decreased by 42.84% and depression concerned mostly light-harvesting complexes (LHCs) (52.71%), than RC of photosystems (16.98%). The depression of carotenoids' synthesis was the least considerable—only by 19.48%. Naked samples of oats reduce content of pigments in flag leaves under stressful soil growth conditions in significantly smaller degree that does them more attractive for breeding of aluminum- and acid-resistant cultivars of this agricultural crop. The researchers conducted in 2015–2018 showed significant differences in structural and quantitative composition of a pigment machinery of flag leaves in naked and covered oats at flowering stage of development. Naked oats contained higher amount of pigment as a part of LHCs, and covered oats—in RC of photosystems: in LHCs—9.53 ± 0.30 and 8.44 ± 0.30 mg/g, in RC—3.57 ± 0.07 and 4.11 ± 0.10 mg/g, respectively. The average content of chlorophyll *b* in leaves of covered samples of oats (7.03 ± 0.25 mg/g of dry matter) was 13.1% lower, then in naked samples (7.95 ± 0.25 mg/g of dry matter). Variability of pigments' content by years was significantly higher for group of covered samples (21.5%), than for naked one (17.9%). At naked oats chlorophyll *a/b* ratio was statistically significant lower, than at covered oats, respectively, 1.73 ± 0.02 and 1.92 ± 0.04. Thus, the content of photosynthetic pigments can be use as a diagnostic indicator of resistance to environmental factors (air temperature, conditions of moistening, existence of toxic ions of metals in the soil) at cultivars and breeding lines of cereal crops and as a diagnostic indicator of belonging to different genetic taxa.

11.1   INTRODUCTION

It is well known that climate changes take part not only in global scale but also in smaller territories such as within area of a one administrative region. These changes are causing to correction in modern cultivar models of agricultural crops or design new one. Existing approaches to design of such models consider scientific foundation in use of some morphological and physiological traits that reflect in the best way plant adaptation to growing conditions and industry demands to yield quality in the exact place of cultivation.[1,2] At the first stage, genetic and physiological and ecological analysis of initial material is done with a glance to exact soil and climatic conditions in region for which the new cultivar is planned to create. More often simple, easy estimated traits are considered such as structure elements of yield.[3–5] Other parameters, such as level of disease or abiotic stress resistance are more difficult to estimate and use in breeding practice. Taking into account the key role of stability in action of photosynthetic machinery in plant adaptation to stress impacts of abiotic nature,[6] structure and functional analysis of leaf pigment complexes must be one of most important direction in screening of collection and breeding specimens,[7,8] as well as wild-growing populations of newly economic valuable plant species (wild berries, medicinal plants, trees etc.).[9,10]

The prevailing form of chlorophyll in terrestrial plants is chlorophyll *a*, which locates both in reaction centers (RC) and in light-harvesting complexes (LHCs) of chloroplasts. Chlorophyll *b* locates in LHC only. Carotenoids are the accessory light-harvesting pigments; they help to dissipate excess energy and so protect photosystem from damage.[11]

During physiological genetic and breeding studies, the next traits of photosynthetic machinery are analyzed:

- total content of chlorophyll; content of chlorophylls *a* and *b*
- total content of carotenoids
- chlorophyll *a*/*b* ratio
- chlorophyll/carotenoids ratio
- pigment content in LHC and RC

Physiological data show that lowering in chlorophyll content is closely related with decrease in plant yield in field conditions.[12,13] Such lowering in pigment content may be linked with structural damage in thylakoid membranes and photosystem II.[14] Keeping of high level of leaves chlorophyll is a desirable trait in breeding for high temperature resistance because it pointed out lesser degree of photoinhibition at stressor action.[8,15,16] Lowering in chlorophyll *a*/*b* ratio may indicate increase of adaptive potential of plants under stress conditions and be indicator of plant resistance.[7,17] Literature data confirm that chlorophyll *a*/*b* ratio indirectly characterizes ratio between RC of photosystems I and II from one hand and size of peripheral antenna of photosystem II.[18]

Statistically significant increase in carotenoids content and total chlorophyll/carotenoids ratio points out activation of defense mechanisms under stress conditions (for example, imbalance in mineral nutrition, low pH, high amount of ions of aluminum, or heavy metals) because defense of photosynthetic machinery under stress is one of the carotenoids function.[19,20]

Therefore analysis of the quantitative content and structure of the pigment complex in plant leaves allows obtaining information on what features of this complex can increase the efficiency and quality of the received yield (grain, fruits, berries, vegetative mass, etc.).

11.2 MATERIALS AND METHODOLOGY

Experiment 1: During 2015–2018 on the sod-podzolic middle clay soil of the Falensky breeding station—branch of the FARC of North-East 20 breeding lines and cultivars of covered oats (*Avena sativa* L.) of a working collection of oats breeding department were tested. Experiment 2: Under the same growing conditions, 10 breeding lines and cultivars of covered oats (*A. sativa* L.) and 7 breeding lines and cultivars of naked oats (*A. nuda* L.) were tested. Names of samples are provided in tables.

Conditions of the growing periods of different research years varied considerably on the hydrothermal mode. In 2015, sufficient moistening and the increased temperature are noted (normal conditions of vegetation). The average yield on samples was 3.3 t/ha. In 2016, during the period from seedlings until flowering, critical for cereals on moisture consumption, the drought of average force is noted. As a result, sharp decrease in average yields to 1.8 t/ha was observed. The growing period of 2017 was characterized by excess moistening and a lack of the sum of effective temperatures, but at the same time the average yield remained at the level of 2015 (3.4 t/ha). In 2018, the deviation of the sum of precipitations from climatic norm made 6.0%, of air temperatures +0.6°C. The sum of effective temperatures was 11.0% higher than perennial average. So, it is possible to characterize 2018 as optimum for growth and development of plants.

For the qualitative and quantitative analysis of pigment structure, 25 individual plants in 4 replication of each genotype were count for the total leaves area, the area of flag and second leaves at a phase of the beginning of flowering. Assessment of content of pigments (chlorophyll *a*, chlorophyll *b*, carotenoids) was carried out with use of the UVmini-1240 spectrophotometer (SHIMADZU Corporation, Japan). Extraction of pigments and calculation of their contents was carried out according to[21] in acetone solutions (100% acetone).

The obtained data were processed statistically with use of software packages of Microsoft Office Excel 2007 and StatSoft Statistica 11. Average values with standard error are given in tables at significance level $p \leq 0.05$.

11.3   RESULTS AND DISCUSSION

Experiment 1: Results of estimation of pigments content in flag leaves of 20 samples of oats which are grown up on acid (pH 3.8, content of aluminum ions up to 23 mg/100 of the soil) and neutral (pH 6.0, without mobile aluminum) soil backgrounds of the Falensky breeding station, are given in Tables 11.1 and 11.2.

TABLE 11.1 Content of Pigments in Flag Leaves of Oats Growing on Neutral Soil Background, mg/g of Dry Matter. Average for 2015–2018.

Breeding line, cultivar	Chlorophyll a	Chlorophyll b	Carotenoids	Chlorophyll a/b ratio	Chlorophyll/carotenoids ratio	Chlorophyll a in LHC, %
2h12o	16.61 ± 0.48	11.19 ± 0.64	3.80 ± 0.05	1.49 ± 0.04	7.30 ± 0.23	80.58 ± 2.29
378h08	13.69 ± 0.55	7.96 ± 0.38	3.62 ± 0.10	1.72 ± 0.02	5.96 ± 0.09	69.64 ± 0.65
397h07	12.14 ± 0.41	6.98 ± 0.32	3.41 ± 0.05	1.74 ± 0.02	5.59 ± 0.14	68.88 ± 0.90
325h12	14.46 ± 0.32	9.96 ± 0.30	3.31 ± 0.06	1.45 ± 0.02	7.38 ± 0.13	82.58 ± 1.00
256h12	16.34 ± 0.46	10.51 ± 0.54	3.85 ± 0.12	1.57 ± 0.04	7.00 ± 0.31	76.96 ± 2.16
188h12	14.65 ± 0.64	9.88 ± 0.62	3.38 ± 0.12	1.49 ± 0.04	7.24 ± 0.22	88.35 ± 1.41
I-4388	9.36 ± 0.47	4.84 ± 0.26	3.03 ± 0.09	1.94 ± 0.03	4.68 ± 0.09	74.89 ± 0.66
I-4592	12.77 ± 0.67	7.24 ± 0.43	3.44 ± 0.20	1.77 ± 0.04	5.82 ± 0.06	67.92 ± 1.34
I-4346	12.67 ± 0.45	7.13 ± 0.32	3.44 ± 0.10	1.78 ± 0.04	5.76 ± 0.13	67.43 ± 1.38
I-4808	15.77 ± 0.54	10.18 ± 0.31	3.75 ± 0.11	1.55 ± 0.02	6.93 ± 0.10	77.56 ± 1.18
I-4815	12.15 ± 0.67	6.76 ± 0.66	3.40 ± 0.08	1.83 ± 0.07	5.55 ± 0.31	66.04 ± 2.86
I-4895	15.09 ± 0.26	10.00 ± 0.30	3.51 ± 0.07	1.51 ± 0.03	7.15 ± 0.19	79.51 ± 1.61
k-3126	15.01 ± 0.24	9.77 ± 0.24	3.41 ± 0.04	1.54 ± 0.02	7.26 ± 0.11	78.09 ± 1.17
k-3622	15.03 ± 0.59	9.28 ± 0.58	3.65 ± 0.09	1.63 ± 0.04	6.66 ± 0.19	73.83 ± 1.96
k-3752	10.60 ± 2.19	6.67 ± 1.38	2.57 ± 0.53	1.59 ± 0.01	6.71 ± 0.05	75.34 ± 0.57
k-3754	14.58 ± 0.30	9.46 ± 0.27	3.45 ± 0.05	1.54 ± 0.02	6.96 ± 0.12	77.82 ± 1.16
k-3862	14.17 ± 0.37	9.13 ± 0.21	3.38 ± 0.11	1.56 ± 0.03	6.91 ± 0.10	77.45 ± 1.27
Boets	14.58 ± 0.56	8.53 ± 0.40	3.94 ± 0.13	1.71 ± 0.02	5.85 ± 0.08	70.11 ± 0.86
15.330	13.12 ± 0.38	6.99 ± 0.28	3.92 ± 0.08	1.88 ± 0.03	5.12 ± 0.08	63.81 ± 1.01
15.331	14.29 ± 0.53	7.80 ± 0.28	3.99 ± 0.15	1.83 ± 0.02	5.55 ± 0.08	65.59 ± 0.85

TABLE 11.2 Content of Pigments in Flag Leaves of Oats Growing on Acid Soil Background, mg/g of Dry Matter. Average for 2015–2018.

Breeding line, cultivar	Chlorophyll a	Chlorophyll b	Carotenoids	Chlorophyll a/b ratio	Chlorophyll/ carotenoids ratio	Chlorophyll a in LHC, %
2h12o	6.75 ± 0.49	3.27 ± 0.25	2.84 ± 0.11	2.07 ± 0.07	3.51 ± 0.15	58.19 ± 1.93
378h08	9.48 ± 0.31	4.61 ± 0.21	3.30 ± 0.09	2.07 ± 0.06	4.28 ± 0.12	58.26 ± 1.53
397h07	6.67 ± 0.43	3.26 ± 0.24	2.48 ± 0.12	2.05 ± 0.07	3.98 ± 0.10	58.71 ± 1.77
325h12	7.99 ± 0.38	4.07 ± 0.23	2.87 ± 0.13	1.97 ± 0.06	4.20 ± 0.09	61.17 ± 1.70
256h12	7.76 ± 0.63	3.79 ± 0.33	3.04 ± 0.11	2.05 ± 0.05	3.78 ± 0.19	58.61 ± 1.27
188h12	8.61 ± 0.48	4.43 ± 0.24	2.97 ± 0.11	1.95 ± 0.03	4.37 ± 0.08	61.72 ± 0.81
1-4388	10.01 ± 0.23	5.30 ± 0.17	3.31 ± 0.06	1.89 ± 0.05	4.62 ± 0.08	63.61 ± 1.76
1-4592	9.87 ± 0.55	5.15 ± 0.34	3.34 ± 0.11	1.92 ± 0.02	4.47 ± 0.12	62.46 ± 0.80
1-4346	5.48 ± 0.23	2.77 ± 0.11	2.42 ± 0.07	1.98 ± 0.04	3.40 ± 0.06	60.78 ± 1.23
1-4808	4.07 ± 0.10	2.03 ± 0.09	1.96 ± 0.04	2.02 ± 0.08	3.12 ± 0.06	59.82 ± 2.04
1-4815	9.18 ± 0.26	4.47 ± 0.12	3.31 ± 0.11	2.05 ± 0.01	4.12 ± 0.05	58.48 ± 0.28
1-4895	5.64 ± 0.29	2.76 ± 0.12	2.36 ± 0.10	2.04 ± 0.02	3.56 ± 0.04	58.71 ± 0.46
k-3126	5.69 ± 0.22	2.68 ± 0.16	2.26 ± 0.06	2.16 ± 0.17	3.71 ± 0.10	56.85 ± 3.47
k-3622	6.40 ± 0.32	3.22 ± 0.18	2.40 ± 0.12	1.99 ± 0.03	4.02 ± 0.11	60.28 ± 1.00
k-3752	8.54 ± 0.22	4.33 ± 0.17	2.82 ± 0.06	1.98 ± 0.04	4.56 ± 0.08	60.70 ± 1.35
k-3754	7.26 ± 0.23	3.52 ± 0.15	2.58 ± 0.09	2.07 ± 0.03	4.18 ± 0.04	58.04 ± 0.99
k-3862	7.97 ± 0.31	3.83 ± 0.17	2.75 ± 0.09	2.08 ± 0.04	4.28 ± 0.08	57.74 ± 1.15
Boets	9.18 ± 0.25	4.53 ± 0.17	2.97 ± 0.07	2.03 ± 0.04	4.52 ± 0.07	59.17 ± 1.24
15330	8.60 ± 0.25	4.39 ± 0.16	3.08 ± 0.06	1.96 ± 0.03	4.22 ± 0.06	61.19 ± 0.90
15331	9.70 ± 0.20	4.76 ± 0.10	3.10 ± 0.06	2.04 ± 0.02	4.66 ± 0.06	58.91 ± 0.68

As it is appearing from data in Tables 11.1 and 11.2, the studied samples differed significantly from each other on the content of single elements of a pigment complex of chloroplasts of flag leaves on both soil backgrounds.

Under conditions of a neutral soil background, the average content of pigments in flag leaves of covered oats was 13.54; 8.23, and 3.49 mg/g of dry matter for chlorophyll *a, b,* and carotenoids, respectively. At change of soil conditions, these average values decreased to level 7.74; 3.86 and 2.81 mg/g of dry matter. Thus, the average level of a depression was for chlorophyll *a*—42.8%; for chlorophyll *b*—53.1%; and for carotenoids—19.5%.

If to consider distribution of pigments by structural parts of photo-systems, then it is possible to note that at average decrease in content of chlorophyll *a* by 42.8%, the depression concerned in higher degree LHCs (52.71%), than the RC of photosystems (16.98%).

Thus, it is possible to assume that under conditions of the acid-soil stress caused by toxic action of ions of aluminum (the major toxic factor of acid sod-podzolic soils) conditions of harvesting of light energy change mostly whereas processes of its transformation into organic substances in the RC are sufficiently protected from a stressor. The content of carotenoids decreases approximately in the same degree, as chlorophyll *a* in the RC that can demonstrate minor changes in dissipation of thermal energy at change of conditions of the soil environment.

However, average values can be used for the comparative analysis of stress influence on pigment complexes of different grain crops. For breeding work, it is much more important to select those samples, which considerably differ from average values for all set of samples. In that case it is possible to note that least depression in chlorophyll *a* content at acidification of the soil concerned such breeding lines as k-3752 (19.4%), I-4815 (24.4%), and at a line I-4388 stimulating effect is noted, though insignificant statistically—the content of a pigment increased by 6.9%.

Change in content of chlorophyll *b* was the smallest for breeding lines I-4592 (decrease by 28.9%), I-4815 (for 33.9%), I-3752 (for 35.0%); as to breeding line I-4388—in this parameter statistically significant depression is not noted too.

According to these data, it is possible to offer for further profound breeding work the breeding lines listed in the previous paragraphs as having the most resistant pigment complexes in the conditions of an aluminum-acid soil background.

Experiment 2: For revealing possibilities of using data on content of pigments in flag leaves at differentiating of two oats species, experiment 2 was conducted. Data on content of pigments are presented in Table 11.3.

As it is shown in Table 11.3, on average for years of a research, covered oats contained 12.54 ± 0.35 mg of chlorophyll *a* in a gram of leaves dry matter at phase of flowering, and naked oats—13.10 ± 0.32 mg. The variability by years of researches was higher at lines of covered oats: the average coefficient of variation of a chlorophyll *a* in leaves made 12.1% whereas for naked samples—9.6%.

Covered breeding lines 378h08, I-4584, and a naked line 683h05 showed the content of a pigment stable by years (variability is less than 5%). Covered lines 168h10 (variation coefficient by years—17.3%), I-4595 (18.3%), and I-4618 (27.6%) reacted to change of cultivation conditions most strongly.

However, it is possible to note distinctions in quantitative distribution of these pigments between the RC of photosystems and their antenna LHCs. In general, naked samples contained statistically significant higher amount of a pigment as a part of LHC of antennas, and covered samples— in the RC of photosystems: in LHC—9.53 ± 0.30 and 8.44 ± 0.30 mg/g, in RC—3.57 ± 0.07 and 4.11 ± 0.10 mg/g, respectively.

Possibly, a high part of chlorophyll *a* in RC of photosystems, where process occurs of transformation of light energy into the chemical one going for synthesis of organic substances, led to the fact that the average grain mass per panicle in covered oats exceeded for 28.5% a similar indicator for naked samples (1.76 ± 0.02 and 1.37 ± 0.02 g accordingly). Differences in 1000-grains mass had approximately the same level—29.4% (38.56 ± 0.32 for covered and 29.70 ± 0.32 g for naked samples).

The chlorophyll *b* is present in leaves of the higher plants only as a part of LHCs of photosystems and, unlike a chlorophyll *a*, does not play a role of primary donor of electrons in the reactionary centers.[11] Its average content in leaves of covered samples of oats during the researches was 7.03 ± 0.25, and at naked samples—7.95 ± 0.25 mg/g of dry matter, that is, is 13.1% higher.

These data can also confirm lower efficiency of transformation of solar energy to organic substances in chloroplasts of flag leaves in naked oats, in comparison with covered samples.

The variability of pigments content by years was significantly higher for group of covered oats (21.5%), than for naked oats (17.9%). Strong reaction to conditions of cultivation had samples of covered oats 168h10

TABLE 11.3 Content of Pigments in Flag Leaves of Covered and Naked Oats Growing on Neutral Soil Background, mg/g of Dry Matter. Average for 2015–2018.

Breeding line, cultivar	Neutral soil background, pH 6.0			Acid soil background, pH 3.8		
	Chlorophyll *a*	Chlorophyll *b*	Carotenoids	Chlorophyll *a*	Chlorophyll *b*	Carotenoids
	Covered oats					
168h10	11.48 ± 0.99	6.38 ± 0.84	2.92 ± 0.26	5.52 ± 0.42	2.34 ± 0.68	1.95 ± 0.21
397h07	11.66 ± 0.32	6.29 ± 0.48	3.05 ± 0.16	6.53 ± 0.30	3.30 ± 0.26	2.26 ± 0.04
2h09	11.80 ± 0.32	6.43 ± 0.37	3.04 ± 0.14	6.03 ± 0.21	2.91 ± 0.89	2.11 ± 0.25
378h08	12.75 ± 0.36	7.46 ± 0.99	3.06 ± 0.16	5.51 ± 0.23	2.60 ± 0.41	1.96 ± 0.07
3h14	12.96 ± 0.33	7.04 ± 0.15	3.34 ± 0.26	6.24 ± 0.31	2.92 ± 0.39	2.25 ± 0.05
I-4584	15.11 ± 0.50	8.95 ± 0.91	3.74 ± 0.26	7.90 ± 0.55	4.14 ± 0.52	2.53 ± 0.35
I-4592	11.48 ± 0.47	6.58 ± 0.58	2.93 ± 0.12	7.08 ± 0.23	3.55 ± 0.97	2.37 ± 0.32
I-4595	12.72 ± 0.33	6.90 ± 0.77	3.35 ± 0.38	6.19 ± 0.23	3.01 ± 0.84	2.18 ± 0.31
I-4618	13.16 ± 0.63	7.25 ± 0.33	3.37 ± 0.71	6.40 ± 0.99	3.04 ± 0.66	2.25 ± 0.30
Argamak	12.33 ± 0.15	7.03 ± 0.36	3.05 ± 0.10	7.89 ± 0.55	3.63 ± 0.78	2.53 ± 0.22
	Naked oats					
1h07	13.19 ± 0.94	7.88 ± 0.18	3.18 ± 0.07	7.00 ± 0.67	3.23 ± 0.57	2.46 ± 0.17
857h05	12.98 ± 0.89	8.09 ± 0.97	3.11 ± 0.17	7.79 ± 0.37	4.15 ± 0.93	2.51 ± 0.34
7h12o	13.70 ± 0.85	8.31 ± 0.15	3.27 ± 0.08	8.54 ± 0.70	4.34 ± 0.68	2.69 ± 0.15
683h05	14.53 ± 0.63	9.05 ± 0.99	3.36 ± 0.07	7.88 ± 0.37	4.10 ± 0.93	2.53 ± 0.24
629h09	11.92 ± 0.37	7.13 ± 0.48	2.92 ± 0.17	9.24 ± 0.68	4.94 ± 0.83	3.31 ± 0.69
14h12o	12.36 ± 0.67	7.21 ± 0.62	3.06 ± 0.15	8.42 ± 0.61	4.22 ± 0.63	2.68 ± 0.09
Vyatsky	13.03 ± 0.26	7.94 ± 0.52	3.14 ± 0.07	8.63 ± 0.29	4.69 ± 0.87	2.67 ± 0.28

(coefficient of a variation is 28.9%) and I-4618 (32.1%). The greatest stability of pigments content among a set of covered oats was shown by a breeding line I-4584 (10.2%), among naked oats—a breeding line 683h05 (11.0%).

Comparison of average content of chlorophyll in flag leaves of oats plants by years of a research showed independence of pigments' accumulation on climatic conditions of growth. It is known that optimum temperature for synthesis of a chlorophyll are 30°C,[22] and considering that the average temperature of July for years of researches was 14.8 (2015), 20.0 (2016), 17.2 (2017), and 20.3°C (2018); it was possible to expect the greatest accumulation of pigments in 2016 and 2018, and the smallest one—in 2015. The second factor—amount of precipitations—is also closely linked with processes of chlorophyll synthesis.[22] The lack of water in leaves leads to destruction of molecules of a chlorophyll; therefore, it was possible to expect significant differences in accumulation of chlorophyll in leaves of plants in 2015 and 2016: The hydrothermal coefficient of July in 2015 made 1.45 (excess moistening), and in 2016—only 0.32 (severe drought). July, 2017 under moistening conditions was close to 2015 (hydrothermal coefficient = 1.32). These discrepancies of theoretically expected results and actually obtained data demand a further research.

Except the total content of different forms of a chlorophyll, such parameter as a mass ratio of a chlorophyll *a*/*b* can give information on ecological resistance of a plant. This parameter is widely used, for example, in ecological researches[23] and is linked with adaptation to habitat illumination conditions.

In the review analysis of influence of external factors on structure of a pigment complex of the higher plants made by,[24] authors came to a conclusion that the very different stressful factors (low or high temperatures, a drought, salinity of a substratum, ozone, carbon dioxide, a season of year) exert rather weak impact on a chlorophyll *a*/*b* ratio; the main impact is exerted by such factor as light. As the studied set of oats samples was grown up in the identical conditions of illumination, distinction in ratio of pigments have to reflect genetic structure of these samples.

Usually, a chlorophyll *a*/*b* ratio in the range of 2.5–4.0 is characteristic of plants.[25] As the light harvesting pigment-protein complex of photosystem I (LHC-I) has the stable ratio of these pigments equal to three, all variability in level of this ratio in the whole leaf is explained by change of structure of a LHC of photosystem II (LHC-II).[26,27]

In general, at naked oats the chlorophyll *a*/*b* ratio in all years of study was statistically significant lower, than at covered oats, respectively 1.73 ± 0.02 and 1.92 ± 0.04. These data can be interpreted in such a way that at naked oats chloroplasts of flag leaves in a stage of flowering have bigger antenna complexes of photosystem II. Low value of parameter can be explained by the prevalence of cloudy weather within a week before sampling. According to data of the website https://yandex.ru/pogoda/101105/month/july, on average for the last 10 years only 11 clear days were observed in July near settlement Falenki. The website http://www.meteo-tv.ru/rossiya/kirovskaya-obl/falenki/weather/climate/ specifies that on average for 1961–2010 in this area it were 12–14 rainy days in July. As the main role of photosystem II is connected with processes of a water photolysis, it is possible to assume that moisture supplying of oats flag leaves was at the high level, and probable even with excess of optimum values, as demanded higher arrival of light energy to photosystem, that is, increase in the sizes of its antenna complexes.

As for influence of conditions of different years, both types of oats had similar pigment complexes in 2015 and 2016 on this parameter. In the conditions of 2017, the ratio of a chlorophyll *a*/*b* in flag leaves of plants was 1.4–1.5 times higher.

Important part of the photosynthetic machinery of leaves of the higher plants are carotenoids: They can catch light energy in those parts of a spectrum which is poorly used by a chlorophyll; in the conditions of a surplus of the absorbed energy carotenoids dissipate it in the form of heat, thus protecting the RC from photodamage.[11] In general, even in optimum conditions of growth and development only about 80% of the absorbed light energy is used in the course of photosynthesis; under the influence of environmental factors and a plant's physiological condition in this part can decrease ever more.[28]

In the studied set of samples, we were not revealed statistically significant differences in the carotenoids content between naked and covered oats; the content of supplementary pigments of photosynthesis in flag leaves made, respectively, 3.15 ± 0.05 and 3.18 ± 0.08 mg/g of dry matter.

Variation of the trait by years of a research at naked oats was at the level of an standard error—only at two breeding lines it exceeded 5% (line 857h05—5.6%, line 629h09—5.7%). Among a set of covered oats, the variability of the trait was higher, reaching 11.5% at a breeding line I-4595 and 21.0% at breeding line I-4618. At other lines and cultivars, the coefficient of a variation fluctuated from 2.5 (a line 3h14) up to 9.1% (a line 168h10).

In ecological and breeding researches of photosynthesis process, such indicator of action of the pigment machinery as mass ratio chlorophyll/ carotenoids is often used. Most often, this ratio is used for evaluation of extent of destruction in chlorophyll molecules because in the course of leaf aging or during reaction to a stress, a chlorophyll collapses earlier then carotenoids owing to what mass ratio of pigments decreases.[29] On average for plants of an open habitat, the ratio of content chlorophyll/carotenoids varies from 4.2 to 5.0, and for growing in a shadow—from 5.5 to 7.0.[26]

In our researches, this ratio was higher at naked oats in all years of observations, than at covered oats: respectively, 6.65 ± 0.08 and 6.09 ± 0.10. It can indicate favorable conditions for synthesis of pigments (absence of stresses and damages of leaf blades) and that flag leaves actively grew and functioned at this time. In 2017, the value of parameter was significantly lower than in 2015 and 2016 by 1.4 times, nevertheless, without going beyond the lower limits of norm for plants of open habitats.

Cluster analysis is another way to identification differences between two groups of genotypes of oats that allow to present these distinctions graphically taking into account the content of all types of pigments at the same time. Distribution of pigments between structural parts of photosystems and degree of their depression at cultivation in stressful conditions in comparison with control was considered. The received results are presented in Figures 11.1–11.3.

In Figure 11.1, the grouping of the studied oats samples on two large clusters at their cultivation on a neutral soil background is shown.

As it is followed from data of Figure 11.1, at cultivation on a neutral soil background the studied genotypes tend to a differentiation though at the same time the covered breeding line I-4584 settles down in the same cluster as majority of naked samples, and naked breeding lines 629h09 and 14h12o are grouped with covered oats.

C, covered genotypes; N, naked genotypes. Numbers following letters correspond to number of breeding line/cultivar in Table 11.3.

At transition to stressful conditions of acid soil field, the studied genotypes are also distributed in two big clusters, but already only one sample of each group unite with samples of other group (Fig. 11. 2).

So, on reaction of a pigment complex to stressful soil factors, the naked breeding line 1h07 behaves like covered genotypes, and the covered breeding line I-4584 is again distributed with group of naked genotypes.

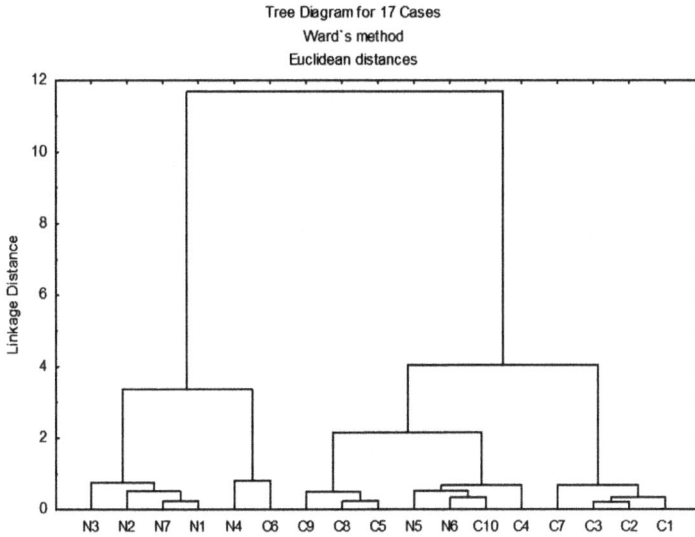

FIGURE 11.1 Distribution of oats genotypes on clusters considering the content of pigments and their distribution by structural parts of photosystems at cultivation on a neutral soil background (average data for 2015–2018).

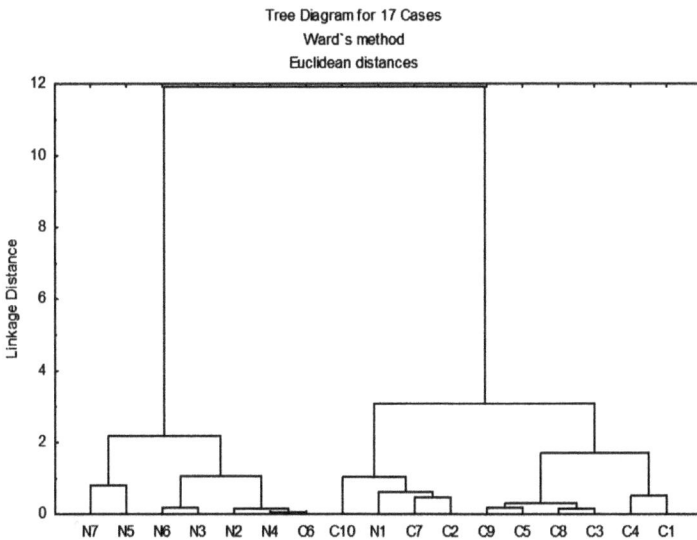

FIGURE 11.2 Distribution of oats genotypes on clusters taking into account the content of pigments and their distribution by structural parts of photosystems at cultivation on an aluminum-acid soil background (average data for 2015–2018) (designations—see Figure 11.1).

Thus, at neutral reaction of soil solution, investigated groups of samples of naked and covered oats generally differ by distribution of chlorophyll between the RC and LHCs of photosystems: naked oats have the higher content of both forms of chlorophyll in LHC, and covered one—on the contrary—contain the most part of chlorophyll in the RC.

There are practically no significant differences in the content of carotenoids, mass ratio chlorophyll *a/b*, and total chlorophyll/carotenoids.

In the conditions of acid reaction of soil solution naked oats have significant advantage over covered oats as on the total keeping of all three groups of pigments, and on the mass content of chlorophyll *a* in both structural parts of photosystems—in the RC, and in LHCs.

The only parameter having higher value at group of covered oats is the mass ratio chlorophyll *a/b*. However, in general, this difference is not beyond statistically significant differences.

As the result of such different reaction to acidity of soil solution, a plant of naked oats reduces the content of pigments in flag leaves under stressful soil conditions of growth in significantly smaller degree than covered one. This fact makes them more attractive in breeding of acid- or aluminum-resistant cultivars of this agricultural crop.

However, when accounting degree of a depression of pigments' synthesis and their redistributions between the RC and antenna LHCs of photosystems of chloroplasts in flag leaves of oats plants, naked genotypes are accurately distributed in one cluster where, except them, they also get two covered genotypes—cultivar Argamak and breeding line I-4592 (Fig. 11.3).

The reasons of inconsistency of results in use of data on the pigments' content by different researchers. Despite high efficiency of initial photophysical and photochemical stages (about 95%), only some percent of solar energy passes into a harvest. Losses are caused by limitation of process at the biochemical and physiological levels as well as incomplete absorption of light. In all plant communities, the received solar radiation is used inefficiently. Among the factors limiting primary production of terrestrial ecosystems there are: (1) the lack of water limiting photosynthesis speed; (2) the shortage of elements of mineral nutrition, which slowing down speed of formation of assimilating tissues and reducing efficiency of photosynthesis; (3) temperature, adverse for growth; (4) pass of the most part of solar radiation beside photosynthesizing organs (because of seasonal abscission of leaves, action of phytophages, and parasites); (5)

low efficiency of photosynthesis in leaves (even under ideal conditions in the most productive agricultural systems it seldom exceed 10% of photosynthetic active radiation). Maximum efficiency of photosynthesis of intensive cereal crop cultivars under ideal conditions is about 3–10%.

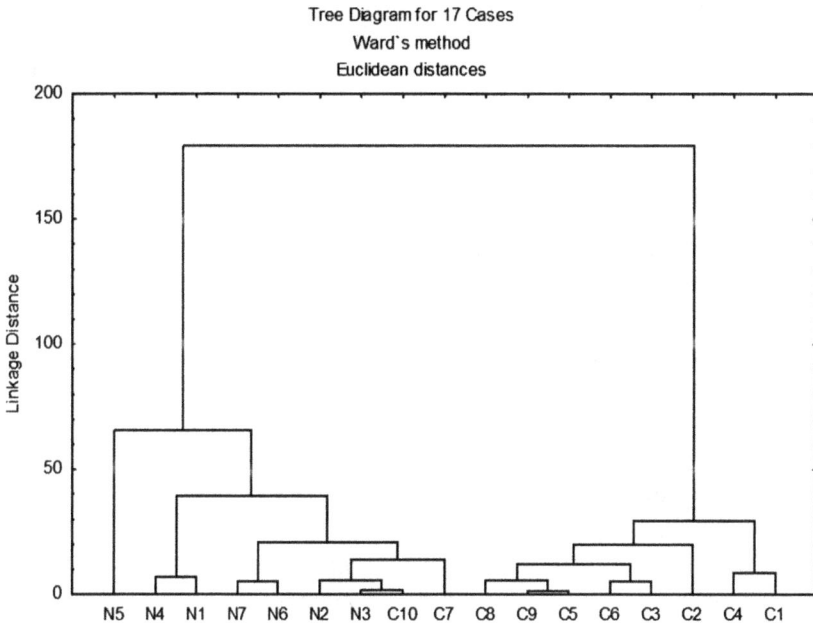

FIGURE 11.3 Distribution of oats genotypes on clusters considering degree of a depression of pigments' content and their distribution between structural parts of photosystems under the influence of an aluminum-acid soil stress (average data for 2015–2018) (designations—see Figure 11.1)

The total amount of the organic substance which is saved up for the separate period of time depends not only on photosynthesis, but also on opposite directed processes of respiration and photorespiration. Besides, the mass of a plant can change in dependence of change of direction of synthetic processes: so, for example, cellulose is 10% lighter, than glucose of which it was formed. At the same time, it is necessary to consider and waste of various parts: leaves, roots, root hairs, a root cap, etc. that especially considerably affects balance of organic substance at wood plants.

The content of pigments in leaf blades of cereal crops is the inherited sign; however, process of synthesis of chlorophyll is under double genetic control: most genes belong to chloroplasts, but about 1/3 of genes governing pigment synthesis is nuclear genes.

Amount of photosynthetic pigments in plant leaves depends on their age state and a phase of development. From the leaves, lower to the top layers, the content of a chlorophyll, as a rule, increases and usually it become as much as possible in a flag leaf before earing stage.

In some conditions, the content of pigments can be more in flag and second leaves concerning the below-located leaves, in others—contra verse. It is necessary to assume that to some extent it demonstrates relative autonomy of the phytomers providing ecological plasticity of a cultivar.

One of the reasons is that changes of pigments content in leaves under stress conditions can be caused both by the accelerated destruction of pigments, and violation of process of their biosynthesis. Besides, in the course of degradation, the chlorophyll *b* can be converted into chlorophyll *a* that will lead to increase in the total content of chlorophyll.

There is no simple, magic gene of resistance to abiotic stresses. Resistance to any abiotic stress is the complex trait including the whole network from a set of gene modules. At the same time, different mechanisms are controlled by different genetic systems. Both in sensitive, and in cultivars, resistant against stresses, increase in chlorophyll content can be registered; however, at each of these group, different processes lead to such increase: at sensitive cultivars—at the expense of a reduction of leaf area; at resistant one—due to strengthening of chlorophyll synthesis.

Influence of growth conditions at different year of cultivation. On average in 4 years of researches only six pair of statistically significant correlations are revealed: the mass ratio of chlorophyll *a/b* was higher at the cultivars having the higher mass of grain per panicle ($r = 0.411$), per plant ($r = 0.485$), and 1000-grain mass ($r = 0.418$). The same parameters of yield structure were statistically higher at cultivars having high part of chlorophyll *a* in the RC of photosystems of a flag leaf ($r = 0.378$; 0.429, and 0.448). In 2015, normal on moistening level, the minimum number of statistically significant pair of correlations in the studied set of oat cultivars is noted between elements of yield structure and elements of pigment machinery: the cultivars having higher amount of spikelets per panicle were characterized by lower content of a total chlorophyll and chlorophyll *b* in a flag leaf ($r = -0.384$ and -0.369), but at the same time the most part of chlorophyll *a* belongs to the RC of photosystems ($r = 0.376$).

The number of pair of statistically significant correlations sharply increased in years, adverse by moisture supply. So, at excess moistening of 2017, the cultivars containing higher amount of chlorophyll *b* in a flag leaf had the smaller mass of grain per plant (r = −0.381). Increase in a part of chlorophyll *a* in the total fund of chlorophyll was observed at cultivars with the raised number of spikelets (r = 0.439) and grains (r = 0.398) in a panicle, higher mass of grain per panicle (r = 0.519), per plants (r = 0.600) and 1000-grain mass (r = 0.370). It was characteristic the higher part of chlorophyll *a* in the RC of photosystems, in which processes of cultivar transformation of the reserved light energy into primary assimilates take place. Correlation coefficients with the mentioned parameters of panicle efficiency were within 0.374–0.537.

Under drought conditions of 2016, statistically significant interrelations between the content of pigments in flag leaves and structural development of reproductive organs were observed: at the cultivars having higher values of signs "panicle length," "mass of a panicle," "number of spikelets per panicle," and "number of grains per panicle" it was increased content of chlorophyll *a* (r = 0.368; 0.540; 0.455, and 0.552, respectively), chlorophyll *b* (r = 0.386; 0.604; 0.409, and 0.589, respectively). The increased content of carotenoids was followed by increase in number of spikelets and grains per panicle (r = 0.403 and 0.389). In a drought condition, however, increase in value of such parameters as mass of a panicle and amount of grains in it was followed by decrease in a part of chlorophyll *a* in the total pool of pigments and as a part of the RC (r = −0.531; −0.461; −0.521, and −0.465, respectively).

There are some reasons for such low number of statistically significant correlations between elements of yield structure of oat plants and parameters of a state of pigment machinery in flag leaves.

First, genotypic variability in elements of yield structure was much higher than in parameters of pigment complex. The variability in content of pigments under normal conditions of moistening was about 10–12% whereas for structural parameters—from 17 to 23%. Decrease in coefficients of a variation in drought conditions of 2016 to 11–18% led to appearance of much more significant correlations. Possibly, indicators of pigments content in a flag leaf are more conservative for species *A. sativa* L. and *A. nuda* L., than structural parameters of generative organs, or it indicates lack of aimed breeding on parameters of development of the pigment machinery of leaves.

Second, rather high degree of resistance of a pigment complex of flag leaves to adverse abiotic growth conditions, than of structural components of reproductive organs. In scientific literature there are indications on the similar facts. It is specified also that stressful environmental conditions of cultivation (except for illumination conditions) make rather weak impact on a ratio of pigments.

Third, in spite of the fact that use of a flag leaf of cereal crops for evaluation of pigments content is explained by their maximum contents in a flag leaf in a flowering stage, the flag leaf brings the greatest, but not the only contribution to synthesis of assimilates. Possibly, there is a need of evaluation of pigments content and their variability for a second leaf of oat plants.

Fourth, in the made experiments, only the content of pigments per unit of leaf dry matter was evaluated. This indicator does not consider the nature of distribution of pigments in a total leaf volume and the total content of pigments in whole leaf. Possibly, use of such parameters in the subsequent researches will allow to reveal closer correlations with the values of elements of panicle productivity.

Absence of aimed oats breeding on change of parameters of a state of pigment machinery in leaves can also assume so far. Possibly, the cultivars attracted in crossings during selection programs in various breeding centers have similar parameters of a structural-functional condition of a pigment complex. It is necessary to notice that similar programs were actively carried out with other cereal crop—wheat. Data on the content of chlorophyll in leaves of oats are most often used in works on a stress physiology as evaluation parameters of genotypes resistance to adverse factors of the cultivation conditions.

11.4   CONCLUSIONS

Obtained data show that under aluminum stress, content of chlorophyll *a* in flag leaf of oats plant depressed by 42.8%; of chlorophyll *b*—by 53.1%, and of carotenoids—by 19.5% in average. The most resistant breeding lines and cultivars had higher content of each pigment. Distribution of pigment between structural parts of photosystems changes significantly: lowering of pigment content in LHCs was 52.71%, in RC of photosystems—16.98%.

Least depression in chlorophyll *a* content at acidification of the soil was in breeding lines k-3752 (19.4%), I-4815 (24.4%). Change in content of chlorophyll *b* was the smallest for breeding lines I-4592 (decrease by

28.9%), I-4815 (for 33.9%), I-3752 (for 35.0%). Breeding line I-4388 did not show any statistical changes in pigment content.

On average for years of a research, covered oats contained 12.54 ± 0.35 mg of chlorophyll *a* in a gram of leaves dry matter at phase of flowering, and naked oats—13.10 ± 0.32 mg. The variability by years of researches was higher at lines of covered oats: the average coefficient of variation of a chlorophyll *a* in leaves made 12.1% whereas for naked samples—9.6%.

Average content of chlorophyll *b* in leaves of covered oats during the researches was 7.03 ± 0.25, and at naked oats—7.95 ± 0.25 mg/g of dry matter, that is, 13.1% higher.

Comparison of average content of chlorophyll in flag leaves of oats plants by years of a research showed independence of pigments' accumulation on climatic conditions of growth.

At neutral reaction of soil solution investigated groups of samples of naked and covered oats generally differ by distribution of chlorophyll between the RC and LHCs of photosystems: naked oats have the higher content of both forms of a chlorophyll in LHC, and covered one—on the contrary—contain the most part of chlorophyll in the RC. There are practically no significant differences in the content of carotenoids, mass ratio chlorophyll *a/b*, and total chlorophyll/carotenoids.

Taking together these data justified that analysis of pigment content in flag leaf can be use as a diagnostic indicator of resistance to environmental factors (air temperature, conditions of moistening, existence of toxic ions of metals in the soil) at cultivars and breeding lines of cereal crops and as a diagnostic indicator of belonging to different genetic taxa.

KEYWORDS

- **oats**
- **chlorophyll**
- **carotenoids**
- **light-harvesting complex**
- **soil**
- **aluminum**
- **abiotic stress**

REFERENCES

1. Kumakov, V. A. *Physiological Foundation of Models of Wheat Cultivars*; Agropromizdat: Moscow, 1985; p 270 (in Russian).
2. Grebennikova, I. G.; Aleynikov, A. F.; Stepochkin, P. I. The Construction of a Spring Triticale Variety Model on the Basis of Modern Information Technologies. *Comput. Technol.* **2016**, *21*, 53–64 (in Russian).
3. Batalova, G. A. *Oats in Volga-Vyatka Region;* Orma Publ.: Kirov, 2013, p 288 (in Russian).
4. Korjakovtseva, L. A.; Volkova, L. V. Ground of Parameters of Model of a High-Yield Variety of Spring Soft Wheat for Conditions of Non-Chernozem Zone of Russia. *Agrar. Sci. Euro-North-East.* **2014**, *6*, 13–18 (in Russian).
5. Shchennikova, I. N. Models of Spring Barley's Varieties for Conditions of Volga-Vyatka Region. *Agrar.Sci. Euro-North-East* **2015**, *6*, 9–13 (in Russian).
6. Kreslavski, V. D.; Carpentier, R.; Klimov, V. V.; Murata, N.; Allakhverdiev, S. I. Molecular Mechanisms of Stress Resistance of Photosynthetic Apparatus. *Biochem. (Moscow) Suppl. Seri. Membr. Cell Biol.* **2007**, *24*(3), 195–217 (in Russian).
7. Nahakpam, S. Chlorophyll Stability: A Better Trait for Grain Yield in Rice under Drought. *Indian J. Ecol.* **2017**, *44*(Special Issue 4), 77–82.
8. Cao, X.; Mondal, S.; Cheng, D.; Wang, C.; Liu, A.; Song, J.; Li, H.; Zhao, Z.; Lin, J. Evaluation of Agronomic and Physiological Traits Associated with High Temperature Stress Tolerance in the Winter Wheat Cultivars. *Acta Physiol. Plant* **2015**, *37*(4), 90.
9. Petridis, A.; van der Kaay, J.; Chrysanthou, E.; McCallum, S.; Graham, J.; Hancock, R. Photosynthetic Limitation as a Factor Influencing Yield in Highbush Blueberries (*Vaccinium corymbosum*) Grown in a Northern European Environment. *J. Exp. Bot.* **2018**, *69*(12), 3069–3080.
10. Li, X.; Xiao, J.; He, B. Chlorophyll Fluorescence Observed by OCO-2 is Strongly Related to Gross Primary Productivity Estimated from Flux Towers in Temperate Forests. *Remote Sens. Environ.* **2018**, *204*, 659–671.
11. Croce, R.; van Amerongen, H. Natural Strategies for Photosynthetic Light Harvesting. *Nature Chem. Biol.* **2014**, *10*, 492–501.
12. Reynolds, M. P.; Balota, M.; Delgado, M. I. B.; Amani, I.; Fischer, R. A. Physiological and Morphological Traits Associated with Spring Wheat Yield Under Hot, Irrigated Conditions. *Austr. J. Plant Physiol.* **1994**, *21*, 717–730.
13. Mohammadi, M.; Karimizadeh, R. A.; Naghavi, M. R. Selection of Bread Wheat Genotypes Against Heat and Drought Tolerance Based on Chlorophyll Content and Stem Reserves. *J. Agric. Social Sci.* **2009**, *5*, 119–122.
14. Ristic, Z.; Bukovnik, U.; Prasad, P. V. V. Correlation Between Heat Stability of Thylakoid Membranes and Loss of Chlorophyll in Winter Wheat under Heat Stress. *Crop Sci.* **2007**, *47*(5), 2067–2075.
15. Talebi, R. Evaluation of Chlorophyll Content and Canopy Temperature as Indicators for Drought Tolerance in Durum Wheat (*Triticum durum* Desf.). *Australian J. Basic Appl. Sci.* **2011**, *5*(11), 1457–1462.
16. Sharifi, P.; Mohammadkhari, N. Effects of Drought Stress on Photosynthesis Factors in Wheat Genotypes During Anthesis. *Cere. Res. Commun.* **2015**, *44*(2), 1-11.

17. Maglovski, M.; Gersi, Z.; Rybansky, L.; Bardacova, M.; Moravcikova, J.; Bujdos, M.; Dobrikova, A.; Apostolova, E.; Kraic, J.; Blehova, A.; Matusikova, I. Effect of Nutrition on Wheat Photosynthetic Pigment Responses to Arsenic Stress. *Pol. J. Environ. Stud.* **2019**, *28*(3), 1821–1829.

18. Eggink, L. L.; Park, H.; Hoober, J. K. The Role of Chlorophyll b in Photosynthesis: Hypothesis. *BMC Plant Biol.* **2001**, *1*, 2.

19. Strzalka, K.; Kostecka-Gugala, A.; Latowski, D. Carotenoids and Environmental Stress in Plants: Significance of Carotenoid-Mediated Modulation of Membrane Physical Properties. *Russ. J. Plant Physiol.* **2003**, *50*(2), 168–172.

20. Joshi, P. N.; Ramaswamy, N. K.; Iyer, R. K.; Nair, J. S.; Pradhan, M. K.; Gartia, S.; Biswal, B.; Biswal, U.C. Partial Protection of Photosynthetic Apparatus from UV-B-Induced Damage by UV-A Radiation. *Environ. Exp. Bot.* **2007**, *59*, 166–172.

21. Lichtenthaler, H. K.; Buschmann, C. Chlorophylls and Carotenoids: Measurement and Characterization by UV-VIS Spectroscopy. In *Current Protocols in Food Analytical Chemistry (CPFA)*; Wrolstad, R. E.; Acree, T. E.; An, H.; Decker, E. A.; Penner, M. H.; Reid, D. S.; Schwartz, S. J.; Shoemaker, C. F.; Sporns, P., Eds.; New York: John Wiley and Sons, 2001; pp F4.3.1–3.8.

22. Nagata, N.; Tanaka, R.; Satoh, S.; Tanaka, A. Identification of a Vinyl Reductase Gene for Chlorophyll Synthesis in Arabidopsis Thaliana and Implications for the Evolution of *Prochlorococcus* Species. *Plant Cell* **2005**, *17*, 233–240.

23. Li, Y.; He, N.; Hou, J.; Xu, L.; Liu, C.; Zhang, J.; Wang, Q.; Zhang, X.; Wu, X. Factors Influencing Leaf Chlorophyll Content in Natural Forests at the Biome Scale. *Front. Ecol. Evol.* **2018**, *6*, 64.

24. Esteban, R.; Barrutia, O.; Artetxe, U.; Fernández-Marín, B.; Hernández, A.; García-Plazaola, J. I. Internal and External Factors Affecting Photosynthetic Pigment Composition in Plants: A Meta-Analytical Approach. *New Phytol.* **2015**, *206*, 268–280.

25. Richardson, A. D.; Duigan, S. P.; Berlyn, G. P. An Evaluation of Noninvasive Methods to Estimate Foliar Chlorophyll Content. *New Phytol.* **2002**, *153*, 185–194.

26. Lichtenthaler, H. K. Chlorophylls and Carotenoids: Pigments of Photosynthetic Biomembranes. *Methods Enzymol.* **1987**, *148*, 350–382.

27. Biswal, A. K.; Pattanayak, G. K.; Pandey, S. S.; Leelavathi, S.; Reddy, V. S.; Govindjee; Tripathy, B. C. Light Intensity-Dependent Modulation of Chlorophyll *b* Biosynthesis and Photosynthesis by Overexpression of Chlorophyllide *a* Oxygenase in Tobacco. *Plant Physiol.* **2012**, *159*, 433–449.

28. Buschmann, C. Variability and Application of the Chlorophyll Fluorescence Emission Ratio Red/Far-Red of Leaves. *Photosynthesis Res.* **2007**, *92*, 261–271.

29. Junker, L. V.; Ensminger, I. Relationship Between Leaf Optical Properties, Chlorophyll Fluorescence and Pigment Changes in Senescing *Acer saccharum* Leaves. *Tree Physiol.* **2016**, *36*, 694–711.

CHAPTER 12

Some Indirect Methods for Predicting the Rooting Ability of Apple Tree (*Malus* spp.) Stem Cuttings

OLGA A. OPALKO and ANATOLY I. OPALKO*

National Dendrological Park "Sofiyivka" of NAS of Ukraine, 12-a Kyivska St.., Uman, Cherkasy region, 20300 Ukraine

Corresponding author. E-mail: opalko_a@ukr.net

ABSTRACT

Possibilities for evaluating the ability for morphogenic regeneration, in particular for predicting the rooting potential of stem cuttings of the number of *Malus* spp. representatives (an apple tree) by indirect methods are discussed. Garden cress—*Lepidium sativum* L., radish—*Raphanus sativus* L. var. *radicola* Pers., cultivar "Chervona z bilym kinchykom" and common bean—*Phaseolus vulgaris* L., "Synel'nykivs'ka 8" have been used for bioindication of *Malus* spp. stem cuttings rooting potential. Radish seeds, garden cress sprouted seeds, and cuttings of common beans have been treated with extracts from cuttings of the studied cultivars, hybrid seedlings, crabapples, and clonal apple rootstocks. Physiological readiness of the donor extract plants for regeneration and rooting potential of stem cuttings of *Malus* spp., treated with these extracts, have been predicted according to radish seeds germinability and intensive growth and develop-ment of primary roots from garden cress sprouted seeds and intensity of adventitious roots formation in stem cuttings of common bean. It has been found that radish seeds and bean cuttings can be used as biotesters with certain premonition, and sprouted seeds of garden cress do not meet the requirements of the regenerative capacity tester of the stem cuttings of apple tree. The fact that there is no significant dependence between the manifestations of the rooting potential of *Malus* spp. stem cuttings and the yield of the cultivars and hybrids from which these cuttings were prepared for rooting gives grounds for refuting the common belief that the rooting ability is inherent only in primitive forms of apple tree.

12.1 INTRODUCTION

The problem of finding and improving the ways of scion-rooting vegetative propagation of woody plants is connected with the efforts to exploit their ability to morphogenic regeneration, in particular with rooting potential of stem cuttings. The rooting ability of stem cuttings depends of the genotype and ontogenetic features (juvenile traits) and physiological state of ortet (the original plant from which a clone is started through rooted cuttings) as well as a number of endogenous and/or exogenous factors. There are many direct and indirect ways of evaluating of plant regeneration potential.[1–4]

12.1.1 *HISTORY OF REGENERATION RESEARCH*

The term "regeneration" appeared in scientific literature in the 28th century as the term used by the following zoologists: René Antoine Ferchault de Réaumur, French entomologist and person of encyclopedic knowledge[5,6] and Abraham Trembley, Swiss naturalist, best known for his studies of the freshwater hydra.[7] It must, however, be admitted that the phenomenon of regeneration had been described by many biologists, but under another terms starting from the times of Ancient Greece and Rome. The earliest striking example of morphogenic regeneration was found in the legend of ancient Greece about the Lernaean Hydra who was being growing two heads for every head chopped off.[5] Among the most ancient scientific evidence on the organ regeneration phenomenon in living organisms after various natural traumas was Aristotle's (384–322 BC) works. He noted that the tails of lizards and snakes could regenerate.[8,9] Pliny the Elder (AD 23–79) also described snakes dropping winter skin in spring (membrane), making them look smooth and young.[10]

   Thomas Hunt Morgan was indicating the functional importance of physiological plant regeneration as a recovery after natural deterioration (changing the root cork-cap cells, and tree cortex trunks, replacing old elements of xylem with new ones, etc.). He described "epimorphosis" as regeneration by the means of proliferation (cell rapid reproduction of the damaged organism part) and a peculiar to lower animals "morphallaxis" as regeneration by the existing body tissues transformation.[11,12] However, we tend to support the propose of the abandonment of "epimorphosis" and "morphallaxis" categories and their replacement by a new unifying principle considering believing that epimorphosis is only one of the versus of morphallaxis.[3,13]

Among the well-known in the last century researchers of the phenomenon of regeneration in plants, first of all, we should name Krenke,[14] whose general ideas are still relevant. Krenke applied quantitative methods in studying the age-related variation of somatic characteristics and formogenesis and regeneration factors. In his papers, the importance of regeneration ability for the success of plant vegetative propagation has been emphasized. It became a good fundamental for further theoretical and applied research.[1,15]

Regeneration processes in plants occur under the influence of many factors. These are first of all phylogenetic features, which in the most concentrated form can be gathered in hereditary characteristics (genotype) of each species, variety, form, or cultivar. On the other hand, the ontogenetic features of a particular plant, its physiological state, as well as endogenous and exogenous factors of chemical (general chemical compounds and releasing substances), physical (wound irritants, ionizing radiation, temperature, moisture supply, photoperiod), and biological nature (phytosanitary state, ontogeny phase) are extremely important. In 1880, the German botanist Breslau Julius von Sachs (1832–1897) proved that the root-forming substance is produced in leaves and is translocated down the stem to promote rooting. Since then, the numbers of rooting cofactors that promote in rooting are found.[16] According to Krenke,[14] the ideas about the hormonal theory of plant ontogeny had been expressed by Charles Darwin (1809–1882) in 1880, but the development and practical application of these ideas came with great delay.[15,17–19]

Diverse localized damages occur on the plants (especially perennial woody plants) during their life period. So in the process of evolution they developed an adaptive mechanism of protection against damage, that is, the ability to recover. All signs of post-damage regeneration can be united into two large groups: morphogenic regeneration, when lost parts and organs recover, and also a new organism from one part of a primary body can be developed; and non-morphogenic post-trauma regeneration which results in healing of all possible wounds.[1,14] At present physiological stress caused by trauma/damage is considered to be an inductor of adaptive response of an organism which promotes regeneration.[1]

The plant cells are totipotent, that allows development beyond the globular embryo stage, in result, they are able to fully implement their genetic program and give a start to a new plant.

Therefore, after some damage to the plant or the introduction of an excised fragment of a tissue or an organ used to initiate an *in vitro* culture an undifferentiated mass of callus cells are formed in a wounded place. According to the traditional view, the new shoots or roots are generated with these dedifferentiated callus cells; that is, cell dedifferentiation occurs on the way to the regeneration of a new plant.[20–22] In contrast, some plant regeneration studies show that plant cells can regenerate damaged tissues, apparently without dedifferentiation.[23] And callus is an organized and differentiated tissue that is generated mainly from a specialized population of primary cells.[3,23–26]

12.1.2 BIOINDICATION AND MAIN ADVANTAGE OF PLANT INDICATORS

Bioindication technique is mainly used to determine biologically significant anthropic ecosystem pollutant loads. That is, bioindication is now one of the methods of environmental research.[27–29] The monitoring of the features of plant growth and development, as well as the behavior of animals, make it possible to fairly assess the biological effects of air, water, and soil pollution, and to predict the spatial distribution of pollutants and their possible accumulation in large areas.[30–34]

It is known that the developmental characteristics (growth rate, flowering process, fruit formation, color and color intensity, etc.) of many plant and animal species may change in response to various environmental irritants, especially in terms of the excession of the evolutionary force and duration of the stimulus influence. Man has long used the peculiarities of the individual plants and animals response to the conditions of growth and development to assess the environment and predict its future changes. The oldest documented evidence of the assessment of the response peculiarities of living organisms to artificially altered conditions include the experiments of the ancient Greek philosopher Aristotle (384–322 BC), who studied the behavior of freshwater fish, which were placed in salt water.[29,35]

The well-known old practice of using canaries in coal mines as gas detectors to assess the potential toxic hazard by carbon monoxide, marsh gas (methane) and other toxic gases before they hurt the miner. These gases are odorless, colorless and equally dangerous to humans and canaries, but birds are much more susceptible to the gas and reacts faster and more noticeably than humans, thereby alerting miners to the presence of poison.[36]

Swedish botanist, physician, and zoologist Carl Linnaeus (Carl von Linné, 1707–1778) proposed interesting approaches to determining time by bioindicating insolation levels, or integrated solar irradiance. He developed a flower clock plan called Horologium Florae. The principle of time determination was based on the results of Linnaeus's long-term observations of the daily dynamics of flowers opening and closing time of many plants. As a result of generalization, he proposed to plant plants of certain studied species depending on the sequence of flowering over the day.[37,38]

Ancient farmers have already known that different plants have certain preferences regarding soil, solar irradiance, precipitation, air and soil temperature, etc. They have used these plants as bioindicators for a very long time but only the second half of the 19th century became a time of current meaning of bioindication.

We should mention the works of the Finnish botanist and entomologist Nylander (1822–1899), who used tinctures of iodine and hypochlorite to determine the taxonomy of lichens. In addition, Nylander published in 1866 in Paris the results of his research on lichens, stating that lichen grows poorly in polluted air, and the good development of many lichen species (and the number of thalli) attests to the purity of the air.[29,39]

The species composition as well as the growth and development features of some plant species can be indicators of the acidity of the soil on which they grow. Regarding the response to soil acidity, plants are divided into:

- Calcifuge, that better grow on acid soils, such as lingonberry (*Vaccinium vitis-idaea* L.), bog blueberry, northern bilberry, or western blueberry (*Vaccinium uliginosum* L.), Labrador tea (*Ledum* L.), dwarf birch (*Betula nana* L.);
- Neutrophilic species (neutral soils living), which include most fruit and vegetables that prefer neutral to slightly acidic pH soils;
- Acidofuge, that are better grown on soils with an alkaline reaction of the medium, such as lavender (*Lavandula* L.), honeysuckle (*Lonicera* L.), lilac (*Syringa vulgaris* L.), mullein or velvet plant (*Verbascum* L.).

In natural arrays, some neutrophilic species tend to expand their niches toward base-rich sites beyond pH 6 than in the more acidic part. For example, the horsetails (*Equisetum arvense* L.) tangle indicate the increased acidity of the soil in the area where they grow. There is evidence that in natural arrays, some neutrophilic species tend to extend their range toward neutral and more alkaline sites with a pH greater than 6, than to more acidic ones niche toward base-rich sites beyond pH 6 than in the more acidic part.[40]

Plants, unlike animals, have much less ability to avoid unwanted external influences by temporarily changing their place of residence or moving to the desired area with appropriate conditions. Therefore, at the population and organismal level, plants are able to respond to ecotope stress mainly by varying individual morphological features, the optimal parameters of which are decisive for natural and artificial selection, as well as due to selection changing the composition of the population. Organisms with a narrow reaction norm and/or those whose reaction norm does not coincide with the desired adaptive response, primarily with an increase in the reproductive rate (wild plants) and an increase in the quantity and quality of the crop (cultivated plants), as well as incapable for morphogenic and non-morphogenic regeneration could be eliminated by natural and artificial selection.[41-43] As a result, plants as bioindicators of the environmental state have several advantages over animals. Observations on the reaction of certain plants (with the well-known limits of the established regulatory deviations) to fluctuations in environmental conditions allow a fairly objective assess the state of the environment. The benefits of indicator plants include their ability to summarize biologically important environmental changes; plants are capable to react on short-term and volatile emissions of toxicants; react on the speed of changes occurring in the environment; indicate where the pollutants accumulate and how they migrate; enable on the early stages to normalize the permissible load on ecosystems.

Among the various ways of assessing the state of the environment, bioindication methods are particularly widely used in forest typology, phytocenology, and to determine the level of air pollution by lichen (lichen indication), moss (moss indication), fungi (myco-bioindication), or other plants.[44-49]

Monitoring the status, growth, and development of bioindicator plants is now recognized as the most expeditious, the fastest and cheapest way of assessing the environment that does not require expensive equipment. Traditional approaches to chemical and physical measurements of long-term ecosystem changes are expensive to observe;[50] herein bioindication can provide easy and cost-effective tools for supervision of environmental and ecosystem integrity.[51,52]

Within the general systemic paradigm of bioindication, the use of allelopathy phenomenon is quite natural. The phenomenon of allelopathy, as an allelochemical interaction between different living organisms, especially plants[53] may be the theoretical basis of applied bioindication. It should be noted that the facts of the negative impact of certain plants on other living organisms, in particular on plants known from ancient times. Aristotle's follower, Greek philosopher Theophrastus (371–281 BC), observed the inhibition of pigweed (*Amaranthus* L.) on alfalfa (*Medicago sativa* L.).

In ancient China about 300 plants pesticidal abilities, including those with allelopathic effects had been well-known nearly 2000 years ago, in 1st–3rd centuries AD.[54]

In 1937, the Austrian professor Hans Molisch offered a scientific explanation for the influence of one plant on another through the release of chemicals. He also was the first who had used the term "allelopathy."[55] However, it was primarily about inhibition, that is, the harmful effects of these chemicals on the target organisms. This is evidenced by the very structure of the compound word allelopathy, in which the first part of *allelo-* (Greek allēlo) means each other, and the second part *-pathy* means the form that occurs in borrowed words from the Greek language in the sense of suffering, feeling, and herein, in compound words of current formation are often used with the meaning of morbid affection, disease (arthropathy; deuteropathy; neuropathy etc.). Later, the beneficial effects of such interactions with secondary metabolites produced by plants, algae, bacteria, and fungi were described. As a result, the semantic meaning of the term allelopathy has expanded to any mutual effects of one organism on survival, growth and development, reproduction, and certain features and properties of other living organisms.

Ukrainian botanist and plant physiologist Grodzinsky (1926–1988) considered allelopathy in this broad sense. In Ukraine and in the republics of the former USSR, he is the father of modern doctrine of allelopathy. The fundamental, albeit somewhat debatable, work of Grodzinsky[56] has helped to formulate our vision of the use of allelopathy phenomenon to bioindicate the physiological state of woody plants, including apple trees (*Malus* spp.) in predicting for rooting potential of stem cuttings.

As general references on history of allelopathy, the reader could find in the works of Willis[57] and to pursue this topic further, the reader should consult Chou[54] and a collection of papers "Allelopathy: current trends and future applications" edited by Cheema and associates.[58]

The above considerations, as well as the value of the genus *Malus* Mill. for horticulture (fruit and ornamental), food industry, and pharmacy give grounds to argue that the search for indirect ways of predicting the regenerative potential of apple trees, in particular the prediction of the rooting capacity of apple tree (*Malus spp.*) cuttings for scion-rooting vegetative propagation is a topical task.

12.1.3 BRIEF HISTORY OF APPLE DOMESTICATION AND TAXONOMY OF THE GENUS MALUS MILL.

The history of the apple plants origin dates back to the Cretaceous period in South-East Asia about 70 million years ago.[59,60] An apple plant that brings together representatives of the current genus *Malus* Mill. (*Rosaceae* Juss.) apparently was known to the ancient man in late Paleolithic-Mesolithic period about 10–15 thousand years ago, as a wild fruit tree. This is evidenced by the remains of apple fruits and their schematic images found by archaeologists in the study of ancient settlements.[61] However, the beginning of meaningful harvesting of apple fruit is associated with early Neolithic period.[62]

In all probability from the East (its primary habitats), the apple tree first came to Palestine, and then to Egypt and Europe as a result of human migration along the Old Silk Road, which linked Western China to the Middle East and the Danube Valley in the Neolithic period and Bronze Age. With regard to the history of domestication of *Malus* spp., it should be considered almost proven that the domestication of apple trees began in the small Tien Shan area on the border of Kazakhstan with China.[63] The earliest written evidence of the apple tree dates from the 4th century BC. These are works of the founder of the scientific botany Theophrastus from Ancient Greece, which had clearly distinguished cultivars and wild apple trees. However, he did not use the name *Malus* to describe some of the varietal characteristics and methods of cultivation that concerned the two known cultivars of the genus *Malus*. He also reported that the gardens of the Panticapeia (now Kerch) had been rich in apple and pear trees.[61]

Two centuries later, in Ancient Rome, Catonis described seven cultivars in his treatise "De agri cultura,"[64] and afterwards Pliny the Elder reported about more than 20 cultivars of apple tree. Pliny the Elder became using the word *malus* to identify apple trees. From the Greeks and Romans, the apple tree culture passed to the Western Europe, where its cultivars were grown mostly in monasteries for a long time. The apple tree hit the territory of present-day Ukraine almost simultaneously with the pear during the times of Kyiv Rus, where Christian monks brought it from Byzantium in the 10th century. The famous gardens of the Kyiv-Mohyla Lavra, laid in 1051 by Antoniy Pecherskyi, as well as the monastery gardens of the Kitaevo Hermitage, Feofaniya, Golosiyevo, and others.[61]

Over the last centuries, in temperate climates of the Northen Hemisphere, apple trees have become a leading fruit crop,[65] with world apple production of 60–66 million tonnes per year, accounting for about 15% of the world's total fruit and berry production.

Due to the ordering of the place of the genus *Malus* in the system Magnoliophyta Cronq., Takht. and W. Zimm. (= Angiospermae Lindl.), in particular in the intra-family intermediate taxa of the family Rosaceae Juss., and clarification of their nomination in accordance with the principle of priority[66] adopted in the botanical nomenclature. Nowadays, for the subfamily uniting former subfamilies Spiraeoideae Endl., Pyroideae Burnett (former Maloideae C. Weber), and Amygdaloideae, namely Amygdaloideae Arn. recognized as a valid name.[67] And, correspondingly, the former tribe Pyreae received a name Maleae Small, and the subtribe Pyrinae—name Malinae Rev. (Table 12.1).

And, eventually, the genus *Malus* is now classified as part of the large subfamily Amygdaloideae, which consumed the former subfamilies Amygdaloideae, Spiraeoideae, and Pyroideae (= Maloideae), tribes—Maleae Small, and subtribe—Malinae Rev.[66–71]

Subtribe Malinae also consists of the following genera: *Amelanchier* Medik. (saskatoon, also known as, serviceberry or shadbush), *Aronia* Medik. (black chokeberry), *Chaenomeles* Lindl. (chaenomeles or Japanese quince), *Cotoneaster* Medik. (cotoneaster), *Crataegus* Tourn. ex L. (hawthorn, thornapple), *Cydonia* (European quince), *Eriobotrya* Lindl. (loquat, also known as, Japanese medlar, Japanese plum, or Chinese plum), *Heteromeles* M. Roem. nom. cons. (toyon, also known as, Christmas berry and California holly), *Mespilus* Bosc ex Spach (medlar), *Photinia* Lindl. (Photinia), *Pseudocydonia* C. K. Schneid. (Chinese quince), *Pyracantha*

M. Roem. (firethorn), *Pyrus* (pear), *Sorbus* (rowan or mountain-ash), and others less known genera and interspecific hybrids.[72,73]

TABLE 12.1   The Position of the Genus *Malus* in the System of Magnoliophyta According to Different Plant Classification Systems

Taxon	Classification systems of plants		
	Engler, 1903 Ref [68]	**Takhtajan, 2009 Ref [69]**	**APG IV, 2016 Ref [66]**
Division	Embryophyta siphonogama	Magnoliophyta	-
Subdivision	Angiospermae		-
Classis	Dicotyledoneae	Magnoliopsida (Dicotyledons)	-
Subclassis	Archichlamydeae	Rosidae	-
Superordo	-	Rosanae	-
Ordo	Rosales	Rosales	Rosales
Subordo	Rosineae	-	-
Familia	Rosaceae	Rosaceae	Rosaceae
Subfamilia	Pomoideae	Pyroideae (Maloideae)	Amygdaloideae
Tribus	-	Maleae	Maleae
Subtribus	-	-	Malinae
Genus	*Malus*	*Malus*	*Malus*

Most *Malus* spp., and its congeners in the subtribe Malinae are diploid ($x = 17$) and cross-pollinated. Herein, most species and genera of the other subfamilies of the *Rosaceae* family are characterized by an appreciably reduced basic chromosome numbers $x = 7$, $x = 8$, or $x = 9$, which gave the cytologists of the early 20th century the ground to assume the polyploid origin of pome fruit plants chromosome number.

English geneticists Cyril D. Darlington and A. A. Moffett assumed, that Pyroideae (current Amygdaloideae) appeared from some *Rosoideae* and is a triple trisomic tetraploid ($x = 7 + 7 + 3 = 17$). Doubling of the number of chromosomes (all seven) with the addition of one more chromosome from three different pairs in some ancient progenitors of *Rosaceae* with haploid set $x = 7$ took place. It is very common chromosome number in the family *Rosaceae*.[67,74]

At the same time with Cyril D. Darlington and A. A. Moffett, American geneticist Karl Sax, on the contrary, believed that Pyroideae (currently

Amygdaloideae) are allopolyploids arising of a result of doubling the number of chromosomes in hybrids between remote progenitors of two distant generic types.[75] According to the researcher, this could be representatives of the subfamily Prunoideae, which has a basic chromosome number x = 8 and the subfamily Spiroideae—with x = 9, which uniting produced x = 17 chromosomes in a common genome of the Pyroideae (current Amygdaloideae). More followers supported the Karl Sax hypothesis; since, the probability of triple trisomy is significantly lower than amphipolyploidy. These views on the origin of the pome fruit genome have been almost unquestioned for over 60 years.[67] However, since the end of the last century, the hypothesis of origin of basic haploid chromosome number of pome fruit plants (x = 17, $2n$ = 34) has become increasingly convincing due to the autopolyploidy with Spiraeeae with x = 9 with subsequent nullisomy.[60] According to this hypothesis, the autotetraploid, in prehistoric times, was formed from an ancestral form of Spiraeeae. The virtual scheme includes the steps of doubling x = 9 to x = 18 (2n = 36) and the nullisomy stage of the newly formed autotetraploid (2n = 36-2 = 34), which provided the genera of subtribe Malinae the present principal number of chromosomes x = 17. It is believed that it is from this ancestral form of Spiraeeae that all the diversity of apple of pome fruit plants developed. The results of molecular genetic analysis of Rosaceae[76–78] showed that on all nuclear and chloroplast cladograms the sister group of apple pome fruit plants consistently had a North American endemic of dry open forests with acid soils, the *Gillenia Moench* genus from the Amygdaloideae subfamily. The Gillenia plants are known as Indian-physic, Fawn's Breath, American ipecac, or Bowman's root. This gave good reason to believe, that the genome of pome fruit plants probably was formed monophyleticly as a result of autopolyploidy of the genus *Gillenia*. That is, the emergence of the common to apple tree main number of chromosome x = 17 is associated with the autopolyploidization of some ancestor of Gillenia.[67]

Most of *Malus* spp. are diploids ($2n$ = $2x$ = 34), are found of triploids from $2n$ = $3x$ = 51 (*M. hupehensis* (Pamp.) Rehder, *M. coronaria* (L.) Mill., *M. ioensis* (Alph.Wood) Britton and *M. toringo* (Siebold) de Vriese) and tetraploid species ($2n$=$4x$=64) (e.g., *M. sargentii* Rehder), and also some species with diploid and tetraploid levels, such as *M. spectabilis* (Aiton) Borkh. and *M. baccata* (L.) Borkh. Diploid apomicts include *M. sikkimensis* (Wenz.) Koehne ex C. K. Schneid. and *M. toringoides* (Rehder) Hughes, but *M. sargentii* Rehder refers to tetraploid apomicts.[79,80]

Accepting evidences of monophyletic nullisomic origin of the pome fruit plants (Maleae), we should explain the facts of mainly bivalent conjugation of their chromosomes by a prolonged evolution, in the process of which during interspecies hybridization and polyploidization within common ancestral group with Gillenia took place. Such course of events is more probable than the gradual formation of a functional diploid from an autoaneuploid that perhaps arose out of an autotetraploid because of its nullisomy.

In the vast majority of scientific and horticultural publications, the scientific name of apple cultivated is *Malus domestica* Borkh. with some minor variations in spelling.[81–83] Despite the widespread use of the grown in orchards of apple cultivars of the preferred scientific name *Malus domestica*, cultivated apple is more correctly name *Malus pumila* Mill., according to accepted International Code of Nomenclature for Cultivated Plants the principle of priority.[84] The name *Malus pumila*, which was the first, indeed publicized, name of a cultivated apple tree, made to abandon the widely used synonym *Malus domestica*. Moreover, discussed during 2010–2014 name *M. domestica* was finally rejected by Nomenclature Committee for Vascular Plants.[72,73] Taking into consideration the complex hybrid origin of cultivated apple, Volodymyr M. Mezhenskyj considers it possible to write the species name M. × *pumila* Mill. (Pro. Sp.), that is, as a hybridogenic taxon, published as a species. The crabapple cultivars are suggested to be attributed to the combined species *Malus* × *gloriosa* Lemoine,[72,73] although not all scientists recognize this name as accepted.[85]

12.2   OBJECTIVES AND METHODS OF THE RESEARCH

These studies were in the plant collections of the National dendrological park "Sofiyivka" of NAS of Ukraine (NDP "Sofiyivka"), and in the orchards of L. P. Symyrenko Research Institute of Pomology of NAAS of Ukraine (RI Pomology) and Uman National University of Horticulture (UNUH) situated in the Central-Dnieper elevated region of Podolsk-Pridneprovsk area of the Forest-Steppe Zone of Ukraine. The area is characterized by temperate-continental climate with unstable humidification and considerable temperature fluctuations. Average many-year amount of precipitation per year is 633.0 mm, its amount being 300–310 mm at +10°C, which corresponds to the precipitation amount in dry southern areas of Ukraine. Average many-year air temperature is +7.4°C.[3]

12.2.1   CULTIVARS, HYBRID SEEDLINGS, CRABAPPLES, AND CLONAL APPLE ROOTSTOCKS STUDIED IN THE EXPERIMENTS

The apple cultivars, namely, "Rosavka," "Symyrenkivets," and "Vnuchka"; 11 hybrid seedlings (Table 12.2); 5 crabapples (*M. baccata* (L.) Borkh, *M. halliana* Koehne, *M. niedzwetzkyana* Dieck., *M. prunifolia* (Willd.) Borkh. var. *Rinki* (Koidz.) Rehd. f. *fastigiata bifera* (Dieck.) Al. Teod. and M. *prunifolia* (Willd.) Borkh. f. *pendula* (Bean) Rehd.); and 2 apple rootstocks (Don 70–456 and 62–396) were involved in the studies.

TABLE 12.2   The Parentages of the Studied Apple Hybrid Seedlings.

Hybrid seedling	Apple parents that were used for crossing	
	Female parent	Male parent
2–6	"Lord Lambournne"	"Golden Spire"
2–7	"Lord Lambournne"	"Golden Spire"
2–7a*	"Lord Lambournne"	"Golden Spire"
6–5	"Calville de Saint-Sauveur"	"Mliyivs'ka krasunia"
6–6	"Calville de Saint-Sauveur"	"Mliyivs'ka krasunia"
6–9	"Calville de Saint-Sauveur"	"Mliyivs'ka krasunia"
6–11	"Calville de Saint-Sauveur"	"Mliyivs'ka krasunia"
1054	24235**	"Rosavka"
1055	24235**	"Rosavka"
1107	24235**	"Rosavka"
1111	24235**	"Rosavka"

Note: *—scion-rooted, **—seedling 24,235 is a seedling of "Mliyivs'ka krasunia"

The studied apple cultivars and hybrid seedlings were originated by Research Institute of Pomology. The crabapples were collected in NDP "Sofiyivka" from a few botanical scientific institutions. The clonal apple dwarf rootstocks were introduced in UNUH from Russia, namely, Don 70–456 was originated by Don Zonal Research Institute of Agriculture of RAAS and 62–396 was originated by Michurinsk State Agrarian University.

The apple cultivars and hybrid seedlings (extract donors) were cultivated in orchard of UNUH. All cultivars and hybrid seedlings were propagated by grafting onto rootstock M 9, but "Rosavka" and seedling 2–7 were studied in 2 variants: grafted and scion-rooted (propagated by layering or stooling). The crabapples were propagated by seeds and the clonal dwarf rootstocks were propagated by layering.

Garden cress—*Lepidium sativum* L., radish—*Raphanus sativus* L. var. *radicola* Pers., cultivar "Chervona z bilym kinchykom" та common bean—*Phaseolus vulgaris* L., "Synel'nykivs'ka 8" were used for prediction of *Malus* spp. stem cuttings rooting potential by indirect methods as plants indicators.[86]

12.2.2 METHODS OF THE RESEARCH

The seed germination bioassays and others biological tests has been made according to the recommendations of Andrei M. Grodzinsky[56] with our modifications. Radish seeds, sprouted seeds of garden cress, and cuttings of common beans were treated with stem cuttings extracts of the studied cultivars, hybrid seedlings, crabapples, and clonal apple rootstocks. All experiments had been made with four-fold replication; each replication was 100 pieces of seeds/cuttings of indicator plants.

Working extracts were prepared in two stages. Initially, cuttings of apple trees were weighed and divided with a conventional garden pruner into 2–4 mm pieces and filled with 10 times the amount of distilled water at a temperature of about 25°C. It was infused in glassware (flasks) for 24 h. Then the tincture was filtered off and the resulting aqueous extract of each variant was diluted with distilled water to a ratio of 1:100. Diluted extracts were filled with sterilized Glass Petri dishes with radish seeds at 4 mL of extract per 100 seeds. Sterilization of Glass Petri dishes was performed on a drying chamber at 130° C for 1 h. The seeds were laid out in Petri dishes between layers of moistened by extracts filter paper: two layers of paper were laid on the bottom of a Petri dish, and one layer of paper covered the seeds. The seeds of the control variants were flooded with distilled water in the same amount. It was germinated at a temperature of 20–25°C. Every 12 h, the number of seedlings germinated on control variants was calculated. For this purpose, preprepared Petri dishes with radish seeds were poured with distilled water (in the control variant) and corresponding extracts from stem cuttings of the studied cultivars, hybrid seedlings, crabapples, and clonal apple rootstocks at 7–8 p.m. and the first calculation of sprouted seeds was performed at 7–8 a.m. in the morning of the following day. When this amount exceeded 50%, sprouted seeds in all variants of the experiment were counted. The obtained results were expressed in relative (of control) percent.

In the garden cress experiment, its seeds were germinated in Petri dishes on moistened with distilled water filter paper in a dark thermostat at 27°C. The germinated seeds with the same root length (2–3 mm) were selected and laid out on Petri dishes in 100 seeds for each variant.

The germinating seeds of garden cress with primary root (radical) filled with diluted extracts from apple tree cuttings. The seeds of the control variant were flooded with distilled water in the same amount. After 24 h, the length of the roots of the garden cress was measured in all variants. The obtained results were expressed as a percentage of the control.

In the common bean experiment, its seeds were sown in plastic containers with soil substrate. The plants were grown in a vegetation structure, and when they reached the phase of two true leaves, they were harvested by cuttings for root regeneration. The common bean cuttings were placed in flasks with diluted extracts from cuttings of evaluated cultivars, hybrid seedlings, crabapples, and clonal apple rootstocks, where they were exposed for 24 h. After treatment, the cuttings were transferred to distilled water flasks. The rooting ability of the extracts was evaluated by the total number of roots, the number of roots with II and III orders branches, and the number of roots of different lengths (0.5–3.0) that developed on cuttings of common beans.

Statistical analysis of experimental data including calculating the least significant difference (LSD) between the treatments, as well as the correlation coefficients (r) between most important features of apple trees, and indices of post-damage regenerative potentials were executed in accordance with Ronald A. Fisher.[87]

12.3 RESULTS AND DISCUSSION

Radish seeds treated by the extracts from stem cuttings of the studied cultivars, hybrid seedlings, and clonal apple rootstocks in the first decade of June were germinated mostly on the second day after soaking. It is the period considered to be the most favorable for rooting the apple trees cuttings for our agroclimatic area.[88] During the first testing period, conducted in the first decade of June, extracts of all fruit forms stimulated germination of radish seeds by 5.5–29.0% (Table 12.3).

The stimulus effect of extracts studied can be divided into three groups. The first group includes extracts from the cuttings of hybrid seedlings 1107 (24235 × "Rosavka"), treatment by which the number of sprouted

TABLE 12.3 Germination of Radish Seeds Depending on the Donor Extracts from the Apple Tree Cuttings and the Testing Period, %.

Extract donor	Testing period			
	First decade of June		Second decade of June	
	%	± of control, %	%	± of control, %
Control, H$_2$O	50.5	0.0	52.5	0.0
Cultivars and hybrid seedlings				
"Vnuchka"	65.5	+ 15.0	65.0	+ 12.5
"Rosavka"	56.0	+ 5.5	7.0	– 45.5
"Rosavka" scion-rooted	55.5	+ 5.0	66.5	+ 14.0
"Symyrenkivets"	70.5	+ 20.0	60.5	+ 8.0
2–6 ("Lord Lambourne" × "Golden Spire")	59.0	+ 8.5	62.0	+ 10.0
2–7 ("Lord Lambourne" × "Golden Spire")	69.0	+ 18.5	71.0	+ 18.5
2–7 ("Lord Lambourne" × "Golden Spire"), scion-rooted	60.0	+ 9.5	60.5	+ 8.0
6–5 ("Calville de Saint-Sauveur" × "Mliivs'ka krasunia")	73.0	+ 22.5	66.5	+ 14.0
6–6 ("Calville de Saint-Sauveur" × "Mliivs'ka krasunia")	66.0	+ 15.5	59.0	+ 6.5
6–9 ("Calville de Saint-Sauveur" × "Mliivs'ka krasunia")	62.0	+ 11.5	60.0	+ 7.5
6–11 ("Calville de Saint-Sauveur" × "Mliivs'ka krasunia")	60.0	+ 9.5	68.0	+ 15.5
1054 (24235 × "Rosavka")	73.0	+ 22.5	64.5	+ 12.0
1055 (24235 × "Rosavka")	64.5	+ 14.0	75.0	+ 22.5
1107 (24235 × "Rosavka")	79.5	+ 29.0	65.0	+ 12.5
1111 (24235 × "Rosavka")	56.0	+ 5.5	56.5	+ 4.0
Clonal apple rootstocks				
Don 70–456	47.0	– 3.5	52.5	0.0
62–396	62.0	+ 11.5	56.0	+ 3.5
LSD$_{05}$	12.1		14.6	

radish seeds exceeded the indicative control variant with water by an average of 29.0%, which is more than double LSD.05 (LSD$_{.05}$ = 12.1%). The second group included extracts of "Vnuchka," "Symyrenkivets," and seedlings 2–7 and 2–7, both from the combination "Lord Lambournne" × "Golden Spire," as well as 6–5 and 6–6, both from the combination "Calville de Saint-Sauveur" × "Mliivs'ka krasunia," and 1054 and 1055 of the combination 24235 × "Rosavka," which exceeded the stimulus effect of control by more than LSD$_{.05}$.

It is noteworthy that the difference between the stimulating effects of extracts from "Rosavka" grafted and "Rosavka" scion-rooted was insignificant. In contrast, the extract from hybrid seedling 2–7 ("Lord Lambournne" × "Golden Spire") scion-rooted was inferior efficacy to the hybrid seedling 2–7 grafted; however, although the difference between them was, as in the previous case, insignificant. The extract from the seedling 2–7 scion-rooted was assigned to the second donor group, as its stimulation effect (+ 18.5%) was significantly greater than in the control variant.

The third group of extracts donors include hybrid seedlings 2–7 ("Lord Lambournne" × "Golden Spire") scion-rooted, as well as 6–9 and 6–11 ("Calville de Saint-Sauveur" × "Mliivs'ka krasunia"), seedlings 1111 (24235 × "Rosavka") and both rootstocks, which deviated from the control variant data within the limits not exceeding LSD$_{05}$.

Analysis of the effect of extracts from cuttings harvested in the second decade of July showed that the germination of radish seeds treated with extracts from "Rosavka" scion-rooted cuttings was larger than the control variant by 14.0%, but the surplus was less than the LSD$_{.05}$ (14.6%).

Instead, the germination of radish seeds treated with the "Rosavka" grafted extract was more than three times less than in distilled water. This inhibition of radish seeds germination of extracted from the "Rosavka" grafted requires reflection and further study, with the involvement of ortets propagated on different rootstocks.

Significant stimulation was provided by extracts from the cuttings of the "Rosavka" scion-rooted, and extracts from seedling 2–7 ("Lord Lambournne" × "Golden Spire") grafted and seedlings 6–5 and 6–11, both from the combination ("Calville de Saint-Sauveur" × "Mliivs'ka krasunia" and 1055 (24235 × "Rosavka"). At the same time, indicators of seedling 6–11 were better than at the previous term of cuttings, as well as "Rosavka" scion-rooted.

In general, the stimulating effects of extracts from the cut in first decade of June stems of most studied cultivars, hybrid seedlings, and clonal apple rootstocks were greater than the effects of extracts from stems cut in the second decade of July. This allows to consider that in the study area (the Central-Dnipro elevated region of the Podilsk-Prydniprovsk area of the Forest-Steppe Zone of Ukraine), the rooting ability of apple tree root (*Malus* spp.) stem cuttings is achieved better in the first decade of June.

Proven efficiency of extracts from stem cuttings of apple cultivars, hybrid seedlings, and clonal apple rootstocks gave reasons for study of the efficiency of extracts from the stem softwood cuttings extracts of crabapples harvested in the first decade of June. In the crabapples extracts experiment, diluted with distilled water, extracts were studied in two concentrations: the first in a ratio of 1:100 and the second in a more concentrated state 1:10. Radish seeds were treated with crabapples stem cuttings extracts according to the method described in the previous experiment (Table 12.4).

TABLE 12.4   Germination of Radish Seeds Depending on the Donor of the Crabapples (*Malus* Mill.) and the Concentration of the Extract.

Extract donor	Number of sprouted seeds		
	%	± of control	% of control
Control, H_2O	51.5	0.0	100.0
Concentration 1:100			
M.baccata (L.) Borkh.	44.5	–7.0	86.4
M. halliana Koehne	53.0	+1.5	102.9
M. niedzwetzkyana Dieck.	59.0	+7.5	114.6
M. prunifolia (Willd.) Borkh. *var. Rinki* (Koidz.) Rehd. *f. fasti-giata bifera* (Dieck.) Al. Teod.	54.0	+2.5	104.8
M. prunifolia (Willd.) Borkh. *f. pendula* (Bean) Rehd.	64.5	+13.0	125.2
Concentration 1:10			
M. baccata (L.) Borkh.	9.0	–42.5	17.5
M. halliana Koehne	6.5	–45.0	12.6
M. niedzwetskyana Dieck.	13.0	–38.5	25.2
M. prunifolia (Willd.) Borkh. *var. Rinki* (Koidz.) Rehd. *f. fasti-giata bifera* (Dieck.) Al. Teod.	15.5	–36.0	30.1
M. prunifolia (Willd.) Borkh. *f. pendula* (Bean) Rehd.	4.5	–47.0	8.7
$LSD_{.05}$	6.9		

Extracts of all crabapples at a concentration of 1:10 inhibited germination of radish seeds. Radish seeds treated with crabapples extracts sprouted in 3.5–11.5 times worse than distilled water treated control seeds.

With regard to the extracts used at the optimum concentration of 1: 100, the studied species according to $LSD_{.05}$ = 6.9% were divided into three groups. The first group of the most effective extracts that significantly increased the radish seeds germination compared to the control variant with water include extracts from stem softwood cuttings of *M. prunifolia* (Willd.) Borkh. *f. pendula* (Bean) Rehd. and *M. niedzwetzkyana* Dieck. The second group, whose values deviated from control within LSD.05, included extracts from *M. halliana* Koehne and *M. prunifolia* (Willd.) Borkh. *var. Rinki* (Koidz.) Rehd. *f. fastigiata bifera* (Dieck) Al. Teod. The third group included an extract from *M. baccata* (L.) Borkh Rehd., which significantly reduced the number of sprouted seeds.

The obtained results make it possible to surmise that some endogenous chemical substances, first of all phytohormones, which are synthesized by crabapples plants growing, may affect the regenerative potency of other plants. According to this influence, we can make indirect forecasting of rooting potential of extract donor plants. The rooting ability of common bean cuttings treated with the stem softwood cuttings extracts of apple cultivars, hybrid seedlings, and clonal apple rootstocks was characterized by much wider variability spectra (Table 12.5) than germination of radish seeds.

The total number of roots per common bean one cutting was highest when using "Rosavka" stem cuttings extracts, hybrids 2–6 and 2–7 ("Lord Lambournne" × "Golden Spire") and both clone rootstocks (Don 70–456 and 62–396). The effectiveness of the remaining donors, with the exception of the "Vnuchka" and the 1055 and 1111 seedlings (both 24235 × Rosavka), also significantly exceeded the stimulating effects of the control with distilled water. In this case, cuttings of beans developed a good root system and in control variants. Extracts from cuttings of hybrid 1111 (24235 × "Rosavka") were significantly inferior to the control variant by the effect of stimulating the number of roots per root, and the efficiency of extraction of the "Vnuchka" and hybrid 1055 (24235 × "Rosavka") differed from the control within $LSD._{05}$ that is, the difference was insignificant.

The extract donor efficiency for rooting ability is also characterized by the development of the root system, the length of adventitious roots, and their branches (the number of secondary and tertiary lateral in breanthing).

TABLE 12.5 The Rooting Potential of Common Bean Cuttings Depending on the Extract Donor of Cultivars, Hybrid Seedlings, and Clonal Apple Rootstocks.

Extract donor	The total number of roots per one cutting, pcs	Percentage of roots (as long as, cm)					Percentage of roots with secondary and tertiary lateral branchings
		<0.5	0.5–1.0	1.0–2.0	2.0–3.0	>3.0	
Control, H$_2$O	76	32.9	19.7	11.8	5.3	30.3	14.5
Cultivars and hybrid seedlings							
"Vnuchka"	77	12.9	3.9	19.5	18.2	45.5	42.8
"Rosavka"	156	16.0	26.9	19.9	25.0	12.2	9.0
"Symyrenkivets"	104	9.7	1.9	16.3	22.1	50.0	31.7
2–6 ("Lord Lambourne" × "Golden Spire")	151	55.0	10.6	14.6	6.6	13.2	11.9
2–7 ("Lord Lambourne" × "Golden Spire")	153	32.6	5.9	19.6	11.8	30.1	22.9
6–5 ("Calville de Saint-Sauveur" × "Mliyivs'ka krasunia")	99	17.2	14.1	17.2	14.1	37.4	22.2
6–6 ("Calville de Saint-Sauveur" × "Mliyivs'ka krasunia")	127	15.0	36.2	12.6	9.4	26.8	26.8
6–9 ("Calville de Saint-Sauveur" × "Mliyivs'ka krasunia")	104	39.5	11.5	16.3	6.7	26.0	26.0
6–11 ("Calville de Saint-Sauveur" × "Mliyivs'ka krasunia")	90	43.2	17.4	5.6	7.8	25.6	18.9
1054 (24235 × "Rosavka")	126	39.7	4.8	11.9	7.9	35.7	35.7
1055 (24235 × "Rosavka")	77	5.2	0.0	6.5	19.5	68.8	58.4
1107 (24235 × "Rosavka")	126	41.3	6.3	15.1	11.9	25.4	22.2
1111 (24235 × "Rosavka")	52	44.3	13.5	19.2	11.5	11.5	19.2
Clonal apple rootstocks							
Don 70–456	141	37.6	5.0	7.8	15.6	34.0	19.9
62–396	160	48.8	17.5	10.6	5.6	17.5	12.5
LSD$_{05}$	7.8						

In the 1111 (24235 × "Rosavka") seedling variant, a least number (52 pcs.) adventitious roots per common bean one cutting were induced. The length of the majority of these roots (57.8%) did not exceed 1.0 cm, and the percentage of roots longer than 3.0 cm was the smallest (11.5%) with a rather low percentage (19.2%) of roots with secondary and tertiary lateral branchings. However, such total inhibition of rooting applies only to extracts from seedlings 1111 (24235 × "Rosavka"). On the other hand, in the variants of bean cuttings, extracts from "Vnuchka" and 1055 seedlings (24235 × "Rosavka"), in which the total number of roots per common bean cuttings were close to the control variants, developed the highest the number of secondary and tertiary lateral branchings. The share of roots in cuttings of common beans more than 3.0 cm long in the variants of extraction with extracts of "Vnuchka" and seedling 1055 (24235 × "Rosavka") was the highest: 45.5% ("Vnuchka") and 68.8% (1055). In these same variants, the number of secondary and tertiary lateral branchings was the largest: 42.8% of "Vnuchka" and 58.4% of seedling 1055 (24235 × "Rosavka"). Therefore, the effectiveness of these two extracts to rooting stimulate in cuttings of beans can be considered quite high. The low indices of the total number of roots per bean root in these variants were offset by the higher indices of root branching, which provided the overall suction functions of the root system storage capacity at a sufficient level.

The grouping of extract donors at the biological effectiveness of their influence on the above mentioned indicator plants made it possible to summarize the data obtained (Table 12.6).

TABLE 12.6 Grouping of Extract Donors at the Effectiveness of their Influence on Radish Seeds and Cuttings of Common Beans.

Indicator of efficiency	Extract donor		
	Most effective	Effective	Less effective
Germination of radish seeds	6–5; 1054; 1107	"Vnuchka"; "Symyrenkivets"; 2–7; 6–6; 1055	"Rosavka"; 2–6; 6–9; 6–11; 1111; Don 70–456; 62–396
The total number of roots per one common bean cutting	"Rosavka"; 2–7; 62–396	"Symyrenkivets"; 2–6; 6–5; 6–6; 6–9; 1054; 1107; Don 70–456	"Vnuchka"; 6–11; 1055; 1111
Percentage of roots with secondary and tertiary lateral branchings on common bean cutting	"Vnuchka"; 1054; 1055	"Symyrenkivets"; 2–7; 6–5; 6–6; 6–9; 6–11; 1107; 1111; Don 70–456	"Rosavka"; 2–6; 62–396

Among the most effective donors, only the 1054 seedling extract was effective when evaluated on two indicators in the absence of none donor that would be effective in three tests. For the effective donor group, extracts from "Symyrenkivets" cuttings and 6–6 seedlings proved effective in three trials and 2–7 seedlings in two trials. However, extracts from this seedling (2–7) were included in the group of the most effective donors in the tester for the total number of roots per bean root. This means that the efficiency of this extract donor is greater than that of the "Symyrenkivets" and 6–6 seedlings. Two indicators confirmed the effectiveness of extracts from cuttings of seedlings 6–9, 1107 and rootstocks Don 70–456. The effectiveness of seedlings 1055 is confirmed by the fact that by the number of branched roots on the cuttings of beans, extracts from cuttings of this seedling was included in the group of the most effective. The ineffective donors on the two testers were extracts from "Rosavka," seedlings 2–6 and 6–11, and rootstocks 62–396.

Experiments with the use of garden cress plants are worth a separate analysis. At both dates of cuttings, most extracts inhibited the growth of garden cress roots (Table 12.7).

TABLE 12.7　Influence of Extracts from Cuttings of Cultivars, Hybrids and Clonal Apple Rootstocks on the Growth of Garden Cress Roots.

Extract donor	Length of the roots of garden cress in dates of cuttings:			
	The second decade of April		The third decade of May	
	mm	± of control	mm	± of control
Control, H_2O	12.9	0	11.1	0
	Cultivars and hybrid seedlings			
"Rosavka"	10.4	-2.5	11.5	+0.4
"Rosavka" (scion-rooted)	11.5	-1.4	10.9	-0.2
2–6 ("Lord Lambournne" × "Golden Spire")	12.3	-0.6	11.3	+0.2
2–7 ("Lord Lambournne" × "Golden Spire")	11.0	-1.9	9.8	-1.3
18/12 ("Lord Lambournne" × "Golden Spire")	10.2	-2.7	9.4	-1.7
	Clonal apple rootstocks			
Don 70–456	10.0	-2.9	11.5	+0.4
62–396	10.9	-2.0	9.6	-1.5
LSD $_{.05}$	1.1		1.3	

When cutting in the second decade of April, the largest inhibition was (more than 2 LSD.$_{05}$) in variants with extracts of cuttings "Rosavka" scion-rooted, seedlings 18/12 ("Rosavka" × "Golden Spire") and rootstock Don 70–456. More than one LSD.05 was inferior to the controls of "Rosavka" root extracts, 2–7 seedlings ("Lord Lambournne" × "Golden Spire"), and rootstocks 62–396. Within LSD.$_{05}$ data of hybrid seedling 2–6 ("Lord Lambournne" × "Golden Spire") differed. When cuttings in the third decade of May, that is, in terms more favorable for regeneration, significantly hindered the growth of the roots of only four donor extracts. These were extracts from cuttings of seedlings 2–7 and 18/12 and rootstocks 62–396. The remaining donors deviated more or less within LSD.$_{05}$.

Comparison of the two periods of extraction does not allow establishing any regularities of their influence on the regrowth of the roots of garden cress, and the difference between the control options and the average indicators in terms of cuttings indicates the absence of a tendency of extracts positive effects even in the growth phase.

It can be assumed that apple tree extracts have in their composition some specific inhibitors of just the garden cress root growth, which neutralize the stimulating effects of endogenous auxins, despite the fact that the same extracts stimulated the growth of common bean roots or germination of radish seeds.

The experiments of studying the effect of extracts from different crabapples cuttings on the regrowth of the roots of garden cress at four terms of cuttings generally confirm the trend found in the analysis of the previously described testing. In most variants, inhibition of root growth under the influence of extracts from crabapples was observed, except for the extract of *M. baccata* (L.) Borkh Rehd. of cuttings divided in the second decade of March (Table 12.8).

At other cuttings dates, the influence of the extract donors differed from the control variant with water within LSD.$_{05}$. During cuttings in the second decade of March, the influence of all donors except *M. baccata* (L.) Borkh Rehd was within LSD.$_{05}$. Cuttings in the first decade of April produced extracts that inhibited the growth of the garden cress roots (except for *M. baccata* (L.) Borkh Rehd.), as it did during cuttings in the second decade of June. When cuttings in the third decade of May extract from cuttings of *M. halliana* Koehne and *M. prunifolia* (Willd.) Borkh. *f. pendula* (Bean) Rehd. inhibited the growth of garden cress roots by more than one LSD.$_{05}$ and more than three LSD.$_{05}$, respectively. The rest variants showed

TABLE 12.8 The Influence of Crabapples Cuttings Extracts on the Growth of Garden Cress Roots.

Crabapple	The length of the roots of garden cress in the dates of cuttings:							
	The second decade of March		The first decade of April		The third decade of May		The second decade of June	
	mm	± of control	mm	± of control	mm	± of control	mm	± of control
Control, H_2O	11.0	0	12.7	0	11.1	0	13.5	0
M. baccata (L.) Borkh.	12.8	+1.8	11.6	-1.1	11.9	+0.8	12.4	-1.1
M. halliana Koehne	10.8	-0.2	10.4	-2.3	9.5	-1.6	10.6	-2.9
M. niedzwetskiana Dieck.	12.1	+1.1	10.7	-2.0	10.0	-1.1	11.2	-2.3
M. prunifolia (Willd.) Borkh. var. Rinki (Koidz.) Rehd. f. fastigiata bifera (Dieck.) Al. Teod.	10.7	-0.3	10.6	-2.1	10.8	-0.3	11.5	-2.0
M. prunifolia (Willd.) Borkh. f. pendula (Bean) Rehd.	10.5	-0.5	10.5	-2.2	7.0	-4.1	11.8	-1.7
LSD_{05}	1.1		1.8		1.4		1.7	

deviations within LSD. $_{05}$. Comparison of averages at different cuttings showed that the effect of extracts from cuttings cut at the beginning of crabapples winter dormancy release differed little from control. Apparently in their composition the number of inhibitors of growth of garden cress roots was smaller in comparison with later terms of cuttings. While cuttings during the transition period, braking increased. This inhibition of the growth of the garden cress roots persisted during the apical growth of shoots, although at these times expected stimulation due to endogenous auxins, but the effect of inhibition continued.

According to the classification of Andrei M. Grodzinsky,[56] the effect of extracts from apple tree cuttings on the regrowth of the garden cress can be attributed to the category of allelopathy. The obtained results are in general agreement with the known facts about the inhibition of the growth of garden cress roots treated by leaf washing-off[89] and extend the observations of our predecessors to extracts from cuttings of apple trees.

Therefore, radish seeds and cuttings can be used to biotest the rooting ability of apple tree (*Malus* spp.) stem cuttings with certain reservations, and the garden cress seeds do not meet the requirements of the apple regeneration ability tester.

The study of regeneration potential is of great scientific importance in itself; however, this problem becomes applicable in the process of analyzing the correlation dependence between regenerative ability and economic traits and properties. At the same time, the greatest interest is the question of how much influence the rooting ability of apple tree stem cuttings on the average yield and the rest of the economic traits of cultivar.

In order to clarify these issues, the correlation coefficients between the apple yield and some other economical traits on the one hand and the regeneration potential on the other hand were calculated. The calculations were made using experimental material obtained in previous experiments.[86] The calculation methodology was classic.[87]

A negative correlation between yield and non-morphogenic post-trauma regenerative capacity in the phase of transition to winter dormancy was established. This is a logical connection. Plants that continue active hormonal functioning in the late-autumn period apparently do not have time to properly prepare for winter conditions. In recent years, winter temperatures have come earlier than usual, which negatively affected the yield of such forms. The relationship between yield and other manifestations of regenerative potential was not significant (Table 12.9).

TABLE 12.9 Correlation Coefficients (*r*) Between Some Features of Apple Trees and their Regenerative Potentials.

Economical traits and properties	Coefficient of correlation between economical traits and regeneration potential					
	Post-trauma regeneration coefficients during the vegetation period				Manifestations of regenerative ability of stem cuttings	
	In the phase of spring awakening	Average seasonal	Maximum	In the phase of transition to winter rest	Callus formation	Rooting
Average yield	-0.08	-0.29	0.42	-0.55*	0.04	-0.21
Yield stability **	0.26	0.44	0.62*	-0.37	0.13	0.24
The percentage share of marketable yield in total yield	0.33	-0.52*	-0.37	-0.73*	0.21	-0.47*
Average fruit weight	-0.36	0.24	0.75*	0.18	-0.23	0.26
Number of fruits per tree	-0.23	-0.17	0.17	-0.11	0.35	-0.57
Taste rating	0.03	0.63*	0.68*	0.33	-0.03	0.10

Note: *—significantly with 95% probability, **—the coefficient of variation of yields over the entire study period.

The absence of a significant correlation between the manifestations of the regenerative potentials of the apple trees stem cuttings and the yield of cultivars and hybrid seedlings from which these cuttings were cut for rooting, gives grounds for refuting the idea that rooting ability is inherent only in primitive forms of apple trees. In addition, taking into consideration the fact that the experiments were conducted with cultivars and hybrids in the process of breeding for rooting, the lack of a significant negative correlation between yield and rooting capacity indicates the success of breeders of L. P. Symyrenko Research Institute of Pomology of NAAS of Ukraine in the efforts to combine yields with rooting ability.

A fairly close correlation ($r = 0.62$) was established between the maximum value of the post-traumatic regeneration coefficient and the yield stability, which indicates the necessity of estimating the regeneration coefficient when performing the general pomological assessment.

The percentage share of marketable yield in total yield is a feature of great importance; however, it depends on the growing conditions even more than the yield. The significantly negative correlation between this characteristic and the average seasonal post-traumatic regeneration coefficient ($r = -0.52$), the same coefficient in the phase of transition to winter dormancy ($r = -0.73$), as well as the rooting ability of cuttings ($r = -0.4$) can be explained by the fact that a fairly homogeneous material was included in the experiment with high percentage share of fruit highest and first grade (92.9–98.3%), and the fact that the physiological processes that accompany regeneration had an additive effect on the fruit crop quality.

The rather close positive correlation between the average weight of one fruit and the maximum rate of post-traumatic regenerative capacity ($r = 0.75$) gives reason to take account of the regenerative potential at the pomological assessment. The absence of a significant negative relationship between the manifestations of the regenerative capacity of stem cuttings, including rooting ability, and the average weight of one fruit, confirms the previous findings about the breeding successes on the combination of yield and top-quality fruit with rooting ability.

In terms of the number of fruits per tree, the connection of this trait with regeneration potential was, for the most part, irrelevant. The negative correlation $r = -0.57$ is only correlated with root power. This phenomenon suggests that selection for the number of fruits was inactive, because a more significant criterion was the sign of fruit size, which traditionally has a negative correlation with the number of fruits on the plant. In our experiments, the correlation coefficient between these traits is $r = -0.52$.

Among some unexpected results, a positive correlation was found between the taste of the fruit and the mid-season ($r = 0.63$) and maximum ($r = 0.68$) post-traumatic regeneration coefficient. This phenomenon requires a deeper analysis with more sampling and more target plant materials.

12.4 CONCLUSION

Current findings suggest that the radish seeds and cuttings of common beans can be used as biotesters with certain reservations, but sprouted seeds of garden cress do not meet the requirements of the tester for predicting the rooting ability of apple tree (*Malus* spp.) stem cuttings. The fact that there is no negative correlation between the majority of economical traits and properties and the manifestations of regenerative ability of stem cuttings, in particular rooting potential of stem cuttings of *Malus* spp. and the yield of the cultivars and hybrid seedlings from which these cuttings were cut for rooting gives grounds for refuting the widespread belief that the stem cuttings rooting ability is inherent only in primitive forms of apple tree. The positive correlation of apple yield stability with the maximum post-trauma coefficient indicates the need to evaluate the regenerative potential of apple cultivars and hybrids in combination with the evaluation of pomological characteristics and properties.

ACKNOWLEDGMENTS

This material is partly based on the work supported by the National Dendrological park "Sofiyivka" of NAS of Ukraine (№ 0112U002032) in compliance with the thematic plan of the research work. The authors are grateful to the corresponding member of NAS of Ukraine DSc in Biology Ivan Kosenko and Ph.D. Liudmyla Zagoruiko for consultation and discussion.

KEYWORDS

- **bioindication**
- **callus**
- **clonal apple rootstocks**
- **crabapples**
- **horticulture**
- **rooting potential**
- **vegetative propagation**
- **woody plants**

REFERENCES

1. Opalko, A. I.; Kucher, N. M.; Opalko, O. A. Method for Evaluation of Regeneration Potential of Pear Cultivars and Species (*Pyrus* L.). In *Ecological Consequences of Increasing Crop Productivity: Plant Breeding and Biotic Diversity;* Opalko, A. I., et al., Eds.; Apple Academic Press: Toronto, New Jersey, 2015; pp 141–154.
2. Opalko, A. I.; Kucher, N. M.; Opalko, O. A. Regeneration Potential of Pear Cultivars and Species (Pyrus L.) of the Collection of The National Dendrological Park "Sofiyevka" of NAS of Ukraine. In *Chemical and Biochemical Technology Materials, Processing, and Reliability*; Varfolomeev, S. D., et al., Eds.; Apple Academic Press: Toronto, New Jersey, 2014; pp 299–311.
3. Opalko, O. A.; Opalko, A. I. Non-Morphogenetic Post-Trauma Regeneration Potential of *Malus halliana* Koehne (Rosaceae Juss.). *Int. J. Plant Reprod. Biol.* **2019,** *11*(2), 167–175. DOI: 10.14787/ijprb.2019 11.2.
4. Perez-Garcia, P.; Moreno-Risueno, M. A. Stem Cells and Plant Regeneration. *Dev. Biol.* **2018,** *442*(N 1), 3–12. DOI: 10.1016/j.ydbio.2018.06.021.
5. Goss, R. J. The Natural History (and Mystery) of Regeneration. In *A History of Regeneration Research: Milestones in the Evolution of a Science*; Dinsmore, C. E., Ed.; Cambridge University Press, 1991; pp 7–23.
6. Réaumur, R. A. F. Sur les diverses reproductions qui se font dans les écrevisses, les omars, les crabes, etc. et entr'autres sur celles de leurs jambes et de leurs écailles. *Mémoires de l'Academie Royale des Sciences*, 1712; pp 223–245.
7. Trembley, A. *Mémoires pour servir á l'histoire d'un genre de polypes d'eau douce, á bras en forme de cornes*; Verbeek: Leiden, 1744, p 410.
8. Aristotle. Generation of Animals. In *Historia Animalium*; Book 1–3., Vol. 1[Trans.: Arthur Leslie Peck]; Harvard University Press: Cambridge, 1965; pp 3–128.
9. Odelberg, S. J. Unraveling the Molecular Basis for Regenerative Cellular Plasticity. *PLoS Biol.* **2004,** *2*, N 8, e232, 1068–1071. DOI: 10.1371/JOURNAL.PBIO.0020232.

10. Williams, D. L. Regenerating Reptile Retinas: A Comparative Approach to Restoring Retinal Ganglion Cell Function. *Eye* **2017,** *31*(N 2), pp 167–172. DOI: 10.1038/ eye.2016.224.

11. Maienschein, J. T. H. Morgan's Regeneration, Epigenesis, and (w)Holism. In *A History of Regeneration Research: Milestones in the Evolution of a Science;* Dinsmore, C. E., Ed.; Cambridge Univ. Press: New York, 2007; pp 133–149.

12. Morgan, Th. H. Regeneration and Liability to Injury. *Sci. New Ser.* **1901,** *14*(N 346), pp 235–248.

13. Agata, K.; Saito, Y.; Nakajima, E. Unifying Principles of Regeneration I: Epimorphosis Versus Morphallaxis. *Dev. Growth Differ.* **2007,** *49*(N 2), 73–78. DOI: 10.1111/j.1440-169X.2007.00919.x.

14. Krenke, N. P. *Regeneration of Plants*; Moscow; Leningrad: USSR Academy of Sciences Press, 1950; p 676 (in Russian).

15. Hartmann, H. T., Kester, D. E. *Plant Propagation: Principles and Practices*; Prentice-Hall Inc.: Englewood Cliffs, 1975; p 622.

16. Melvin, N. W. Vegetative propagation. *Temperate-Zone Pomology: Physiology and Culture*; Timber Press: Portland, 1993; pp 116–129.

17. Bruinsma, J. Exogenus and Endogenus Growth-Regulating Substances in Studies on Plant Morphogenesis. *Perspect. Exp. Biol.* **1976,** *2*, 69–76.

18. Cottrell, J. E.; Dale, J. E. The Effects of Photoperiod and Treatment with Gibberellic Acid on the Concentration of Soluble Carbohydrates in the Shoot Apex of Spring Barley. *New Phytol.* **1986,** *102*(N 3), 365–373. DOI: 10.1111/j.1469-8137.1986.tb00814.x.

19. Kefeli, V., Kalevitch, M. V. *Natural Growth Inhibitors and Phytohormones in Plants and Environment*; Springer Science & Business Media, 2013; p 324.

20. Kunakh, V. A. *Biotechnology of Medicinal Plants. Genetic, Physiological and Biochemical Bases*; Logos: Kyiv, 2005; p 730. (in Ukrainian).

21. Pais, M. S. Somatic Embryogenesis Induction in Woody Species: The Future After OMIC's Data Assessment. *Front. Plant Sci.* **2019,** *10*(Art. 240), pp 1–18. DOI: 10.3389/ fpls.2019.00240.

22. Tvorogova, V. E.; Lutova, L. A. Genetic Regulation of Zygotic Embryogenesis in Angiosperm Plants. *Russ. J. Plant Physiol.* **2018,** *65*, 1–14. DOI: 10.1134/S10214437 18010107.

23. Sena, G.; Wang, X.; Liu, H. Y.; Hofhuis, H.; Birnbaum, K. D. Organ Regeneration Does Not Require a Functional Stem Cell Niche in Plants. *Nature* **2009,** *457*(7233), 1150–1153. DOI: 10.1038/nature07597.

24. Atta, R.; Laurens, L.; Boucheron-Dubuisson, E.; Guivarc'h, A.; Carnero, E.; Giraudat-Pautot, V.; Chriqui, D. Pluripotency of Arabidopsis Xylem Pericycle Underlies Shoot Regeneration from Root and Hypocotyl Explants Grown *in vitro. Plant J.* **2009,** *57*(N 4.), 626–644. DOI: 10.1111/j.1365-313X.2008.03715.x.

25. Sugimoto, K.; Jiao, Y.; Meyerowitz, E. M. Arabidopsis Regeneration from Multiple Tissues Occurs via a Root Development Pathway. *Dev. Cell.* **2010,** *18*(N 3), 463–471. DOI: 10.1016/j.devcel.2010.02.004.

26. Sugimoto, K.; Gordon, S. P.; Meyerowitz, E. M. Regeneration in Plants and Animals: Dedifferentiation, Transdifferentiation, or Just Differentiation? *Trends Cell Biol.* **2011,** *21*(N 4), 212–218. DOI: 10.1016/j.tcb.2010.12.004.

27. Asif, N.; Malik, M.; Chaudhry, F. N. A Review of on Environmental Pollution Bioindicators. *Pollution* **2018**, *4*(N 1), 111–118. DOI: 10.22059/poll.2017.237440.296.

28. Stöcker, G. Zu einigen theoretischen und methodischen Aspekten der Bioindikation. In *Methodischen und theoretischen Grundlagen der Bioindication.* Teil, 1.; Schubert, R., Schuh J., Eds.; Martin-Luther Universität: Halle-Wittenberg. Wissen-schaftliche Beiträge, 1980; pp 10–21.

29. Wiłkomirski, B. History of Bioindication (*Historia bioindykacji*). *Monitoring Środowiska Przyrodniczego* **2013**, *14*, 137–142.

30. Bahday, T. V.; Panas, N. E.; Antonyak, H. L.; Bubys, O. E. Biomonitoring Environmental Assessment of Natural Reservoirs. *Sci. Messenger Lviv Natl. Univ. Vet. Med. Biotechnol. Named After S. Z. Gzhytsky.* **2016**, *18*(N 1(65)), Ser. 3, 190–194 (in Ukraine).

31. Bertrand, L.; Monferrán, M. V.; Mouneyrac, C.; Amé, M. V. Native Crustacean Species as a Bioindicator of Freshwater Ecosystem Pollution: A Multivariate and Integrative Study of Multi-Biomarker Response in Active River Monitoring. *Chemosphere* **2018**, *206*, 265–277. DOI: 10.1016/j.chemosphere.2018.05.002.

32. Havlicek, E. Soil Biodiversity and Bioindication: From Complex Thinking to Simple Acting. *Eur. J. Soil Biol.* **2012**, *49*, 80–84. DOI: 10.1016/j.ejsobi.2012.01.009.

33. Santos, R. S. D.; Barreto-Garcia, P. A. B.; Scoriza, R. N. Mycorrhizal Fungi and Litter as Indicators of the Edge Effect in a Fragment Seasonal Forest. *Ciência Florestal* **2018**, *28*(N 1), 324–335. DOI: 10.5902/1980509831603.

34. Zhukov, O. V.; Kunah, O. M.; Dubinina, Y. Y.; Ganzha, D. S. Diversity and Phytoindication Ability of Plant Community. *Biol. Bull. Bogdan Chmelnitskiy Melitop. State Pedagog. Univ.* **2017**, *7*(N 4), 81–99. DOI: 10.15421/2017_90 (in Ukraine).

35. Cairns, J.; Pratt, J. R. A History of Biological Monitoring Using Benthic Macroinvertebrates. In *Freshwater Biomonitoring and Benthic Macroinvertebrates;* Rosenberg, D. M., Resh, V. H., Eds.; Chapman & Hall: NY, 1993; pp 10–27.

36. Singham, M. The Canary in the Mine. *Phi Delta Kappan* **1998**, *80*(N. 1), 9–15.

37. Gardiner, B. G. Linnaeus' Floral Clock. *Linnean* **1987**, *3*(N. 1), 26–29.

38. Tomalin, M. Ecological Horology. In *Romantic Ecocriticism: Origins and Legacies;* Hall, D. W., Ed.; Lanham, MD: Lexington Books, 2017; pp 21–42.

39. Nylander, W. Les lichens du Jardin du Luxembourg. *Bulletin de la Société Botanique de France* **1866**, *13*, 364–372.

40. Reinecke, J.; Wulf, M.; Baeten, L.; Brunet, J.; Decocq, G.; De Frenne, P.; Naaf, T. Acido-and Neutrophilic Temperate Forest Plants Display Distinct Shifts in Ecological pH Niche Across North-Western Europe. *Ecography* **2016**, *39*(N 12), 1164–1175. DOI: 10.1111/ecog.02051.

41. Gehring, Ch.; Irving, H. R. Peptides and the Regulation of Plant Homeostasis. In *Plant Signaling Peptides;* Irving, H. R., Gehring, C., Eds.; Springer: Berlin; Heidelberg, 2012; 183–198. DOI: 10.1007/978-3-642-27603-3_10.

42. Guex, J. Environmental Stress and Atavism in Ammonoid Evolution. *Eclogae geologicae helvetiae* **2001**, *94*(3), 321–328. DOI: 10.5169/seals-168897.

43. Wheeler, J. I.; Irving, H. R. Plant Peptide Signaling: An Evolutionary Adaptation. In *Plant Signaling Peptides;* Irving, H. R., Gehring, C., Eds.; Springer: Berlin; Heidelberg, 2012; pp 1–23. DOI: 10.1007/978-3-642-27603-3_1.

44. Blagnytė, R.; Paliulis, D. Research into Heavy Metals Pollution of Atmosphere Applying Moss as Bioindicator: A Literature Review. *Environ. Res. Eng. Manag.* **2010,** *54*(N 4), 26–33. DOI: 10.5755/j01.erem.54.4.93.

45. Henderson, A. Literature on Air Pollution and Lichens XLVIII. *Lichenologist* **1999,** *31*(N 1), 111–119. DOI: 10.1006/lich.1998.0183.

46. McMurray, J. A.; Roberts, D. W.; Geiser, L. H. Epiphytic Lichen Indication of Nitrogen Deposition and Climate in the Northern Rocky Mountains, USA. *Ecol. Indic.* **2015,** *49*, 154–161. DOI: 10.1016/j.ecolind.2014.10.015.

47. Onete, M.; Pop, O. G.; Gruia, R. Plants as Indicators of Environmental Conditions of Urban Spaces from Central Parks of Bucharest. *Environ. Eng. Manag. J. (EEMJ)* **2010,** *9*(N 12), 1637–1645. DOI: 10.30638/eemj.2010.225.

48. Parmar, T. K.; Rawtani, D.; Agrawal, Y. K. Bioindicators: The Natural Indicator of Environmental Pollution. *Front. Life Sci.* **2016,** *9*(N 2), 110–118. DOI: 10.1080/21553769.2016.1162753.

49. Smith, R. I. L. Vascular Plants as Bioindicators of Regional Warming in Antarctica. *Oecologia* **1994,** *99*(N 3–4), 322–328. DOI: 10.1007/BF00627745.

50. Metcalfe-Smith, J. L. Biological Water-Quality Assessment of Rivers: Use of Macro-invertebrate Communities. In *The Rivers Handbook: Hydrological and Ecological Principles,* Vol. 2; Calow, P. P., Petts, G. E., Eds.; John Wiley & Sons, 2009; pp 144–170. DOI: 10.1002/9781444313871.

51. Neumann, M.; Baumeister, J.; Liess, M.; Schulz, R. An Expert System to Estimate the Pesticide Contamination of Small Streams Using Benthic Macroinvertebrates as Bioindicators. II. The knowledge base of LIMPACT. *Ecol. Indic.* **2003,** *2*(N 4), 391–401. DOI: 10.1016/S1470-160X(03)00025-6.

52. Pander, J.; Geist, J. Ecological Indicators for Stream Restoration Success. *Ecol. Indic.* **2013,** *30*, 106–118.

53. Gniazdowska, A.; Bogatek, R. Allelopathic Interactions Between Plants. Multi Site Action of Allelochemicals. *Acta Physiol. Plant.* **2005,** *27*(N 3B), 395–407. DOI: 10.1007/s11738-005-0017-3.

54. Chou, C. H. Introduction to Allelopathy. *Allelopathy*; Reigosa, M. J., Pedrol, N.; González, L., Eds.; Springer: Dordrecht, 2006; pp 1–9. DOI: 10.1007/1-4020-4280-9_1.

55. Molish, H. *The Influence of One Plant on Another: Allelopathy*; Narwal, S. S.; Fleur, Trs. Linda J. La, Bari Mallik, M. A., Eds.; Scientific Publishers (India): Jodhpur, 2001; pp 23–132.

56. Grodzinsky, A. M. *Fundamentals of Chemical Interaction of Plants*; Scientific Opinion: Kyiv, 1973; p 207 (in Ukraine).

57. Willis, R. J. *The History of Allelopathy*; Springer: Dordrecht, 2007. DOI: 10.1007/978-1-4020-4093-1.

58. Cheema, Z. A.; Farooq, M.; Wahid, A, Eds.; *Allelopathy: Current Trends and Future Applications.* Springer: Berlin, Heidelberg, 2013, p 517.

59. Langenfelds, T. V. *Apple-Trees—Morphological Evolution, Phylogeny, Geography and Systematic of the Genus*; Zinatne: Latvia, Riga, 1991; pp 234 (in Russian).

60. Peil, A.; Kellerhals, M.; Höfer, M.; Flachowsky, H. Apple Breeding—from the Origin to Genetic Engineering. *Fruit Veg. Cereal Sci. Biotechnol.* **2011,** *5*, 118–138.

61. Opalko, O. A.; Konopelko, A. V.; Opalko, A. I. An Apple (*Malus* Mill.) in History and Culture of the Ukrainian and other Ethnces. *Sib. J. For. Sci.* **2019,** *N 4*, 18–35. DOI: 10.15372/SJFS20190403. (in Russian).

62. Reiss, M. *Apple*; Reaktion Books: London, 2015; p 224.

63. Janik, E. *Apple: A Global History*; Reaktion Books: London, 2011; p 110.

64. Catonis, M. P. *De agri cvltvra liber. Post Henricvm Keil itervm edidit Georgivs Goetz*; Tevbneri: Lipsiae, 1922; pp 20–74.

65. Ignatov, A.; Bodishevskaya, A. *Malus. Wild Crop Relatives: Genomic and Breeding Resources, Temperate Fruits;* Kole, C., Ed.; Springer: Berlin, Heidelberg, 2011; pp 45–64. DOI: 10.1007/978-3-642-16057-8_3.

66. APG IV. An Update of the Angiosperm Phylogeny Group Classification for the Orders and Families of Flowering Plants: The Angiosperm Phylogeny Group. *Bot. J. Linn. Soc.* **2016,** *181*(1), 1–20. DOI: 10.1111/boj.12385.

67. Opalko, A. I.; Andrienko, O. D.; Opalko, O. A. Phylogenetic Connections Between Representatives of the Genus *Amelanchier* Medik. *Temperate Crop Science and Breeding: Ecological and Genetic Study;* Bekuzarova, S. A., Bome, N. A., Opalko, A. I., et al., Eds.; Apple Academic Press: Oakville, Waretown, 2016; pp 201–232.

68. Engler, A. Syllabus der Pflanzenfamilien. Eine Übersicht über das gesamte Pflanzensystem mit Berücksichtigung der Medicinal- und Nutzpflanzen nebst einer Übersieht über die Florenreiehe und Florengebiete der Erde zum Gebrauch bei Vorlesungen und Studien über specielle und medicinisch-pharraaceutische Botanik; Verlag von Gebrüder Borntraeger: Berlin, 1903; pp 29–237.

69. Takhtajan, A. L. *Flowering Plants*, 2nd ed.; Springer Science+Business Media: NY, 2009; p 871.

70. Reveal, J. L. An Outline of a Classification Scheme for Extant Flowering Plants. *Phytoneuron* **2012,** *2012–2037*, 221.

71. Turland, N. J.; Wiersema, J. H.; Barrie, F. R.; Greuter, W.; Hawksworth, D. L.; Herendeen, P. S., Knapp, S.; Smith, G. F, Eds.; *International Code of Nomenclature for Algae, Fungi, and Plants (Shenzhen Code) Adopted by the Nineteenth International Botanical Congress Shenzhen, China, July 2017*. Regnum Vegetable 159; Koeltz Botanical Books: Glashütten, 2018. DOI: 10.12705/Code.2018.

72. Mezhenskyj, V. M. On Streamlining the Ukrainian Names of Plants. Information 5. Species Names for Pome Fruit Crops. *Plant Var. Stud. Prot.* **2015,** *N 3–4*(28–29), 4–11. DOI: 10.21498/2518-1017.

73. Mezhenskyj, V. M. On Streamlining the Ukrainian Names of Plants. Information 6. Names of Some Subtribe Malinae Reveal Taxa (Information 6). *Plant Var. Stud. Prot.* **2015,** *N 1*(30), 5–11. DOI: 10.21498/2518-1017.1(30).2016.61699.

74. Darlington, C. D.; Moffett, A. A. Primary and Secondary Chromosome Balance in *Pyrus. J. Genet.* **1930,** *22*(N 2), 129–151.

75. Sax, K. The Origin and Relationships of Pomoideae. *J. Arnold Arbor.* **1931,** *12*(N 1), 3–22.

76. Campbell, C. S.; Evans, R. C.; Morgan, D. R.; Dickinson, T. A.; Arsenault, M. P. Phylogeny of Subtribe Pyrinae (Formerly the Maloideae, Rosaceae): Limited Resolution of a Complex Evolutionary History. *Plant Syst. Evol.* **2007,** *266*(N. 1–2), 119–145. DOI: 10.1007/s00606-007-0545-y.

77. Dickinson, T. A.; Lo, E. Y. Y.; Talent, N. Polyploidy, Reproductive Biology, and Rosaceae: Understanding Evolution and Making Classifications. *Plant Syst. Evol.* **2007**, *266*(N 1–2), 59–78.

78. Potter, D.; Eriksson, T.; Evans, R. C.; Oh, S. H.; Smedmark, J. E. E.; Morgan, D. R.; Campbell, C. S. Phylogeny and Classification of Rosaceae. *Plant Syst. Evol.* **2007**, *266*(N. 1–2), 5–43. DOI: 10.1007/s00606-007-0539-9.

79. Tatum, T. C.; Stepanovic, S.; Biradar, D. P.; Rayburn, A. L.; Korban, S. S. Variation in Nuclear DNA Content in Malus Species and Cultivated Apples. *Genome* **2005**, *48*(N 5), 924–930.

80. Way, R. D.; Aldwinckle, H. S.; Lamb, R. C.; Rejman, A.; Sansavini, S.; Shen, T.; Yoshida, Y. Apples (Malus). *Acta Hortic.* **1990**, *290*, 1–62. DOI: 10.17660/ActaHortic.1991.290.1.

81. Eccher, G.; Ferrero, S.; Populin, F.; Colombo, L.; Botton, A. Apple (Malus domestica L. Borkh) as an Emerging Model for Fruit Development. *Plant Biosyst. Int. J. Deal. Aspects Plant Biol.* 2014. Vol. 148, N 1. pp. 157–168. DOI: 10.1080/11263504.2013.870254.

82. Gross, B. L.; Henk, A. D.; Richards, C. M.; Fazio, G.; Volk, G. M. Genetic Diversity in *Malus* × *domestica* (Rosaceae) Through Time in Response to Domestication. *Am. J. Bot.* **2014**, *101*(N. 10), 1770–1779. DOI: 10.3732/ajb.1400297.

83. Hancock, J. F.; Luby, J. J.; Brown, S. K.; Lobos, G. A. Apples. In *Temperate Fruit Crop Breeding: Germplasm to Genomics*; 2008; pp 1–32. DOI: 10.1007/978-1-4020-6907-9-1.

84. Brickell, C. D. International Code of Nomenclature for Cultivated Plants, 9th ed.; In *Scripta Horticulturae*; Commission Chairman, Ed.; 2016, N 18; p 190.

85. *Malus* × *gloriosa* Lemoine. *The Plant List is a Working List of all Known Plant Species.* Version 1.1. September 2013. URL: http://www.theplantlist.org/tpl1.1/record/rjp-39428 (accessed Dec 30, 2018).

86. Balabak, A. F.; Opalko, O. A. Indirect Methods for Predicting the Rooting Capacity of Apple Tree Cuttings. *Collect. Sci. Pap. Uman State Agrar. Acad.* **2001**, *53*, 70–77 (in Ukraine).

87. Fisher, R. A. *Statistical Methods for Research Workers*; Cosmo Publications: New Delhi, 2006; p 354.

88. Chupryniuk, V. Ya.; Homeniuk, V. I. Propagation of Apple Trees by Softwood Stem Cuttings. *Collection of Scientific Papers of L. P. Symyrenko Research Institute of Pomology of NAAS of Ukraine and Uman State Agrarian Academy;* 2000; pp 41–46 (in Ukraine).

89. Moroz, P. A. *Allelopathy in Orchards*; Naukova dumka: Kyiv, 1990; p 208 (in Ukrainian).

Glossary

Agrocenosis
It is an artificial ecosystem created by man in the course of economic activity.

Agrotechnology
The technology of agriculture as the methods or machinery needed for efficient production.

Amylase
An enzyme that catalyzes the hydrolysis of starch (Latin *amylum*), glycogen, and related polysaccharides to oligosaccharides, maltose, or glucose; any of several digestive enzymes that break down starches and one of several enzymes that are produced in the pancreas. Specific amylase proteins are designated by different Greek letters (α-amylase EC 3.2.1.1; β-amylase EC 3.2.1.2; and γ-amylase or glucan 1,4-α-glucosidase 3.2.1.3.)

Angiosperm Phylogeny Group (APG)
An international group of systematic botanists who collaborate to establish a consensus on the taxonomy of flowering plants (angiosperms) by molecular techniques.

Angiosperm Phylogeny Group IV
It is the fourth version of a modern system of plant taxonomy for flowering plants (angiosperms) based on analysis of sequences of chloroplast markers, and as nuclear genes, and the features or characteristics of the plants being developed by the Angiosperm Phylogeny Group.

Artemia
The brine shrimp (Crustacea: Branchiopoda) is a widespread and dominant crustacean, species present in many hypersaline environments. It is filter-feeding organism, which is vital in order ecosystem functioning and the impact of natural and anthropogenic stressors, because the adult brine shrimp accumulated chemicals from the water. The brine shrimp is recommended as test organism for the evaluation of the ecological status of the natural water and for the marine and hypersaline water.

Asexual reproduction
A type of reproduction by which offspring are produced from a single parent as budding, fission, or spore formation, rather than through fertilization (the union of gametes), and inherit the genes of that parent only. This is propagation through multiplication of vegetative parts is the only method for the *in vivo* propagation

of certain cultivated plants, as they do not produce viable seeds, for example, banana, garlic. grape, fig, chrysanthemum, etc., asexual reproduction also used for propagation of some cultivars, especially allogamous (cross-pollinated) plants in which seed propagation does not ensure the availability of cultivar features in seed progeny. To save cultivar purity, vegetative propagation is essential to be used.

Automorphic soil
Any soil whose properties result primarily from climatic action and from vegetation as the main soil-forming factors and correlates with zonal soil.

Bacillus
A genus of gram-positive rod-shaped bacteria of the Firmicutes phylum, with 266 species. The genus *Bacillus* is a diverse group of spore-forming bacteria. *Bacillus* species are important in both the medical and bioprotective aspects of microbiology.

Bacillus subtilis
The spore-forming bacteria of the genus *Bacillus* of the family Bacillaceae. Are able to process plant and animal waste, using substrates such as starch, carbohydrates, proteins, agar, cellulose, etc.; they are able to synthesize numerous compounds with antimicrobial potential (peptide and lipopeptide antibiotics, bacteriocins). Widely used in the composition of drugs (probiotics) that improve the digestive system and increase the immunity of animals.

Bacillus subtilis CABI
The strain of *Bacillus subtilis* (Ehrenberg) Cohn.

Biogeocenosis
The term biogeocenosis was first used in 1945 in a publication by Russian geobotanist Vladimir Sukachev in the sense of a interrelated complex of living and inert components associated with each other by material and energy exchange in a particular unit of space; one of the most complex systems in nature.

Bio-indication
A time-dependent, sensitive registration of anthropogenic or anthropogenically altered environmental factors by distinguished dimensions of biological objects and biological systems under defined circumstances.

Bog (fen)
The mire that due to its location relative to the surrounding landscape obtains its water mainly from precipitation and not influenced by groundwater; Sphagnum-dominated vegetation (ombrotrophic).

Calyptra (also calypter)
A term proposed by Van Tieghem and Douliot (1988; in: Ann. Des Sci. Nat.: Bot. VIII) for that portion of the root cap in lateral roots which belongs strictly to the root system, and a thickened membrane of parenchymal cells that protects the growing root of vascular plants.

Carotenoids
Is a class of mainly yellow, orange, or red fat-soluble pigments, including carotene, which gives color to parts of plants, such as carrot, ripe tomatoes, autumn leaves, etc.

Celsius
Is the degree or degrees Celsius relating to a temperature scale that registers the freezing point of water as 0° and the boiling point as 100° under normal atmospheric pressure: the formula for converting a Celsius temperature to Fahrenheit is °F = °C + 32.

Centimetre or centimeter
Is a unit of length in the metric system, which is equal to one-hundredth of a meter.

Chemical mutagenesis
A method of the induction of mutations by the use of chemical agents that interact with nucleic acids. In 1946, the mutations were chemically induced in *Drosophila* by Joseph A. Rapoport (in Russia) using formaldehyde as well as Charlotte Auerbach and John M. Robson (in Great Britain) using mustard gas. The mutagenicity of alkylating agents was determined by Joseph A. Rapoport, and the chemical mutagenesis method was introduced to applied genetics and breeding.

Chemiluminescence
The light emitted as the result of a chemical reaction. It's also known, less commonly, as chemiluminescence. Light isn't necessarily the only form of energy released by a chemiluminescent reaction. Heat may also be produced, making the reaction exothermic.

Chlorophyll
That is any of several green pigments found in the mesosomes of cyanobacteria, as well as in chloroplasts of algae and plants. Chlorophyll is essential for photosynthesis, allowing plants to absorb the energy of light. A correlation was revealed between the content of chlorophyll and the intensity of photosynthesis.
The chlorophyll content is specific for the leaves of each species and plant variety and varies significantly depending on lighting, mineral nutrition, leaf age, and nitrogen content in the soil and other conditions. The chlorophyll content is estimated in a laboratory way. The method is subtle but difficult to perform in the field. An instrumental method for determining the total chlorophyll content in a field has been invented (see SPAD 502).

Clonal propagation
A vegetative (asexual) propagation from a single cell or plant by multiplication of genetically identical copies of individual plants.

Clone
A group of genetically identical cells or individuals, derived from a common ancestor named the ortet (original plant), by asexual mitotic division; in

horticulture or agriculture, a group of individuals originally taken from a single specimen and maintained in cultivation by vegetative propagation.

CMCase

An enzyme isolated from fungi and bacteria named the carboxymethyl cellulase (EC 3.2.1.4). It catalyzes the endohydrolysis of 1,4-β-glucosidic linkages in cellulose, lichenin, and cereal β-glucans.

Crabapple

The umbrella term for plants from the genus *Malus* Mill. (Rosaceae Juss.) exhibiting usually small acidic fruit (diameters <5 cm). There is an abundance of crabapple germplasm resources. More than 30 natural species have been recorded in the classical taxonomic systems, with consistent groupings and more than hundreds of ornamental crabapple cultivars exist. Many cultivars originated from selective breeding or were accidentally discovered, and thus their genetic background and phylogenetic relationships are uncertain.

Cutting

A section of a plant that is removed and used for propagation; cuttings may consist of a whole or part of a stem (foliaceous or non-leafy), leaf, bulb, or root; a root cutting consists of root only; other cuttings have no roots at the time they are made and inserted; as opposed to division, a kind of propagation that consists of part of the crown of a plant or of its above-grond portion and roots; several types of cuttings can be taken from the parent stock, which depends on the point on the parent stock where the cutting is taken; four major categories of cuttings are *stem cuttings*, *leaf cuttings*, *leaf-bud cuttings*, and *root cuttings*.

Cysts

The *Artemia* dormant eggs have led to extensive use of *Artemia* in aquaculture. The cysts may be stored for long periods and hatched on demand to provide a convenient form of live feed (nuplias) for larval fish and crustaceans.

Diploid chromosome number

Refers to having complete the number of sets of chromosomes found in a somatic cell; diploid chromosome number refers to having complete the number of sets of chromosomes found in a somatic cell for diploid $2n = 2x$, for tetraploid (four sets of chromosomes), $2n = 4x$, and for hexaploid (six sets), $2n = 6x$, etc.

Drone

An aircraft that does not have a pilot and is controlled by someone on the ground.

Ecosystem

Refers to individual concepts of ecology. The word itself stands for "ecological system." The term was proposed by ecologist A. Tensley in 1935. Ecosystem combines several concepts: biocenosis—a community of living organisms; biotope—the habitat of these organisms; and types of relationships of organisms in the specified habitat. The metabolism occurs between these organisms in a given biotope. That is, in essence, an ecosystem is a combination of components of animate and inanimate nature, between which energy is exchanged. Thanks to this, it is possible to create the conditions necessary for permanent life. The basis of any ecosystem on our planet is the energy of sunlight. To classify ecosystems, scientists chose one feature—the habitat, since this is what determines the climatic, bioenergetic, and biological features.

Flax
Linum usitatissimum known as flaxseed is a member of the genus *Linum* in the family Linaceae. It is a food and fiber culture grown in cooler regions of the world.

Germinability
The degree of potential for germination.

Germination
The beginning of growth of a seed, spore, or other structure, usually following a period of dormancy and generally in response to the return of favorable external conditions.

β-Glucanase
An enzyme that hydrolyzes β-glucans for the breakdown of hemicellulose, xylan, and cellulose used in the production of alcohol, beer, in the pulp and paper, and textile industries.

Haploid number of chromosomes
Refers to having exactly half the number of sets of chromosomes found in a somatic cell and is different from "x" which means the monoploid number, that is, the number of chromosomes in a single (nonhomologous) set.

Hardwood cuttings
The cuttings are prepared from shoots that grew the previous summer (the wood is firm and does not bend easily); they are cut in winter or early spring while the plant is still dormant.

Hatching process
The hatchery of the larvae (nauplia) from *Artemia* cysts (eggs).

Heavy metals
These are metals with relatively high densities, atomic weights, or atomic numbers. The criteria vary depending on the author and context: in metallurgy, a heavy metal may be defined on the basis of density, whereas in physics the distinguishing criterion might be atomic number, and a chemist or biologist would likely be more concerned with chemical behavior. A density of more than 5 g/cm^3 is sometimes quoted as a commonly used criterion and is used in the body of this article.

Herbaceous cuttings
The cuttings are taken from succulent, herbaceous plants; this type of cutting roots fast but is next to nothing use in horticulture and forestry practice.

Holocene

The current period of geologic time. Another term that is sometimes used is the Anthropocene Epoch, because its primary characteristic is the global changes caused by human activity. The Holocene Epoch began 12,000–11,500 years ago at the close of the Paleolithic Ice Age and continues through today. The Holocene and the preceding Pleistocene together form the Quaternary period encompasses the last ~2.6 Ma during which time Earth's climate was strongly influenced by bipolar glaciation and the genus *Homo* first appeared and evolved.

The Holocene can be subdivided into five time intervals, or chronozones, based on climatic fluctuations: Preboreal (10–9 ka BP), Boreal (9–8 ka BP), Atlantic (8–5 ka BP), Subboreal (5–2.5 ka BP), and Subatlantic (2.5 ka BP–present). The abbreviation "ka" means "kilo-annum" before present (era).

Homomorphism

In biology is similarity of external form or appearance but not of structure or origin.

Landscape

Any transport corridor which integrate with ecological system and any natural society becomes a driver of temporal heterogeneity. Along the transport corridor, the soil structure, and the ecological balance of vegetation are disturbed.

Leaf-bud cuttings

The cuttings are not used extensively in horticulture and forestry applications; the leaf-bud cutting includes the leaf itself, petiole, and a small piece of stem with the axyal bud; this form of cutting propagation is useful when material is scarce, because the same amount of stock will produce twice as many new plants as that of stem cuttings.

Leaf cuttings

The cuttings are not used extensively in horticulture and forestry applications; this form of propagation utilizes the leaf to promote new plant growth; a root and shoot will form and develop from the leaf cutting into a new plant.

Least significant difference

Between the treatment variants at the 5% level for significance (significantly with 95% probability).

Before Linum usitatissimum L., see Flax

do interval

Lipases

Any enzyme that catalyzes the hydrolysis of fats (lipids). They are the most used enzymes in synthetic organic chemistry, catalyzing the hydrolysis of carboxylic acid esters in an aqueous medium or the reverse reaction in organic solvents.

MAP kinase (MAPK)

Mitogen-activated protein kinase is a type of protein kinase that is specific to the amino acids—serine and threonine (i.e., a serine/threonine-specific protein kinase).

MAPKs are involved in directing cellular responses to a diverse array of stimuli, such as mitogens, osmotic stress, heat shock, and proinflammatory cytokines. They regulate cell functions including proliferation, gene expression, differentiation, mitosis, cell survival, and apoptosis. It is specific to the amino acids—serine and threonine-specific protein kinase. MAPKs are involved in directing cellular responses to a diverse array of stimuli, such as mitogens, osmotic stress, heat shock, and proinflammatory cytokines. They regulate cell functions including proliferation, gene expression, differentiation, mitosis, cell survival, and apoptosis.

Minerotrophic, see Peatland

Mire
It's peatlands where peat is currently being formed.

Mitosis
The process of cell nuclear division by which two daughter nuclei are produced, each identical to the parent nucleus; before mitosis begins each chromosome replicates to two sister chromatids; these then separate during mitosis so that one duplicate goes into each daughter nucleus that results in two daughter cells each having the same number and kind of chromosomes as the parent nucleus, typical of ordinary tissue growth.

Mitotic Index (MI)
The fraction of cells undergoing mitosis in a given sample.

Multicopter
A mechanically simple aerial vehicle whose motion is controlled by speeding or slowing multiple downward thrusting motor/propeller units. Multicopters can be divided into different types of multicopters: quadcopters, hexacopters, or even octocopters. The names indicate the number of rotor arms and propellers. Quadro stands for four, so a quadcopter drone (or quadrocopter drone) is an aircraft equipped with four rotors. The rotor arms are arranged in a square or rectangle and thus allow a stable and steady ascent and flying of the drone. While the word "hexa" in hexacopter stands for six rotors and an octocopter is equipped with eight rotors. However, the four-rotor design is the most common type among multicopter.

Mutagenesis
The process of obtaining hereditary changes in living cells or organisms through changes in the genetic structure. Mutagenesis may be natural, arising regardless of human activity, and artificial, induced when exposed to the chromosome apparatus using mutagens.

Mutagens
The agents leading to receiving of a mutant genotype (or genotypes).

Mutational selection
Of plants is carried out as a result of exposure to chemical, or physical mutagens on seeds, shoots or shredded leaves, and subsequent selection in generations of

samples with desirable traits and their reproduction. Unlike genetic engineering, mutational selection excludes the possibility of targeted genetic changes, since this kind of mutation is random.

Nauplia
It is *Artemia* larvae.

NC group
Is experimental group of dry cysts in glass tubes, which were irradiated with magnetic field near north pole during 24 h, then they were incubated in sterile marine water for 48 h.

Nitrogen (N)
A naturally occurring element (inert gas) that comprises about 80% of the Earth's atmosphere. Nitrogen is essential for growth and reproduction in both plants and animals. It is found in amino acids that make up proteins, in nucleic acids, that comprise the hereditary material and life's blueprint for all cells.

Nitrogen fertilizers
This is one of the most common categories of fertilizers produced out of nitrogen (N) chemical combinations. Nitrogen is absorbed by the plant roots in two forms: nitrate form (NO_3) and ammonical form (NH_4). Most of the crop plants prefer nitrogen in nitrate form. Ammonical form of nitrogen is, however, easily covertible into nitrate form. Most of the fertilizers contain nitrogen in these two available forms. Urea, however, contains nitrogen in amide form but this form of nitrogen is swiftly converted by soil microorganisms into ammonical form and then into nitrate form.

NT group
Is experimental group of cysts in glass tubes in sterile marine water with 35‰ salinity which were exposed to near north pole of magnetic field during the entire hatching period (48 h).

Nucleolus
A clearly defined, often spherical area of the eukaryotic nucleus, composed of densely packed fibrils and granules, that is associated with the nucleolus organizer; its composition is similar to that of chromatin, except that it is very rich in RNA and protein; it is the site of the preribosomal RNA synthesis and processing and of ribosomal particle assembly; the assembly of ribosomes starts in the nucleolus but is completed in the cytoplasm.

Nucleus
The double-membrane-bound organelle containing the chromosomes that is found in most non-dividing eukaryotic cells; it disappears temporarily during cell division; within the nucleus several independent approaches point to the compartmentalization of particular activities such as transcription, RNA processing, and replication; chromosomes are revealed to occupy defined domains

and to represent highly differentiated structures; the numerous activities that use DNA and RNA as a template occur with a defined spatial and temporal relationship (e.g., compartmentalization of nuclear functions of storage, transmission and sale of hereditary information with protein synthesis).

NW group
Is experimental group of cysts incubated in sterile marine water (salinity 35‰) in glass tubes, treated with static magnetic field 50 mT near north pole of the magnet during 24 h.

Optical density
An indicator determined on a spectrophotometer at a wavelength of 600 nm. For a bacterial culture, this is an indicator of the number of cells (cell concentration) in the medium.

Ortet
The original plant from which a clone is started through rooted cuttings (clone head), grafting, tissue culture, or other means of vegetative propagation (e.g., the original plus tree used to start a grafted clone for inclusion in a seed orchard).

Palynology
The science that studies plant pollen, spores, and certain microscopic plankton organisms (collectively termed palynomorphs) in both living and fossil form. Botanists use living pollen and spores (actuopalynology) in the study of plant relationships and evolution, while geologists (palynologists) may use fossil pollen and spores (paleopalynology) to study past environments, stratigraphy (the analysis of strata or layered rock), historical geology, and paleontology. Palynology is a useful tool in many applications, including a survey of atmospheric pollen and spore production and dispersal (aerobiology), in the study of human allergies, the archaeological excavation of shipwrecks, and detailed analysis of animal diets.

***Para*-aminobenzoic acid (*p*-aminobenzoic acid—PABA)**
It is 4-aminino-2-hydrooxybezoic acid with molecular mass 137.1; an organic compound with the formula $H_2NC_6H_4CO_2H$; a white-gray crystalline substance, only slightly soluble in water, soluble in hot water (80–90°C), well soluble in benzene, ethanol, and acetic acid; it is classified as nontoxic vitamin-like compound of group B, known also as vitamin H_1 or vitamin B_{10}; microorganisms use it as a precursor of folic acid synthesis; it participates in the synthesis of purines and pyrimidines, ultimately in the synthesis of DNA and RNA. PABA as chemical compound is known since 1863, but its high biological activity in low concentrations was first discovered in 1939 with well-known geneticist J. A. Rapoport in experiments with *Drosophila*. He showed that the positive effect on living systems is based on the previously unknown phenomenon of its interaction with ferments. This interaction results in the restoration of the ferments activity decreased in some cases at the genetic level (e.g., because of excess of recessive

gene) or because of damaging environmental factors. In subsequent studies, the ranges of PABA suitable for different objects were determined. It was proved that PABA is a promoter of phenotypic activity and increases immunity; it has virucidal and antimicrobial action, showed dioxidic functions. There are data about PABA effect decreasing harmful mutagens action, PABA positively effect on all characters determining yield structure and increasing adaptive plant properties including the resistance to a series of diseases.

Peatland
An organic wetland ecosystem with or without vegetation with a naturally accumulated peat layer at the surface. They are wetlands whose soils consist almost entirely of organic matter derived from the remains of dead and decaying plant material. In general, peat forms when the amount of photosynthetically produced organic matter exceeds the loss of organic matter through decomposition. Peatlands are of two types: fens and bogs. Fens are peatlands that are minerotrophic, while bogs are ombrotrophic.

Phosphazines
White crystalline powder; molecular mass of 137.1 g; contains two groups of ethyleneimine and a pyrimidine. Phosphemid possesses a cytogenetic effect (shown on plant and animal cells), causes high variability of genetically significant traits in the first generation with the induction of mutations in subsequent generations after processing seeds of grain crops (wheat, barley, and flax).

Phosphemide (Phosphemidum—Lat.)
(Di-(ethyleneimide)-pyrimidyl-2-amidophosphoric acid, Phosphemidum Lat.) is included in the group.

Photometry
The measurement of the intensity of visible light (quantification methods); the branch of physics concerned with such measurements.

Phytocenosis
A historically formed association of plants on homogeneous territory. It is the main part of biocenosis and ecosystem. The smallest taxon in the classification of associations is vegetable association.

Probiotics
A living microorganism that, when administered in adequate amounts, confer a health benefit on the host, can improve the functioning of the gastrointestinal tract, normalize the composition, and increase the biological activity of normal intestinal microflora.

Proteases (also called peptidases)
Proteolytic enzymes from the class of hydrolases that degrades proteins by hydrolysis of peptide bonds. Proteases catalyze (increases the rate) of proteolysis, the breakdown of proteins into smaller polypeptides or single amino acids. This

category is divided into two classes: endopeptidases and exopeptidases. Endopeptidases break bonds of nonterminal amino acids (within the protein) while exopeptidases break bonds of terminal amino acids (the end of the protein).

Proteinases
Enzymes that promote decomposition and synthesis of peptide bonds between amino acid residues of the protein molecule are carried out; proteases include pepsin, trypsin, papaya, etc. Proteinase is synonymous with endopeptidase.

Quadrocopter (also called quadcopter)
This is the multicopter that lifted and propelled by four rotors.

Radicle
The part of a plant embryo that develops into the primary root.

Revegetation
Is the process of transplanting and restoring the soil of disturbed land. Crop production helps prevent soil erosion and increases the ability of the soil to absorb more water during significant rainfall events and in the aggregate.

Root cuttings
The cuttings, which are used in horticulture and forestry propagation, should be taken from the young plant stock during the winter and spring months to ensure that they are saturated with stored foods; this time frame also prevents cutting during the time the parent plant is rapidly expanding shoot growth; cutting during active expansion will take food stores away from the root system.

Rootstock (syn. rhizome)
The bottom or supporting root used in horticulture to receive a scion in grafting; most temperate-zone fruit trees are propagated by asexual methods of grafting or budding to preserve the characteristics of the aerial portion, or scion, of the plant; in some cases, the scion cultivar of the plant cannot be reproduced by seed from adventitious roots on cuttings, and so propagation by grafting onto a rootstock is necessary; rootstocks are also used for other purposes, such as size control, disease resistance, or winter hardiness, etc.

SC group
Is experimental group of dry cysts in glass tubes, which were irradiated with magnetic field near south pole during 24 h, then they were incubated in sterile marine water for 48 h.

Scion
A portion of a shoot or a bud on one plant that is grafted onto a stock of another.

Semihardwood cuttings
The stem cuttings are taken from the current season's growth after the wood has matured (of mid-July to early fall for most plants).

Slopes
That are lateral surfaces along the transport corridors (roads, railways, embankments, etc.) with the natural earth's surface and road embankment.

Smallest significant difference
The smallest significant difference is observed between treatment options at a significance level of 5% (significantly with a probability of 95%).

Softwood cuttings
The cuttings obtained from soft succulent neoplasms of woody plants at the beginning of their maturation.

SPAD 502
It is the chlorophyll counter SPAD 502 (Minolta Camera Co, Ltd., Tokyo, Japan). The relationship between the chlorophyll content in leaf cells detected by the SPAD 502 counter and signs of productivity and quality plants was revealed in some plant species. Determination of chlorophyll content in the leaves is carried out directly in the field. The relationship between the chlorophyll content in leaf cells detected by the SPAD 502 counter and signs of plant productivity and quality was detected in some plant species. The method has been tested in Russia and Germany.

Static magnetic fields
Constant fields, which do not change in intensity or direction over time, in contrast to low and high-frequency alternating fields. They have a frequency of 0 Hz. They exert an attracting force on metallic objects containing, for example, iron, nickel, or cobalt, and so magnets are commonly used for this purpose. In nature, the geomagnetic field of the earth exerts a force from south to north that allows. The strength of a static magnetic flux density is expressed in tesla (T) or in some countries in gauss (G). The strength of the natural geomagnetic field varies from about 30 to 70 µT (1 µT is 10-6 T).

Stem
The major structural part of the vascular plant that bears buds, leaves, and flowers; it forms the central axis of the plant and often provides mechanical support. The stem also contains vascular tissues that transport water, minerals, and food throughout the plant.

Stem apex
The top or tip of a stem.

Stem cell
An undifferentiated cell that can divide to form some hereditary cells that continue to function as stem cells, and some cells that are destined to differentiate (become specialized); immature cells, as opposed to somatic cells, that capable of developing into various (specialized) cell types.

Stem cuttings
The cuttings are severed twigs that have been placed into a growing medium and encouraged to develop roots; stem cuttings are broken down into subclasses consisting of hardwood, semihardwood, softwood, and herbaceous cuttings (from the current season's growth; taken during the winter season; semihardwood cuttings are produced from woody, etc.).

ST group
Is experimental group of cysts in glass tubes in sterile marine water with 35‰ salinity which were exposed to near south pole of magnetic field during the entire hatching period (48 h).

SW group
Is experimental group of cysts incubated in sterile marine water (salinity 35‰) in glass tubes, treated with static magnetic field 50 mT near south pole of the magnet during 24 h.

Tilling
"Targeting induced local damage in genomes" is reverse genetics method (from genotype to phenotype) developed.

U per 1 million cells
A unit of enzymatic activity. The enzyme unit, or international unit for enzyme (symbol U, sometimes also IU), is a unit of enzyme's catalytic activity; 1 U (μmol/min) is defined as the amount of the enzyme that catalyzes the conversion of one micromole of substrate per minute under the specified conditions of the assay method. Relative values (U per 1 million cells) are calculated in terms of the number of cells to exclude the density factor of the culture.

Unmanned Aerial Vehicle to Estimate Nitrogen Status of Field Crops
A mechanically aerial vehicle equipped with a digital commercial camera and a multispectral sensor for comparing the spectral reflectance through the acquisition of Normalized Difference Vegetation Index on different field crops via Unmanned Aerial Vehicle with onboard a digital camera and by a ground-based instrument to test the variation in nitrogen levels.

USDA Plant Hardiness Zone Map
The 2012 USDA Plant Hardiness Zone Map is the standard by which gardeners and growers can determine which plants are most likely to thrive at a location. The map is based on the average annual minimum winter temperature, divided into 10°F zones.

Vegetation index
A measure of the amount of photosynthetic active biomass, which is calculated by points as the ratio of the difference between the photometry indexes in the near-infrared and red.

Vegetative propagation

A reproductive process that is asexual and so does not involve a recombination of genetic material (propagation by stolons, rhizomes, tubers, tillers, bulbs, bulbils, or corms, e.g., cloning of potato or strawberry).

Vegetative shoot

The aerial portion of a plant, composing the stem and leaves; new or young growth that arises from some portion of a plant.

Xylanase (endo-1,4-β-xylanase)

A glycoside hydrolase enzyme (EC 3.2.1.8) that degrade the linear polysaccharide xylan into xylose by catalyzing the hydrolysis of the glycosidic linkage (β-1,4) of xylosides that degrade the linear polysaccharide xylan into xylose, thus breaking down hemicellulose, one of the major components of plant cell walls. Generally, xylanases are synthesized by microbes and filamentous fungi, and the xylanase family is very diverse.

Index

For Product Safety Concerns and Information please contact our EU
representative GPSR@taylorandfrancis.com
Taylor & Francis Verlag GmbH, Kaufingerstraße 24, 80331 München, Germany